Aquaculture and the Environment in the United States

Edited by

J. R. Tomasso
Clemson University
Clemson, South Carolina, USA

Published by

U.S. Aquaculture Society,
A Chapter of the World Aquaculture Society
Baton Rouge, Louisiana USA

LaDon Swann, Mississippi-Alabama Sea-Grant Consortium
Managing Book Editor
U.S. Aquaculture Society, A Chapter of the World Aquaculture Society

ISBN: 1-888807-09-1

How to cite this volume:

Tomasso, J. R. (editor). 2002. Aquaculture and the Environment in the United States. U. S. Aquaculture Society, A Chapter of the World Aquaculture Society, Baton Rouge, Louisiana, USA.

How to obtain copies of this volume:

Send request to: The World Aquaculture Society
                        143 J. M. Parker Coliseum
                        Baton Rouge, Louisiana, 70803
                        United States
                        phone: +1-225-388-3137
                        online orders: http://www.was.org

TABLE OF CONTENTS

Preface

The rapid expansion of aquaculture around the world coupled with a general increase in the appreciation of quality environments has led to questions about the impact of aquaculture on the environment. These questions apply to the United States as well as the rest of the world. In support of the public-policy debate on these issues within the United States, the U.S. Aquaculture Society (a Chapter of the World Aquaculture Society) has sponsored this book. It is hoped the book will serve as a primer on the relationship between U.S. aquaculture activities and the environment. The authors and I have endeavored to present technical overviews of specific topics. It is not our purpose to take positions in public-policy debate; rather, we hope to provide a technical basis for the debate. Opinions and interpretations expressed in the book are those of the authors and not of the U.S. Aquaculture Society, the World Aquaculture Society or the editor.

The authors were selected because of the recognition they have achieved in their respective fields. Consequently, they were given a great deal of latitude in how they approached their subject. This has resulted in some overlapping topics among some of the chapters. Indeed, in a few cases two authors interpreted the same information in different ways. I did not try to help the authors find a common ground because I think these differences in interpretation point out the complexity and uncertainty inherent in the field of environmental impact assessment.

Any project such as this requires the assistance of a number of people, and I would like to recognize them. The members of Board of Directors of the U.S. Aquaculture Society spent a great deal of time evaluating the merits of the original proposal for this book and committed the necessary financial resources. LaDon Swann, as the Society's Managing Editor for books, had the responsibility of making accepted manuscripts into a book. He was assisted by Pat Edwards, Kay Bruening and the Mississippi-Alabama Sea Grant Consortium. Shawn Young assisted me, and the South Carolina Agricultural Experiment Station provided salary support for Shawn and me while we committed time to this project. Finally, I would like to thank the following reviewers: Chris Bridger, Paul Brown, Gary Carmichael, Brett Dumbauld, Rebecca Dunning, Arnie Eversole, Gary Fornshell, Charles E. Helsley, Connie Keeler-Foster, Michael Masser, Nick Parker, John Plumb, Michael Rust, Bill Simco, Jim Tidwell, Harry Westers, and Jim Winton.

Joe Tomasso
December 2001

LIST OF CONTRIBUTORS

Vicki S. Blazer
Research Fishery Biologist
United States Geological Survey
National Fish Health Research Laboratory
Leetown Science Center
1700 Leetoen Road
Kearneysville, West Virginia 25430 USA
vicki_blazer@usgs.gov

Claude E. Boyd
Professor & Butler/Cunningham Eminent Scholar in Agriculture and the Environment
Department of Fisheries and Allied Aquacultures
Auburn University
Auburn, Alabama 36849-5419 USA
ceboyd@acesag.auburn.edu

Shulin Chen
Associate Professor
Department of Biological Systems Engineering
Washington State University
Pullman, Washington 99194 USA
chens@wsu.edu

Carole R. Engle
Professor & Director
Aquaculture/Fisheries Center
Mail Stop 4912
University of Arkansas at Pine Bluff
1200 N. University Drive
Pine Bluff, Arkansas 71601 USA
cengle@uaex.edu

Gary Fornshell
Extension Educator
University of Idaho
Twin Falls County Extension Office
246 Third Avenue East
Twin Falls, Idaho 83301 USA
gafornsh@uidaho.edu

Delbert M. Gatlin III
Professor
Department of Wildlife & Fisheries Sciences
Texas A&M University
College Station, Texas 77843-2258 USA
d-gatlin@tamu.edu

Ronald W. Hardy
Professor & Director
Hagerman Fish Culture Experiment Station
University of Idaho
Hagerman, Idaho 83332 USA

John A. Hargreaves
Associate Professor
Department of Wildlife and Fisheries
Mississippi State University
Box 9690
Mississippi State, Mississippi 39762 USA
jhargreaves@CFR.MsState.Edu

Reginal M. Harrell
Professor
Wye Research and Education Center
University of Maryland
P.O. Box 169
Queenstown, Maryland 21658 USA
rh116@umail.umd.edu

Jeffrey M. Hinshaw
Associate Professor & Extension Specialist
North Carolina State University
455 Research Drive
Fletcher, North Carolina 28732 USA
Jeff_Hinshaw@ncsu.edu

Scott LaPatra
Director of Research & Development
Clear Springs Food, Inc.
Research Division
P.O. Box 712
Buhl, Idaho 83316 USA
scottl@clearsprings.com

Tom Losordo
Professor
Department of Zoology
North Carolina State University
Raleigh, North Carolina 27695 USA
tlosordo@unity.ncsu.edu

Ron Malone
Chevron USA Professor
Department of Civil and Environmental Engineering
Louisiana State University
Baton Rouge, Louisiana 70803 USA
rmalone@unix1.sncc.lsu.edu

Christopher A. Myrick
Assistant Professor
Department of Fishery and Wildlife Biology
Colorado State University
Fort Collins, Colorado 80523-1474 USA
camyrick@cnr.colostate.edu

Paul G. Olin
Interim Associate Director
University of California Sea Grant Extension
2604 Ventura Avenue
Santa Rosa, California 95403 USA
pgolin@ucdavis.edu

Steve Summerfelt
Research Engineer
Freshwater Institute
P.O. Box 1889
Shepherdstown, West Virginia 25443 USA
s.summerfelt@freshwaterinstitute.org

Robert R. Stickney
Director
Texas Sea Grant College Program
2700 Earl Rudder Freeway S.
Suite 1800
College Station, Texas 77845 USA
stickney@tamu.edu

Joseph R. Tomasso, Jr
Professor
Department of Aquaculture, Fisheries & Wildlife
Clemson University
Clemson, South Carolina 29634 USA
jtmss@clemson.edu

Craig S. Tucker
Aquaculture Research Leader & Director, Southern
Regional Aquaculture Center
Thad Cochran National Warmwater Aquaculture
Center
Mississippi State University
P.O. Box 197
Stoneville, Mississippi 38776 USA
ctucker@drec.msstate.edu

Diego Valderrama
Research Associate
Aquaculture/Fisheries Center
Mail Stop 4912
University of Arkansas at Pine Bluff
Pine Bluff, Arkansas 71601 USA
dvalderrama@uaex.edu

Global Aquaculture Production with an Emphasis on the United States

J. R. TOMASSO

Department of Aquaculture, Fisheries and Wildlife, Clemson University

Clemson, South Carolina 29634 USA

ABSTRACT

World aquaculture production of fish, mollusks and crustaceans increased from 11 million MT in 1987 to 31 million MT in 1998. The value of the aquaculture products in 1998 was US$41 billion. China is the leading producer of aquaculture products, accounting for an estimated 67.4% of world production. In contrast, the United States accounted for an estimated 1.4% of world production and just under US$1 billion in farm sales in 1998. Within the U.S., nearly two-thirds of production takes place in the southern region, and ponds are the most common production system. Fish are the most commonly produced category of aquaculture products in the U.S. The demand for aquaculture products throughout the world will probably continue to increase due to an increasing human population, an increase in the per capita consumption of seafood, and the reaching of sustainable capture limits from fisheries.

Aquaculture production has consistently and rapidly increased during the past few decades. The growth has occurred throughout most of the world, including the United States (U.S.). I will summarize here the species or groups of species being cultured, where they are being cultured, monetary value, and the kinds of production systems being used. Aquaculture production in the U.S. will be emphasized, but world production will be included for context.

World Aquaculture Production

The Food and Agriculture Organization of the United Nations (UNFAO) regularly publishes detailed figures on global aquaculture production in both its traditional publications (e.g., UNFAO 1997, 1998a, 1998b) and its web site (www.fao.org). Two summaries and analyses of the UNFAO reports (New 1997, 1999) provide a good overview of world aquaculture production. Those wishing to pursue global production information beyond this paper may find the summaries and analyses much easier to follow than the UNFAO reports.

World aquaculture production of food fish[1] (excluding seaweed production) has increased from 11 million MT in 1987 to 31 million MT in 1998. The value of foodfish produced through aquaculture in 1998 was estimated be US$47 billion. About two-thirds of the production was in China (Table 1). The U.S. accounted for about 1.4% of world production. Table 1 ranks the top-ten aquaculture-producing nations in 1998 by production weight. Weight and value rankings are not always the same due to differential values of products.

Freshwater fishes were by far the leading aquaculture product in 1996 (Table 2). All fishes account for 49% of world aquaculture production by weight. Mollusks, seaweeds, and crustaceans account for 25%, 23% and 4%, respectively. These percentages are similar to 1987 figures, indicating that the four major sectors are all growing at similar rates.

---

[1] The term "food fish" is used by FAO and in this section to include fish, mollusks and crustaceans consumed by humans. In the USDA statistics used in the following section, food fish refers only to fish consumed by humans.

TABLE 1. *Countries with highest aquaculture production (excluding seaweeds) in 1998 (various UNFAO sources; Michael New, personal communication). Countries are ranked by production (million MT) and percent of world production (Percent). Value is in billions of US$.*

	Production	Percent	Value
China	20.8	67.4	21.7
India	2.0	6.6	2.2
Japan	0.8	2.5	3.1
Indonesia	0.7	2.3	2.1
Bangladesh	0.6	1.9	1.5
Thailand	0.6	1.8	1.8
Viet-Nam	0.5	1.7	1.3
USA	0.4	1.4	0.8
Norway	0.4	1.3	1.1
Republic of Korea	0.3	1.1	0.5

TABLE 2. *Global aquaculture production (million MT) of selected groups of species (New 1999; various UNFAO sources). Numbers may not exactly agree with text due to rounding.*

	1987	1996
Freshwater fish	5.9	14.4
Diadromous fish	0.8	1.7
Marine fish	0.3	0.6
Marine shrimp	0.5	0.9
Other crustaceans	0.1	0.2
Oysters	1.3	3.1
Clams, cockles and arkshells	0.5	1.8
Mussels	1.0	1.2
Other mollusks	0.3	2.4
Brown seaweeds	1.8	4.6
Red seaweeds	0.8	1.7
Other seaweeds	0.2	1.5
Total	13.5	34.1

The growth of world aquaculture production appears to be stimulated by three factors: a growing human population, an increase in the per capita consumption of seafood[2], and the reaching of sustainable limits of production from the capture fisheries by the world's fishing fleets. Fig. 1 shows the relationship between capture fisheries and aquaculture over time for food fish. The values for capture fisheries exclude landings of fishes used to produce fish oil and fish meal which have accounted for about a third of recent landings (Tacon 1997/1998). Food fish landings appear to have stabilized at about 60 million MT per year.

Since 1988, the continuing increase in food fish availability (Fig.1) has been predominantly due to the expansion of aquaculture production (New 1997). A recent analysis (New 1997) considered the increasing world population, changing demographics, and a projected 1% annual increase in per capita seafood consumption, and concluded that food fish production through aquaculture will have to equal capture fisheries production (assuming stabilization at 60 million MT) by 2015 (Fig. 2) for seafood supplies to meet demand.

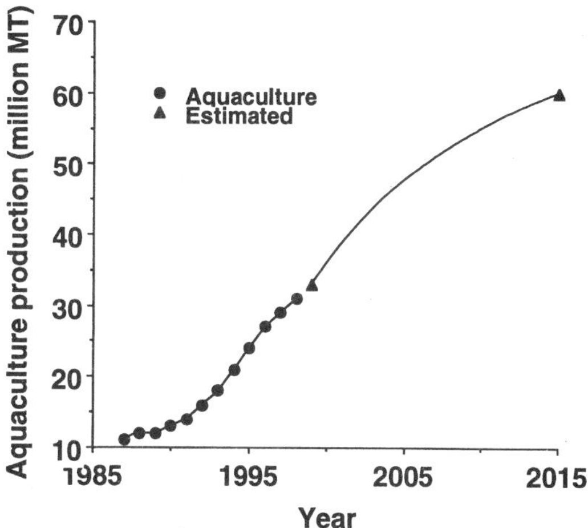

FIGURE 2. *World aquaculture production (excluding seaweeds) from 1987 to 1998 (New 1999, and various UNFAO sources), and an estimate of future production to meet world seafood needs (New 1997).*

Aquaculture Production in the United States

The U.S. Department of Agriculture (USDA) recently completed a "Census of Aquaculture" (USDA 2000, also available at www.nass.usda.gov/census/) which describes the state of the industry in 1998. I will summarize several of the key points here.

Aquaculture products sold by farmers in the U.S. were valued at $978 million in 1998. This is slightly higher than the approximately $800 million in sales of U.S. farmers estimated by UNFAO (Table 1) and is probably due to the USDA census including farm-to-farm sales of stockers. Farm sales of U.S. aquaculture products is one of the smallest measured components of U.S. agriculture (Table 3) where farm sales approached $200 billion in 1997

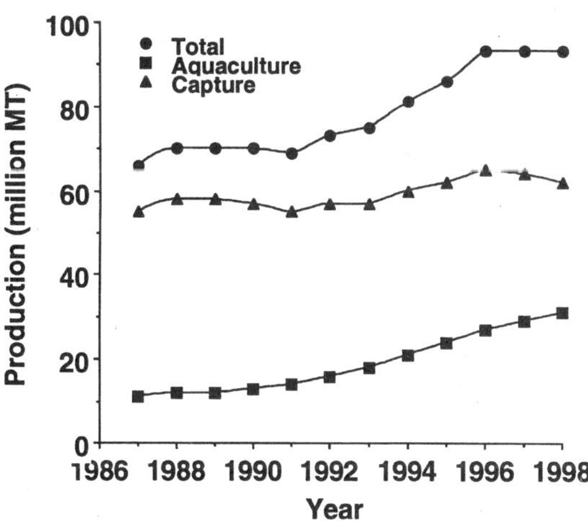

FIGURE 1. *World aquaculture production (excluding seaweeds), foodfish captured from world fisheries, and total food fish available from both sources. Data were compiled from Tacon (1997/1998), New (1999), and various UNFAO sources.*

---

[2] The term "seafood" as it is used here refers to fish, mollusks, and crustaceans consumed by humans, regardless of freshwater or saltwater origin.

TABLE 3. *Farm sales (bllions US$) in the U.S. of selected product categories (USDA 1999).*

Cattle and calves	41
Poultry and poultry products	23
Dairy products	19
Corn for grain	19
Soybeans	16
Hogs and pigs	14
Fruits and nuts	13
Nursery and greenhouse crops	11
Vegetables, sweet corn and melons	9
Wheat	7
Cotton	6
Tobacco	3
Aquaculture products	1

(USDA 1999, also available at www.nass.usda.gov/census/).

By value, nearly two-thirds (65%) of the production took place in the southern region (Table 4), followed by the western (17%) and northeastern regions (13%). Food fish was the largest category of production (71% by value), followed by mollusks (9%), ornamentals (7%), baitfish (4%), and crustaceans (4%). By value, catfish was the largest crop produced (Table 5). Production of channel catfish has consistently increased during the past 30 years (Fig. 3).

More farms use ponds than any other kind of production system in all of the regions of the country (Table 6). Flow-through raceways or tanks are the second most commonly used production system. More farms use groundwater as a water source than any other source, followed by on-farm surface water.

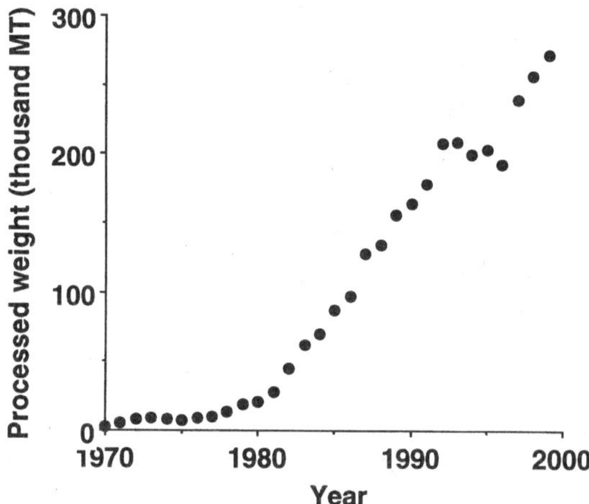

FIGURE 3. *Catfish processing in the United States from 1970 to 2000 (various USDA sources).*

TABLE 4. *Farm-gate sales (X US$1,000) of aquaculture products in the U.S. by region during 1998 (USDA 2000). The percentage of each region's sales accounted for by main product categories is also presented. For assignment of states to regions, see USDA (2000).*

	Sales	Main product categories
Northeastern	127,393	67% food fish[a], 21% mollusks
Southern	637,394	74% food fish, 9% ornamentals
North Central	30,120	60% food fish, 21% baitfish
Western[b]	166,565	69% food fish, 22% mollusks
Hawaii	16,541	33% crustaceans, 6% food fish

[a]The term food fish in USDA (2000) and in this table refers only to finfish for human consumption.

[b]Includes Alaska.

TABLE 5. *Number of farms involved in aquaculture, production (X 1000 kg), and value (X US$1,000) of selected aquaculture products in the U.S. during 1998 (USDA 2000). Numbers for production and value are for the farms that are the final point of sale and do not include sales of stockers to growout operations. Production figures are not included for products that are not sold by weight. Products are ranked by value.*

	Farms	Production	Value
Catfish	1,227	259,973	420,898
Salmon	45	50,267	103,583
Trout	466	26,744	62,608
Tropical fish	192		57,145
Clams	360		50,076
Oysters	207		36,323
Tilapia	137	5,260	23,107
Hybrid striped bass	66	3,944	21,177
Golden shiners	123		18,094
Shrimp	42	1,917	12,188
Softshell crabs	218		10,297
Crawfish	563	7,937	[a]
Feeder goldfish	34		9,302
Fathead minnows	164		7,425
Ornamental goldfish	65		6,749
Turtles	56		4,757
Largemouth bass	136		4,450
Koi	115		3,942
Mussels	29		3,180
Bluegill	129		1,790
Carp	39	755	1,300

[a]Value not provided, ranking is estimated.

Outlook for Aquaculture in the United States

It appears that demand for aquaculture products will continue to increase in the U.S. The annual per capita consumption of seafood in the U.S. peaked in 1994 at 6.9 kg, and decreased to 6.6 kg by 1997 (USDC 1999). However the population is expected to increase from an estimated 270 million in 1998 to 310 million in 2015 (USDC 1999), and will more than make up for any decrease in per capita consumption.

The U.S. dependency on imported seafood may also be a factor in the future growth of aquaculture in the U.S. In 1980 the U.S. imported 2 million MT of seafood. Imports increased to 3 million MT by 1997 (USDC 1999). The 1997 value is over twice the UNFAO estimate for U.S. aquaculture production in 1998 (Table 2).

TABLE 6. *Number of aquaculture farms in the U.S. using various production systems and water sources in 1998 by region (USDA 2000). For assignment of states to regions, see USDA (2000). Northeastern region=NE, Southern region=S, North Central region=NC, Western region=W, Hawaii=H.*

	NE	S	NC	W	H
PRODUCTION SYSTEMS					
Ponds	195	2,131	292	209	51
Flow-through raceways or tanks	98	280	90	123	26
Cages[a]	55	40	6	14	2
Net pens[a]	16	8	4	20	2
Closed recirculating systems	65	145	61	30	27
Prepared bottoms[b]	74	203	2	59	0
Other methods	71	111	2	47	0
SOURCES OF WATER					
Groundwater	146	1,411	195	155	18
On-farm surface water[c]	139	946	209	149	11
Off-farm water[d]	17	70	19	50	36
Saltwater	167	500	0	118	30

[a] By USDA definition, cages are normally placed in lakes and rivers, net pens usually placed in protected bays or inlets.

[b] Bottom in tidal areas that has had material added to improve habitat for shellfish.

[c] A surface water supply not controlled by a water supply organization.

[d] Water that is provided by a water supply organization.

Recently American consumers have demanded more fresh seafood rather than canned or cured. The proportion of seafood from the domestic-capture fishery that was processed into fresh or frozen product has increased from 72% in 1980 to 95% in 1997 (USDC 1999). If the trend toward fresh seafood continues, domestic aquaculture will have to provide it, given the generally prohibitive cost of importing fresh seafood and the plateauing of capture-fishery production.

Acknowledgments

I would like to thank Michael New and Albert Tacon for helping decipher UNFAO statistics. Arnie Eversole reviewed the manuscript and provided helpful suggestions.

Literature Cited

New, M. B. 1997. Aquaculture and the capture fisheries. World Aquaculture 29(2):11–30.

New, M. B. 1999. Global aquaculture: Current trends and challenges for the 21st century. World Aquaculture 30(1):8–13, 63–79.

Tacon, A. G. J. 1997/1998. Global trends in aquaculture and aquafeed production 1984–1995. International Aquafeed Directory 1997/1998:5–37.

UNFAO (United Nations Food and Agriculture Organization). 1997. Review of the state of world aquaculture. FAO Fisheries Circular no. 886, rev.1. FAO, Rome, Italy.

UNFAO (United Nations Food and

Agriculture Organization). 1998a. Aquaculture production statistics 1987–1996. FAO Fisheries Circular no. 815, rev. 10. FAO, Rome, Italy.

UNFAO (United Nations Food and Agriculture Organization). 1998b. Fisheries statistics: Capture production, volume 82. FAO Fisheries Series no. 50. FAO, Rome, Italy.

USDA (United States Department of Agriculture). 1999. National Agricultural Statistics Service, 1997 census of agriculture. Washington, D.C., USA.

USDA (United States Department of Agriculture). 2000. National Agricultural Statistics Service, 1997 census of agriculture, census of aquaculture (1998), volume 3, Special studies, part 3. Washington, D.C., USA.

USDC (United States Department of Commerce). 1999. Statistical abstract of the United States, 119th edition. U.S. Census Bureau, Washington, D.C., USA.

Water Budgets for Aquaculture Production

JOHN A. HARGREAVES

Department of Wildlife and Fisheries, Mississippi State University,

Box 9690, Mississippi State, Mississippi 39762 USA

CLAUDE E. BOYD

Department of Fisheries and Allied Aquacultures, Auburn University

Auburn, Alabama 36849-5419 USA

CRAIG S. TUCKER

National Warmwater Aquaculture Center, Mississippi State University

P. O. Box 197, Stoneville, Mississippi 38776 USA

ABSTRACT

Water budgets can be used to identify and describe the magnitude of inflows, water storage, and outflows to aquaculture production systems. Accounting for water in aquaculture has two significant implications: first, water is a resource that must be used wisely to assure sustainable development; second, water budgets can also be used to estimate effluent volume and develop options for managing discharged water. For aquaculture ponds, inflows include precipitation, runoff, and regulated inflows from ground and surface water sources. Outflows include evaporation, seepage, and overflow. The magnitude of water budget terms for ponds depends on hydrological pond type (levee or watershed) and production phase (brood-fish, fingerling, food-fish). Long-term water budgets are strongly affected by the interval between pond drainings. Although fingerling ponds represent 13% of the total pond area in catfish production, ponds are drained annually and therefore contribute disproportionately (30%) to overall discharge volume. Knowledge of the magnitude of terms in water budgets facilitates identification and prioritization of management options for water conservation in pond aquaculture, which reduces the need to pump groundwater and decreases the volume of water discharged. Water conservation can be achieved by reducing outflows (minimizing seepage, reducing water exchange, reducing draining or drawdown frequency) or by manipulating storage volume. Maintaining water storage capacity in ponds allows capture of rainfall or runoff, rather than allowing it to overflow, thereby reducing future need for groundwater. Maintaining a water storage capacity of 20 to 30 cm can reduce groundwater volume pumped and effluent discharged by 40 to 60% compared to ponds managed without water storage capacity. Maintaining water storage capacity is a simple and effective method of improving the environmental performance of pond aquaculture. Compared to pond aquaculture, opportunities for water conservation in raceways are extremely limited. However, in contrast to pond aquaculture, water use in raceway culture is non-consumptive. Water efficiency in raceway culture can be improved through serial reuse. Improving efficiency of water use in aquaculture will contribute to continued and sustainable development of aquaculture industries.

Water is an obvious and essential input for aquaculture. It is also an essential resource for other forms of agriculture, and indeed, is essential for all life. For many years, aquaculture was a small, diffuse endeavor practiced mainly on a small-scale, artisanal basis in developing countries. At that time, water seemed abundant, at least on a global scale, and demands on water resources by aquaculture were insignificant. Few people voiced concern that aquaculture was either consumptively using water or degrading the quality of water after it was used. In general, it was felt that the benefits of aquaculture—supplying food to needy people and serving as an engine for local economic development—far outweighed the social and economic cost of using water to grow that food. However, over the last 20 years, world population has expanded at a phenomenal rate, placing increased demands on the earth's finite supply of water. Expansion in population and increased awareness of the limitations of clean water have occurred at the very time when aquaculture was changing from a small-scale activity to a major food-producing industry in many countries. In several instances, these coincident phenomena have set aquaculture in conflict with other users of water.

Water budgets account for the quantities of water that flow into and out of an aquaculture production system. Accounting for water in aquaculture has two significant implications. First, as alluded to above, water is a precious resource and must be used wisely to assure sustainable development. Conservative water use in aquaculture is therefore a social requirement, and knowledge of water budgets can lead to development of management practices that make water use more efficient. Water budgets can also be used to estimate the volume of water discharged from different types of aquaculture facilities, which in turn can be used to assign priorities to options for management of discharged water. This is important because water discharged from any food production facility may impact the water body receiving the effluent, and the extent of that impact is greatly influenced by effluent volume.

These two issues—conservative water use and minimizing effluent volume—are intimately linked. Water that is discharged from a facility is a "loss" process on one side of the water budget equation. By decreasing the magnitude of that loss, less water will be needed on the other ("input") side of the

equation. It should be pointed out, however, that efficiency of water use must be viewed on a larger, watershed-level scale, rather than on a limited, single-farm or single-pond scale. If water is suitable for further use after it has been used first in aquaculture, then it can be a resource to support other economic activities, rather than be a true waste.

The requirement for water by aquaculture varies depending upon the type of culture system (Table 1). Raceways non-consumptively use large volumes of water per unit fish produced as the culture system is configured and managed such that a continuous flow of water is critical for water quality maintenance, particularly oxygen supply. Raceways are usually configured so a given parcel of water is used several times as it flows through a series of raceways. However, depending on the quality of the source water and without water treatment to add oxygen or remove products of fish metabolism, there is a limit to the number of raceway segments that can use the same parcel of water. Therefore, in most raceway facilities, water flows through the series of raceways once and is then discharged from the facility. Typical number of uses in Idaho average four to six, while in North Carolina, where the water is more acidic and softer, up to 20 uses is common. In contrast, many recirculating systems are operated with very low water exchange rates, and the water within each production unit is reused many times. Water quality is maintained by mechanical and biological treatment processes that maintain water quality within tolerance limits of the fish crop. Water use in ponds is similar to that in recirculating systems in that water quality is maintained by mechanical and biological processes. However, in recirculating systems, treatment takes place in discrete units, such as screens or settling basins to remove solids and biological filters to remove ammonia, whereas in ponds, the same processes occur naturally as unconfined parts

TABLE 1. *Volume of water used per unit production for different aquaculture systems and selected crops.*

Culture system	Water use (m³/kg production)	Reference
Ponds		
Clarias batrachus		
static, intensive ponds	0.05–0.20	Phillips et al. 1991
Channel catfish		
undrained levee ponds	1.25–1.75	Yoo and Boyd 1994
watershed ponds	6.5–10	Yoo and Boyd 1994
Penaeid shrimp		
	11–55	Phillips et al. 1991
semi-intensive, 5% water exch.	50–100	Yoo and Boyd 1994
intensive, 20% water exch.	40–80	Yoo and Boyd 1994
Raceways		
Trout		
average production	98	G. Fornshell, pers. comm.
unaerated raceways	83–117	Yoo and Boyd 1994
aerated raceways	17–42	Yoo and Boyd 1994
Other agricultural commodities		
Cotton	0.09–0.45	Phillips et al. 1991
Beef	0.042	Phillips et al. 1991
Pork	0.054	Phillips et al. 1991

of the pond ecosystem. Net pen or cage culture is more difficult to classify based on water use than other types of aquaculture systems. Net pens are placed in multiple-use water bodies and therefore do not consume water in the same way as other production systems. However, maintenance of adequate water quality within the production unit depends on the exchange of water from outside the unit, so from that standpoint, water use in net pens is most similar to that in raceways.

In this chapter we will discuss water budgets for aquaculture systems, with the goal of pointing out how water can be used more efficiently and how effluent volume can be reduced. As commercial culture of channel catfish and rainbow trout are the two most important aquaculture industries in the United States, this discussion will focus on water budgets for ponds and raceways. Greater consideration will be given to water budgets for ponds because they are more complex than those for raceways and therefore offer greater opportunities for improvement of water use efficiency.

The Pond-Raised Catfish Industry

Production of channel catfish in earthen ponds is the largest aquaculture enterprise in the United States, with over 287,400 mt of fish processed in 2000 (USDA/NASS 2001a). This represents about half of the total national aquaculture production for all species. The value of food fish sales in 2000 was $468.8 million (USDA/NASS 2001a). Although commercial catfish aquaculture was reported from 35 states in 1998, nearly all production occurs in 13 states in the southeastern United States and two states (Missouri, California) outside that region. Production of food-sized fish from four states (Mississippi, Alabama, Arkansas, and Louisiana) accounts for 97% of the total domestic catfish production. Three counties in Mississippi (Humphreys, Sunflower, and Leflore) account for 39% of the total national pond area. Industry demographics are summarized in the 1998 Census of Aquaculture (USDA/NASS 2000).

Within the southeastern United States, the two major catfish-producing areas are: 1) a relatively well-defined geographical area of the lower Mississippi River valley that includes northwest Mississippi, southeast Arkansas, and northeast Louisiana, and 2) a less well-defined area of west-central Alabama and east-central Mississippi. Ponds located in the Mississippi River valley account for about 78% of the total land area devoted to catfish aquaculture in the United States and about the same proportion of the total production of food-sized catfish. Resources, pond type, and production practices are relatively uniform throughout this area and are described by Tucker (1996, 2000). In contrast, catfish farming in the area of west-central Alabama and east-central Mississippi is much more diverse. Farms vary greatly in size, a mixture of pond types and water sources are used, and catfish may be one of several crops on a given farm (Boyd et al. 2000).

Collectively, the two areas account for about 95% of the total land area used for catfish aquaculture and about the same percentage of food-sized fish production.

Geographic Variation in Pond Facilities and Water Supplies

Lower Mississippi River Valley

The flat topography and alluvial clay soils of the lower Mississippi River valley and the availability of a high-yielding groundwater source lying at a shallow depth are ideally suited for construction of levee ponds. Levee ponds (also called embankment ponds) are built by removing soil from the area that will be the pond bottom and forming it into levees around the pond perimeter. Most ponds are rectangular with a 3:1 to 5:1 ratio of length to width, and average pond size is between 3.2 and 6 ha of water surface area. For ease of harvest, minimum pond depth is no less than 1 m and no greater than about 1.5 m. The height of the levee above normal water level (freeboard) is 30 to 60 cm. Drains are extended through the levee at the deepest end of the pond. Drains are fitted with either an alfalfa valve on the outside of the pond or a swivel-joint on the inside to allow adjustment of water level in the pond.

Wells are drilled after levees are constructed. The number of ponds served from one well depends on pond size and well pumping capacity. Typically, a well is drilled at the intersection of four levee ponds. The minimum pumping capacity should be about 140 L/min per ha of water surface. That rate is sufficient to replace water losses from about twice the maximum daily pond evaporation rate in the southeastern United States and allows for maintenance of pond levels during periods of drought with some excess to meet moderate seepage losses.

After ponds are filled initially, water levels are maintained by inputs of precipitation and pumped ground water. Rainfall significantly reduces the requirement for pumped water after ponds are filled. The need for pumped water is further reduced because farmers routinely manage ponds to capture rainfall rather than allowing it to overflow. This water conservation practice also results in reduced pond effluent volume and reduced discharge of nutrients and organic matter (see Chapter 3).

West-central Alabama and East-central Mississippi

About 75% of the commercial catfish ponds in west-central Alabama are watershed ponds that use mainly rainfall and storm runoff for filling and maintaining water levels. The remaining 25% of the ponds are levee ponds and are filled with pumped surface water from adjacent streams or with groundwater from wells. Some ponds are a combination of the two types in that they may have two or three embankments, but are capable of receiving storm runoff from the watershed adjacent to one or more sides of the pond.

The average size of ponds in west-central Alabama and east-central Mississippi is 4 to 5 ha with an average maximum depth of 2 m at the standpipe and a minimum of 1 m at the shallow end. Most ponds, including watershed ponds, are constructed so that fish can be conveniently harvested with a seine. Ponds are shallow and smooth-bottomed so that they do not need to be drained to facilitate harvest. However, ponds are completely drained for maintenance approximately once every 15 yr. The height of levee or dam above normal water stage (storage and freeboard) ranges from 30 to 60 cm on levee ponds to 1 m on watershed ponds. Drains are located at the deepest end of the pond. These drains are mostly smooth-steel pipe with a valve and riser inside the pond to control the water level and provide drainage. Most ponds in west-central Alabama and east-central Mississippi have been designed using USDA-NRCS soil conservation practice standards.

After initial filling, pond levels are maintained with precipitation and pumped water, depending on availability. As most of the ponds in production are watershed ponds, they are designed to maintain a minimum water stage height with average rainfall.

A major difference between catfish farming in west-central Alabama and east-central Mississippi is the reduced reliance on groundwater supplies in east Mississippi. About half the ponds in east Mississippi are watershed ponds that use rainfall and runoff for filling and maintaining water levels. The remainder are levee ponds or hybrid watershed-levee ponds, but nearly all farms obtain water from nearby streams or other surface water supplies rather than from groundwater.

Pond Water Management for Different Production Phases

Broodfish Ponds

Broodfish ponds represent 1.2% of the total pond area devoted to catfish production. Some farmers harvest and drain broodfish ponds every autumn to replace low-performance breeders and to adjust broodfish standing crop and sex ratios. Most broodfish ponds, however, are drained only every 2 to 5 yr and broodfish are inspected after seining the fish without draining the pond or are simply not inspected each year. Therefore, water budgets for broodfish ponds will depend on the frequency and timing of pond draining.

Fingerling Ponds

Nursery ponds represent about 13% of the total pond area in catfish production (USDA/NASS 2001a). Water use in nursery ponds is much greater than in broodfish or food-fish production ponds because nursery ponds are drained annually. Fingerlings are usually harvested by seining the pond several times over 1 to 3 mo. After as many fish as possible have been removed by seining, the pond is drained so that all remaining fish can be removed.

In contrast to nursery ponds used to raise fry to fingerlings, some ponds are used to produce "stocker" size fish. Small (5–10 cm) fingerlings are stocked at 185,000 to 250,000 fish/ha and grown out to stocker size (20–25 cm), which are then transferred to food-fish ponds. Stocker ponds may be in production for several years before they are partially drained.

In west-central Alabama and east Mississippi, fingerling producers are at a disadvantage because the temperature of groundwater supplies for hatcheries is too low. In addition, water supplies for watershed nursery ponds supplied by rainfall and runoff are too unreliable for the highly ordered annual production cycle needed in fingerling production. Therefore, most fingerling production occurs in the lower Mississippi River valley, and fingerlings are transported to the food-fish production areas to the east.

Food-fish Production Ponds

Nearly all food-fish production ponds are managed with a multiple-batch cropping system known as "understocking," in which more than one year-class of fish is present after the first year of production. Initially, the pond is stocked with a single year-class of fingerlings. The faster-growing individuals of about 0.5 kg and larger are selectively harvested using a large-mesh seine, and fingerlings are added ("understocked") to replace the harvested fish plus any losses suffered during production. The process of selective harvest and understocking continues for multiple years without draining the pond. Producers in east-central Mississippi practice a variation of this cropping system in which as many fish as possible are harvested after a growing season without draining the pond. The pond is subsequently re-stocked with fingerlings and production continues.

Food-fish production in levee ponds are usually drained only when ponds need to be renovated or when there is need to adjust fish inventory by completely harvesting the crop. Commercial ponds remain in production for 3 to over 10 yr without being drained, and the average time between pond drainings is over 6 yr (USDA/APHIS 1997). However, many watershed ponds have deep areas near the dam that prevent normal fish harvest by seining. These ponds must be partially drained periodically so that fish can be concentrated into an area of relatively shallow water near the dam to facilitate harvest. Ponds of this type are common in west-central Alabama and east-central Mississippi but constitute a small proportion of the total pond area used for catfish farming.

Water Sources for Catfish Ponds

The main water sources for catfish ponds are precipitation, runoff from watersheds, and regulated additions of water from groundwater aquifers or surface water streams and reservoirs. Precipitation and runoff are the most important water sources for watershed ponds; precipitation and regulated inflows from groundwater are the primary water sources for levee ponds.

Precipitation

Precipitation falling directly into the pond can be a significant contributor of water to aquaculture ponds in humid climates. The increase in pond water level from direct precipitation is equal to rainfall. The frequency distribution of precipitation events is approximately log-normally distributed: there are many more precipitation events of low amount than of high amount. Based on a 28-yr precipitation record for Stoneville, Mississippi, 50% of daily rainfall events are less than 0.5 cm (range: 0.35–0.84 cm) and 90% of daily rainfall events are less than 3.1 cm (range: 2.4–4.2 cm) (Fig. 1).

In the lower Mississippi River valley, average annual precipitation is about 127 cm.

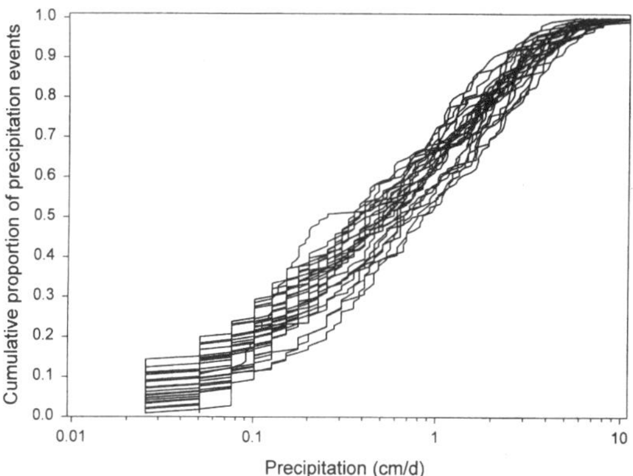

FIGURE 1. *Cumulative distribution of total daily precipitation events. Each line represents data from one year of a 26-yr climate record (1962–1987) for Stoneville, Mississippi (EarthInfo, Inc. 1989).*

Over the long term, average annual rainfall is approximately uniformly distributed through the year (Fig. 2). However, long-term average rainfall obscures seasonal variability and rainfall extremes. The range of monthly rainfall among years is great, particularly during winter and early spring. Variability in rainfall is greater in the summer than in the spring. October is the driest month of the year in the southeastern United States.

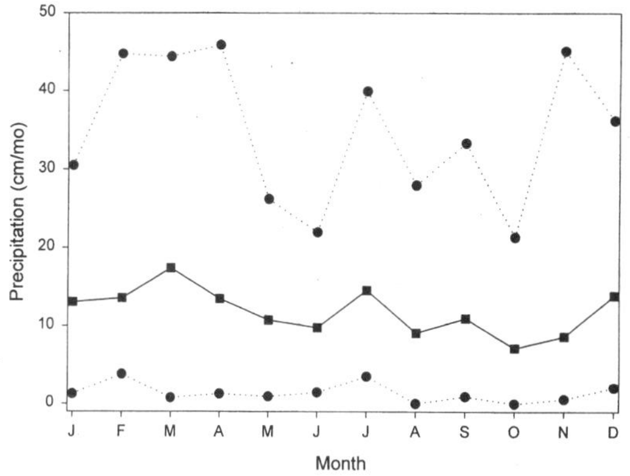

FIGURE 2. *Minimum, mean, and maximum monthly precipitation for Auburn, Alabama. Data from Southeast Agricultural Weather Service Center, Auburn University, Alabama (Yoo and Boyd 1994).*

In the southeastern United States, the primary mechanisms of precipitation vary seasonally. In the winter and spring, rainfall is associated with the west-to-east movement of strong cold fronts that separate cold, dry polar air from warm, moist tropical air. Rainfall amounts per event in the winter and spring are greater than in the summer. In the summer, stable high pressure centered over the western Atlantic Ocean is associated with warm and humid air. The high pressure area directs tropical storms around the edges of the high. Rainfall also occurs as the result of sporadic and widespread convectional showers and storms when the subtropical high weakens.

Providing storage volume in ponds to capture rainfall (or runoff) will minimize the need for pumped water, particularly during the summer, and reduce effluent volume released during pond overflow. Many producers make some effort to capture this free water source. Boyd and Gross (2000) recommend maintaining a water storage capacity equal to the normal, maximum daily precipitation.

Runoff

Once the capacity of soil to hold moisture (infiltration capacity) is exceeded, water will begin to flow across the land surface. Watershed ponds are constructed to capture overland flow from the surrounding watershed, especially if the availability of ground or surface water resources is not reliable or predictable. The volume of runoff will depend on watershed characteristics, particularly slope, type and extent of vegetative cover, soil type, and antecedent moisture. The volume of runoff is also a function of rainfall amount and intensity.

There are several methods for estimating the proportion of rainfall that is converted to runoff (Yoo and Boyd 1994). A water accounting method can be used to estimate monthly runoff. Adding the average rainfall for a month to the initial soil moisture equals the total available moisture. Potential evapotranspiration for the period is subtracted from the total available moisture to yield the remaining measurable moisture. Runoff is equal to the difference between the remaining measurable moisture and the final soil moisture. If the remaining measurable moisture exceeds the available moisture when the soil is fully saturated (approximately 0.83 mm water/cm soil depth), then runoff will occur.

Across the United States, runoff averages 28% of precipitation. In the southeast United States, overland flow is about 21 to 29% of rainfall from November to March, but very little runoff occurs from April to October (Yoo and Boyd 1994). In Alabama, the proportion of rainfall that becomes runoff in small watersheds with intermittent streams (16%) is lower than that for large watersheds with permanent streams (36%) (Boyd and Tucker 1998). The total annual runoff from various soil areas of Alabama ranges from 46 to 65 cm (Yoo and Boyd 1994).

Approximately 6.5 to 9.8 ha of watershed is recommended to support 10,000 m^3 of water storage in the southeastern United States (USDA/SCS1982). In a survey of channel catfish farms in west Alabama, average pond depth was 1.67 m and the ratio of watershed area to pond surface area was 6.32 (Boyd et al. 2000). Therefore, Alabama catfish ponds were constructed at less than the recommended watershed-to-pond area (3.78 ha of watershed supporting 10,000 m^3 of water storage). Many ponds in west Alabama and east Mississippi are constructed as a combination of watershed and levee ponds and many catfish farms in this area have wells to supplement inflow from runoff.

Fluctuations in water level in watershed ponds will exceed those in levee ponds because runoff is a much more important source of water for watershed ponds than levee ponds. During droughts, water levels in ponds supplied only by runoff may decline to dangerously low levels. The contribution of runoff to pond volume can be many times greater in watershed ponds than levee ponds because levee ponds have a watershed area that is only slightly greater than the pond surface area. For example, a typical commercial catfish pond in the Mississippi River valley is constructed on 8.1 ha of land and has a pond surface area of 6.9 ha (watershed area = 14.8% of the total area).

Groundwater

Water for filling and maintaining water level in catfish ponds in the lower Mississippi River valley is pumped from the Mississippi River alluvial aquifer. The aquifer is shallow and consists of layers of sand and gravel overlain by clay. Locally, a layer of impervious clay under the aquifer separates it from deeper water-bearing strata. Depth to water generally ranges from 6 to 30 m; the thickness of the aquifer ranges from < 15 to about 60 m. Wells withdrawing water from the aquifer range from 15 to over 45 m in depth. The high transmissivity and hydraulic conductivity of the coarse alluvium allow high pumping rates: typical well yields range from 1.9 to 15 m^3/min.

The aquifer is recharged from lateral movement of water from the Mississippi River and its deeply incised tributaries in areas of the aquifer adjacent to these water bodies. Recharge also occurs through the sandy soils of loess bluff hills to the east of the alluvial valley. The clay layer lying between the aquifer and the surface precludes significant recharge by vertical movement of precipitation falling on the land surface. The confined nature of the aquifer also reduces the opportunity for pollution of the water.

Wells used for filling ponds in west-central Alabama tend to be relatively deep compared to wells supplying water to catfish ponds in the lower Mississippi River valley. Wells in west Alabama range from 33 to 425 m deep with most being around 35 to 55 m deep. Casing diameters rarely exceed 15 cm, and pumping capacity is limited to 1.5 to 2.3 m³/min. The main aquifers are the Eutaw, Gordo, and Coker Formations, and alluvial and terrace deposits of the Quaternary age. Parts of the Coker and the underlying lower Cretaceous and some areas of the Eutaw formation come relatively close to the surface and may be free flowing. The high chloride content of this water, which protects fish from nitrite toxicosis, makes it highly desirable to catfish culturists.

Surface Waters

Water is pumped into catfish ponds from streams and rivers to supplement groundwater and runoff sources, particularly in west-central Alabama and east Mississippi. However, surface waters are undesirable from many standpoints. Generally, the seasonal availability of surface waters does not correspond to demand. Peak stream discharge occurs during winter to early spring, yet the demand for surface water, particularly as a supplemental water source, is maximum during the summer when stream discharge is near the annual minimum. The seasonal and year-to-year availability of surface waters is extremely variable. In extremely dry years, insufficient surface water may be available to supplement other water sources. In extremely wet years, flooding of streams and rivers may damage pond levees and drainage structures. Overtopping of levees by flood waters may result in the escape of the fish stock. Surface waters are also undesirable because they are sources of contamination by wild fish, potential disease pathogens, agricultural pesticides, fertilizers and sediment. Finally, surface waters are considered Waters of the State and are therefore subject to potential regulatory constraints that may limit withdrawals for pond aquaculture.

Water Losses from Catfish Ponds

Water loss from catfish ponds results from evaporation, seepage, and overflow that occurs when inflows exceed pond water storage capacity. Evaporation rates are controlled by climatic forces beyond the control of the culturist. However, seepage can be reduced by good site selection and pond construction practices, and overflow can be minimized by maintaining water storage capacity.

Evaporation

Evaporation from ponds is the loss of water from the pond surface to the atmosphere. The rate of evaporation depends on air temperature and relative humidity, water temperature, and wind speed. Over the long run, water temperature has the greatest effect on evaporation, so evaporation rates are maximum during summer and minimum during winter (Fig. 3).

As a practical matter, pond evaporation is difficult to measure directly because decreases in pond water stage are due to the combined effects of evaporation and seepage, which cannot be easily differentiated. Accordingly, pond evaporation rates are estimated indirectly by multiplying the evaporation rate determined with a class A evaporation pan by 0.81 (Boyd 1985a). Measurements from class A evaporation pans (sometimes simply called "pan evaporation") can be obtained from weather monitoring stations located throughout the world. Average monthly pan evaporation in the southeast United States ranges from about 2.8 cm/mo (0.9 mm/d) in January to 19.7 cm/mo (6.6 mm/d) in June (Boyd and Tucker 1998). Over the period 1962 to 1986, maximum daily pond evaporation rates in the lower Mississippi River valley during June ranged from 8.9 to 11.4 mm/d (Pote et al. 1988).

Seepage

Ideally, ponds should be constructed in soil

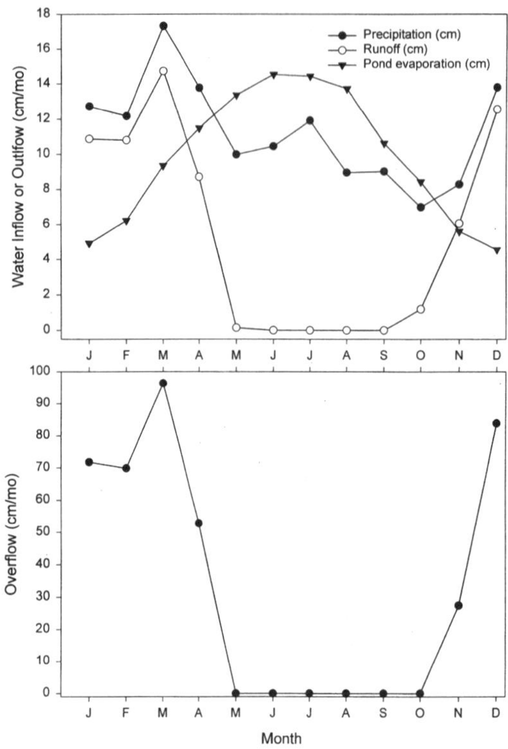

FIGURE 3. *Average monthly precipitation, runoff, and evaporation (top panel) and estimated overflow (bottom panel) from watershed ponds in west-central Alabama. Data from Boyd et al. 2000.*

that is impervious to the movement of water. In practice, however, water moves into or out of all ponds unless they are lined with a watertight material, such as plastic. Although water may move into ponds from the surrounding soil when local water tables are near the soil surface, the net movement of water through most pond soils is generally outwards, and this loss is called seepage or infiltration. Seepage is difficult to measure because soils are not uniformly impervious across the pond bottom and evaporative losses are difficult to partition from seepage losses. Generally, seepage is estimated by subtracting pond evaporation (estimated from measurement of pan evaporation) from total losses measured by decrease in water stage.

Levee ponds in the Mississippi River valley are constructed in montmorillonite clay soils of the Sharkey or Alligator series. These soils are dense and swell when wet, becoming nearly impervious to water flow. Seepage from rice fields constructed in these soils is 0.381 mm/d (Pringle and Pennington 1991). Limited measurements from ponds constructed in the lower Mississippi River valley indicate that seepage is less than 1.3 mm/d (Pote et al. 1988). Seepage from ponds constructed in the Black Belt soil edaphic region of east-central Mississippi and west-central Alabama is 1.5 mm/d (Parsons 1949). In general, ponds in the major catfish-producing areas have low seepage rates according to Boyd (1986), who classified seepage rates as low (0–5 mm/d), moderate (6–10 mm/d), high (11–15 mm/d) and extreme (>15 mm/d).

Excessive seepage is undesirable because it increases the need for pumped water. Also, excessive seepage can pose a threat to the pristine quality of underground water supplies because uncharged substances or anions, such as nitrate, can percolate through soils with the seepage water. Seepage also has a profound effect on the amount of water lost as overflow. In ponds with low seepage rates, a proportionally greater volume of water is discharged as overflow. In ponds with high seepage rates, water is lost to infiltration and overflow rates are low.

Overflow

Overflow will occur when pond inflows exceed pond storage capacity. The volume of overflow from levee ponds is much less than that from watershed ponds because of the much greater contribution of runoff to the water budget of watershed ponds (Table 2). Most pond overflow occurs during winter months when precipitation exceeds evaporation. Very little overflow occurs during the dry summer months when evaporation often exceeds precipitation (Fig. 3). Pond overflow volume

TABLE 2. *Estimated pond overflows (cm/yr) from channel catfish ponds.*

| Pond type | Year | Management scenario | | Reference |
		No storage	Storage	
Levee[a]	Dry (1986)	63.9	0.0	Tucker et al. 1996
	Average	104.3	32.6	
	Wet (1979)	153.9	61.1	
Levee[b]	Average		8.0	Boyd et al. 2000
Watershed	Average		402.2	

[a] Levee ponds in the lower Mississippi River valley with pond water level managed with no water storage potential or pond water level managed to maintain a minimum 7.5-cm water storage potential. Overflows were based on a 30-yr (1961–1990) climatological record for Stoneville, Mississippi, and were calculated for the driest year (1986), the 30-year average, and the wettest year (1979).

[b] Levee and watershed ponds in west Alabama with pond water level managed with 10 cm storage potential. Overflows were based on the average of a 65-yr (1924–1989) climatological record (1924–1989) from the Southeast Agricultural Weather Service Center, Auburn University, Alabama.

can be reduced by maintaining storage capacity for rainfall and runoff.

The effect of watershed ponds on the moderation of peak stream flows following storm events is one unheralded positive impact of watershed ponds on local hydrology. When rainfall and runoff do not exceed storage capacity, water is conserved within the watershed and local groundwater supplies may be augmented through seepage. When rainfall and runoff exceed storage capacity, the pond may serve as a temporary stormwater detention basin, moderating peak discharge from the watershed. Watershed runoff is not instantaneous, but is released from the pond at a lower rate and for a longer duration than would occur in the absence of a watershed pond. Thus, moderation of peak flows by ponds following storm events may prevent downstream flooding.

Pond Water Budgets

Water budgets for ponds can be prepared by measuring or modeling the rates of water inflows and outflows (Table 3). The change in water level is simply the difference between water inflows and outflows. Water budgets are influenced mainly by climatological variables, particularly temperature and rainfall. Most pond water budgets, including those discussed below, have been developed to estimate discharge volume because of interest in the possible impacts of effluents on downstream ecosystems. However, the general hydrological equation can easily be rearranged and solved for estimates of water use, rather than overflow. An example of water budgets developed to estimate water use is provided by Pote et al. (1988).

The following general hydrological budget equations can be used to obtain an estimate of

TABLE 3. *Examples of annual water budgets for levee and watershed channel catfish ponds. Water inflows and outflows are in cm.*

Pond type	Water inflows				Water outflows		
	Rainfall	Runoff	Spring	Groundwater	Evaporation	Seepage	Overflow
Levee[a]	60.2	2.3		78.0–283.0	77.0	58.7–260.0	2.3–10.4
Levee[b]	155.5	29.8		108.7	92.5	185.0	11.0
Levee[c]	132.2	–		89.2	122.8	92.5	9.4
Watershed[d]	147.8	238.8	661.2		118.1	242.6	819.2
Watershed[e]	142.3	176.9			112.3	109.5	17.2
Watershed[f]	135.4	410.8			117.2	54.7	402.2

monthly overflow (Boyd et al. 2000):

Watershed ponds

$$WD_f = WD_i + P + RO - (E + S)$$
$$OF = WD_f - PD$$

Levee ponds

$$WD_f = WD_i + P - (E + S)$$
$$OF = WD_f - PD$$

where

WD_i = water depth in pond at beginning of month,

WD_f = water depth that would be in pond at end of month if it was possible to prevent overflow,

P = precipitation falling directly into pond,

RO = runoff from watershed into pond,

E = evaporation from pond surface,

S = net seepage out of pond,

PD = average pond depth (based on elevation of overflow intake structure), and

OF = overflow.

Water budgets have been prepared for levee ponds (Boyd 1982; Pote et al. 1988) and watershed ponds (Shelton and Boyd 1993; Boyd et al. 2000). Shelton and Boyd (1993) developed water budgets for four watershed ponds in East Alabama. Two of the ponds were on pasture watershed and two had mixed pine-hardwood watersheds. The ponds had watershed to pond area ratios of 5.2 to17.8, which are typical of ponds in the Piedmont and Coastal plain regions of the Southeast. In a year (1982–1983) with higher-than-average rainfall (173 cm; normal rainfall is 142 cm), discharge was equal to 2.5 times pond volume. In a year (1983–1984) with near-average rainfall, discharge averaged 1.4 times pond volume. In both years, discharge varied greatly over time because rainfall in the Southeast varies seasonally. Between May and November of both years, precipitation was not adequate to replace losses from evaporation and seepage. Over that period, pond water levels steadily declined and little discharge was measured.

Most discharge occurred in the winter and early spring: 89% of the total annual pond discharge occurred from December through April in the wet year of the study and 72% in the year with normal rainfall.

In another study of Alabama catfish ponds, Boyd et al. (2000) estimated effluent volumes using climatological data, hydrological models, and survey information about water management on catfish farms in west-central Alabama. The 25 farms included in the survey had a total pond water surface area of 2,378 ha (approximately 27% of the state total pond area). Of the total, 1.11% of the pond area was in fry nursery ponds, 9.55% was in fingerling production, and 89.34% was in food-fish production. All ponds on six farms were levee ponds. On five of the farms with levee ponds, wells were the water source; the other farm obtained water from a stream. The remaining 19 farms had watershed ponds or a hybrid between watershed and levee ponds. It is a common practice to construct ponds so that one overflows into another. So, where a combination of watershed and levee ponds were used, the watershed ponds overflowed into levee ponds. The average ratio of watershed area to pond water surface area was 6.32 (range = 0.52–50; SD = 11.7). As about 10 ha of water is required to maintain the water level in a 1-ha pond in west-central Alabama (Yoo and Boyd 1994), only six farms were operated with runoff as the sole source of water. On 13 farms with a combination of water and levee-type ponds, wells had been developed to supplement the water supply.

The following assumptions applied to the water budget prepared by Boyd et al. (2000):

- Levee ponds do not receive significant runoff from their limited watershed areas.
- Average pond depth was 168 cm (survey data).
- Water was added to ponds from wells or streams to maintain the water level 10 cm below the top of the overflow structures during dry weather (survey data).
- Net seepage from ponds was 0.15 cm/d (Yoo and Boyd 1994).

Precipitation data were obtained from the Southeast Agricultural Weather Service for Marion Junction, Alabama, and pond surface evaporation and runoff were calculated from climatological data using standard methods (Yoo and Boyd 1994).

Solution of the two overflow equations above provided the values of OF in Table 4 and Fig. 3. Negligible runoff enters levee ponds, and rain falling into ponds seldom exceeds evaporation plus seepage out of ponds on an annual basis. Thus, levee ponds have very little annual overflow (OF = 8.0 cm). Watershed ponds have large amounts of annual overflow (OF = 402.2 cm) because of runoff from watersheds that occurs from late fall to mid spring—the cooler, wetter part of the year in the Southeast.

Long-term Water Budgets

The water budgets described above consider water use and overflow for ponds that are full initially. The discharge of water during pond draining and the need for water to refill ponds was not considered in those budgets, but must be included in long-term water budgets. Long-term water budgets for culture ponds vary with the phase of production because the interval between pond drainings is different for broodfish ponds (drained every 1 to 5 yr), nursery/fingerling ponds (drained annually), and food-fish ponds (drained every 3 to 10 yr). It must be emphasized that only a small percentage of catfish ponds are drained for fish harvest.

Long-term water budgets also depend on the type of pond. Levee ponds represent over 90% of the ponds used for catfish culture (USDA/APHIS 1997) and are usually harvested without draining for several years.

Final fish harvest occurs only when ponds need to be drained for renovation or adjustments in fish inventory are deemed necessary. At that time, an attempt is made to harvest all fish by repeatedly seining the pond. Only when the farmer is confident that further seining will yield few fish is the pond completely drained. For levee ponds, there is no runoff input, so water levels must be restored after drawdown by water from wells or streams. Therefore, no adjustment of OF is necessary. The amounts of overflow expected annually are provided in Table 4.

In contrast, watershed ponds may be partially or completely drawn down to facilitate fish harvest. Water level drawdown reduces the amount of overflow below the OF values calculated using the general hydrological budget equations because more of the water entering in rainfall or runoff will be retained in ponds to replace the water that was intentionally discharged during drawdown. Thus, OF values from above must be adjusted to account for the average drawdown schedule for nursery/fingerling and food-fish ponds. For nursery/fingerling ponds that are normally drained annually, the adjusted annual overflow

(OF_{adj}) is:

$$OF_{adj} = OF-PD$$

In watershed ponds used for food-fish production, partial drawdown occurs about every 6 yr and complete drawdown occurs about every 15 yr. Thus, OF was averaged over a 15-yr period to adjust for drawdown. In 15 yr, there will be two drawdowns of 50% of the pond volume and one drawdown of 100% of the pond volume (total = 2 pond volumes). Thus, the OF_{adj} is:

$$OF_{adj} = [(OF \times 15) - (2 \times PD)] / 15$$

The drawdown amount was estimated assuming that ponds are completely full when drawdown was initiated. Nursery/fingerling ponds will yield drawdown effluent equal to pond volume each year, and food-fish pond water levels will be drawn down an amount equal to two pond volumes every 15 yr on average (Table 4).

The discharge values in Table 4 can be combined with survey data on pond use to estimate the proportion of total pond discharge attributable to various factors. For example, overflow caused by excess precipitation represents 88.1% of the total effluent from

TABLE 4. *Estimates of adjusted overflow (OF_{adj}), drawdown (DD), and total effluent (TE) discharged annually from channel catfish production ponds in west-central Alabama.*

Type of pond	Water discharged (cm)		
	OF_{adj}	DD	TE
Levee ponds			
Fry and fingerling	8.0	167.0	175.0
Food fish	7.0	22.3	30.3
Watershed ponds			
Fry and fingerling	235.0	167.0	402.0
Food fish	379.7	22.3	402.0

west-central Alabama catfish ponds. Fry and fingerling ponds generate 47.4% of the drawdown effluent although they represent only about 11% of pond area.

Opportunities for Water Conservation in Catfish Ponds

Conserving water and reducing the volume of effluent from ponds are intimately linked because water use and water discharge are simply two sides of the general hydrological equation:

Inflows = outflows ± change in storage

Based on this equation, conservative water use becomes a matter of either reducing outflows or manipulating storage volume. Reducing outflows can be accomplished by minimizing seepage losses, reducing or eliminating water exchange ("flushing"), and reducing drawdown or draining frequency.

The need for pumped water can be reduced by manipulating storage volume so that maximum use is made of any rainfall or runoff entering the ponds. That is, if some storage volume is always maintained, rainfall or runoff is captured rather than allowed to overflow. Increasing storage capacity is simply accomplished by maintaining water level below the top of the drainage structure. When water is intentionally added, ponds are not filled completely. Manipulating pond storage volume not only reduces the need for pumped water, it obviously decreases overflow volume. In fact, manipulating storage volume is probably the most cost-effective practice available for reducing the discharge of nutrients and organic matter from ponds.

Reducing Seepage

Seepage can be a major loss of water, especially when considered over long periods. Some ponds built on pervious sandy soils seep so badly that they require constant addition of pumped waters during non-rainy periods.

Reducing seepage is a site-selection problem because it is difficult and expensive to reduce seepage in existing ponds. In fact, it is common practice to evaluate sites for pond aquaculture primarily based on the ability of soils to retain water when wet.

Reduce or Eliminate Water Exchange

Water exchange, or flushing, is the practice of running large volumes of water into and out of the pond in an effort to dilute the concentration of substances in the pond. Prior to about 1985, pumped water was applied liberally to many catfish ponds as a presumed panacea for water quality and fish health problems. Catfish farmers believed that "flushing" the pond with pumped groundwater would substantially improve environmental conditions and benefit the fish population. However, research (McGee and Boyd 1983) and practical experience have demonstrated that flushing of commercial catfish ponds at rates allowed by typical well yields (less than 5% of total pond volume per day is generally not beneficial.

Excessive use of water was also reduced when detailed studies of the hydrology of the Mississippi River alluvial aquifer conducted in the 1980s revealed that increased demand for groundwater, aggravated by drought, was causing alarming decreases in the water table for that aquifer. Those studies also served as an impetus to develop conservative water use practices in catfish culture. The development of water conservation techniques was perceived to be in the best interest of fish farmers because adequate groundwater supplies are essential for the continued sustainable development of the aquaculture industry in the lower Mississippi River valley. In addition, water conservation techniques also have the major impact on reduction of mass discharge from aquaculture ponds (Chapter 3). Virtually all catfish ponds in the Southeast are now managed as essentially static systems with insignificant water

TABLE 5. *Estimates of average total water use and water discharge by the channel catfish industry in the United States.*

		Volume ($\times 10^8$ m^3)		
Pond type	Area (ha)[a]	Annual drawdown[b]	Groundwater use or overflow volume[c]	Total groundwater use or effluent volume[d]
Food-fish	63,478	1.29	2.07	3.36
Fingerling	9,947	1.22	0.32	1.54
Broodfish	2,272	0.09	0.07	0.16
TOTAL	75,697			5.06

[a] Water surface area as of 1 January 2001 (USDA/NASS 2001a).

[b] Assumes 1.22 m (4 ft) average pond depth. Calculation of drawdown was based on a pond draining frequency of once every 6 yr for food-fish ponds, once every 1 yr for fingerling ponds and once every 3 yr for broodfish ponds.

[c] Groundwater use represents volume added to replace evaporation and seepage losses while maintaining capacity to store rainwater. Assumes an overflow from ponds with water storage of 32.6 cm/yr (Tucker et al. 1996). In ponds with low seepage, groundwater use is approximately equal to overflow volume.

[d] Total groundwater use is the sum of water used to fill ponds initially and water used to replace evaporation and seepage; total effluent volume is the sum of water discharged during pond drawdown and overflow.

exchange except during periods of unusually high precipitation. Natural biological activity and mechanical aeration maintain adequate environmental conditions for culture, and pumped water is now used only to fill ponds and to replace evaporation and seepage losses.

Reducing Drawdown or Draining Frequency

Water needed to fill ponds is a large part of the water budget, so the frequency of pond draining and refilling has a major impact on water use. As a corollary, the frequency of pond drawdown and draining also has a major impact on the volume of water discharged from ponds. The need for pumped water and the volume of water discharged from catfish ponds is directly related to water use practices for the different catfish production phases. Effluent volumes are relatively high for nursery ponds because they are drained between crops each year. Total effluent volume thus consists of the amount discharged at pond draining (an average of about 122 cm) plus any discharge occurring during production (which will depend on climatological conditions and water use practices as described below). Although fingerling ponds are 13.4% of the total area in catfish production, they contribute 46.7% of the total volume discharged from ponds (that is subsequently refilled from groundwater) during drawdown (Table 5). Annual discharge from drawdown is 12,265 m^3/ha for fingerling ponds, 3,961 m^3/ha for broodfish ponds, and 2,032 m^3/ha for food-fish production ponds based on the assumptions of drawdown frequency outlined in Table 5.

As fingerling ponds represent an inordinate share of the water discharged from catfish ponds, considerable scope for water conservation in channel catfish aquaculture may be derived by improving water management in fingerling ponds. However, reducing drawdown frequency is difficult given current fingerling production practices. Research on the modification of fingerling production methods that will permit water reuse or the production of multiple fingerling crops without draining is needed. As the greatest fraction of catfish ponds are dedicated to food-fish production, managing ponds with the multiple-batch cropping system in which fish are harvested without draining for many years of continuous production will minimize the proportion of water discharged during pond draining.

Maintaining Storage Capacity for Rainfall and Runoff

Effluent volume and the need for pumped groundwater can be reduced if some storage capacity is maintained in ponds to capture rainfall. By this water management (called a "drop-fill") scheme, water is not added to a pond until the water level falls to a certain level below the top of the drainage structure. Then, the pond is not refilled completely, but water is added to allow the maintenance of some capacity to capture and store rainfall. By capturing as much rainfall as possible, the future need for pumped water is offset and the loss of rainfall through overflow is minimized.

The impact of managing ponds to capture rainfall on pond discharge can be modeled using a hydrological equation with appropriate climatological data. Such a model was developed by Pote et al. (1988) and further evaluated by Tucker et al. (1996) and Hargreaves et al. (2001) for a hypothetical levee pond in the lower Mississippi River valley.

Water discharged from levee ponds as a result of overflow only was calculated using the following hydrological budget equation:

$$OF_d = WD_{d-1} - WD_d - P_d - 0.8E_d - S + GW_d$$

where

OF_d = overflow (cm) on day d,

WD_{d-1} = pond water level (cm) at the end of day d-1,

WD_d = pond water level (cm) on day d,

P_d = precipitation (cm) on day d,

E_d = pan evaporation (cm) on day d,

S = daily seepage loss (cm), and

GW_d = ground water pumped into pond (cm) on day d.

Daily observations for P and E from 1961–1990 obtained from the National Weather Service Cooperative Observation System for Stoneville, Mississippi, were used as inputs to the model. Pan evaporation was multiplied by 0.8 to estimate pond evaporation and a value of 0.381 mm/d was assumed for S. The value for S was obtained from seepage losses measured in northwest Mississippi rice fields constructed on the Alligator and Sharkey series of thermic Chromic Epiaquert soils (Pringle and Pennington 1991), which are the soil types most commonly used for catfish pond construction in that region.

Overflow losses were determined for two pond water management scenarios. One scenario assumed that ground water was pumped into the pond at the end of each day to replace E and S. In other words, no storage capacity was maintained in the pond and any P in excess of E + S was lost as overflow through the drain. This scenario estimated the maximum pond overflow that could occur under a given set of climatic conditions. The other scenario is a management option designed to reduce the need for pumped water and reduce overflow volume by allowing for storage of much of the annual rainfall. In that scenario, pond water level was allowed to fluctuate with climatic conditions until pond water levels dropped to 15 cm below the overflow structure due to E + S in excess of P. At that point, ground water was added to raise the water level to 7.6 cm below the top of the drain, representing the potential storage capacity for subsequent rainfall. A water storage potential of 7.6 cm was chosen because climatological records showed that there is a 90% chance that total precipitation will be less than 7.6 cm for any week in northwest Mississippi. Overflow thus occurred only during unusual precipitation events. Overflow losses were estimated using weather data for the 30-yr (1961–1990) average, the year of record with the most precipitation (1979), and the year of record with the least precipitation (1986).

The results of the modeling exercise show that seasonal variation in pond overflow volume is related directly to annual climatic conditions because precipitation and evaporation are major components of the hydrological equation. On average, modeled pond overflow volumes were highest in the

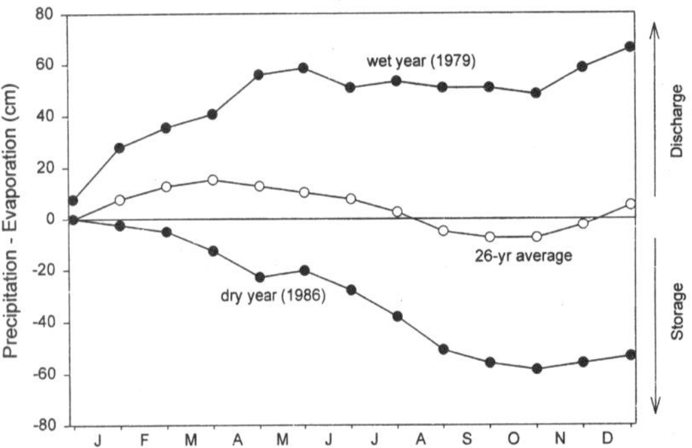

FIGURE 4. *Cumulative monthly precipitation minus evaporation for levee ponds in the lower Mississippi River alluvial valley. Data for 26-yr average from 1962–1987 (Pote et al. 1988).*

winter and spring when rainfall was highest and pond evaporation rates were lowest, except during the driest year on record (1986) when rainfall was evenly distributed among seasons (Fig. 4). The model showed that managing water levels in ponds to maintain a minimum surplus storage potential of 7.6 cm of water greatly reduced overflow volumes compared to ponds managed without surplus storage. In an average year, managing pond water levels to maintain water storage potential reduced the predicted annual overflow to about 30% of that from ponds managed without surplus storage (Table 2). The reduction in pond overflow volume was greatest in the summer: the

predicted overflow in the average summer was only about 8% of that from ponds managed without storage, and the model indicated that there would have been no overflow from ponds managed to maintain storage potential in exceptionally dry summers.

A wider range of pond storage capacities and fill amounts than in the model evaluated by Tucker et al. (1996) was considered by Hargreaves et al. (2001). Average estimated annual water usage was 66 cm in ponds with 5 cm of water storage capacity, 38.1 cm with 15.2 cm of water storage capacity, and 15–18 cm with 46 cm of storage capacity. Average estimated annual effluent discharge was 74 cm from ponds with 5 cm of water storage capacity, 46 cm from ponds with 15.2 cm of water storage capacity, and 20–25 cm from ponds with 46 cm of storage capacity. Based on the model results, maintaining a water storage capacity between 20 to 30 cm can reduce groundwater volume pumped and effluent volume discharged by 40 to 60%.

Ponds discharge more water than is pumped from the ground because the seasons of maximum water demand and greatest rainfall do not coincide. Most water is pumped during summer, when evaporation rates are high, but most water is discharged during winter, when rainfall tends to be more frequent and intense. During winter, ponds are nearly full or overflowing in most years.

Seepage had a strong effect on both the volumes of effluent discharged and groundwater pumped. Ponds with a relatively high seepage rate were predicted to have lower rates of effluent release and higher rates of groundwater use compared to ponds with lower seepage. The high seepage rate diverted water from discharge as effluent to discharge as seepage, but increased the frequency of refills and thus the requirement for groundwater. For ponds with low seepage rates, predictions were opposite: discharge increased and groundwater use decreased.

The results of the model presented here represent average groundwater use and effluent discharged for weather conditions over the 28-yr period. Of course, annual rainfall patterns can range from very dry years to very wet years. During dry years, more water than average will be pumped, and during wet years, very little water may be pumped. Thus, it may be more important to consider extreme years represented in the climate record, rather than the average year.

The reduction in groundwater use has a significant water conservation effect that can be translated into cost savings associated with reduced pumping (Fig. 4). The greater the storage capacity maintained in the pond, the greater the cost savings associated with pumping, the greater the conservation of groundwater resources, and the lower the water volume discharged.

Selection and management of the capacity of a pond to store rainfall will be constrained within several limits associated with practical considerations. First, ponds must be maintained at a pond depth that will allow seining, usually 1.22 m. Maintaining a 1.22-m operational depth with a drop-fill water management scheme is made more difficult by sedimentation that occurs as a result of erosion of pond levees through time. Sedimentation in older ponds can be substantial, severely reducing operational depth, suggesting that the water storage capacity selected for old ponds should be no more than 15 to 20 cm. Increasing the water storage capacity increases the area of exposed, bare-soil levee that is subject to erosion. However, during winter, when strong winds occur during the passage of cold fronts, ponds are usually full to the level of vegetated levee.

Selection of a drop-fill scheme will also be constrained by operation of wells. Most wells are placed to supply water to four ponds. A typical well supplying 3.78 m³/min can add about 16 cm of water to a 6.9-ha pond in 2 d.

Assuming that the well is sequentially operated for each of the four ponds serviced by the well, then 8 d will be required to add 16 cm of water to the four ponds. Thus, adding water should be limited to 16 cm or less. Ponds with lower capacity wells should limit the fill during a pumping event to 10 cm.

Linked-pond Modules

Cathcart et al. (1999) recently described the results of a hydrological model simulation that evaluated the effects of an innovative pond management strategy on effluent volume and groundwater use. The strategy employs production ponds that are linked to one another to allow water flow from one pond to the other, rather than being discharged, in the event of overflow. In several model scenarios, the depth of production ponds was increased by 30 to 90 cm to increase water storage capacity. These ponds were designated production and storage (P+S) ponds. In one configuration, one P+S pond was linked to one additional production pond. In another configuration, one P+S pond was linked to three additional production ponds. When the production ponds overflow, water is stored in the P+S pond rather than discharged. The increased depth of the P+S pond allows the storage of a greater proportion of rainfall and reduces the need for pumped groundwater. In addition, water can be pumped from the P+S pond to provide water to production ponds.

Simulated water level fluctuation in P+S ponds was much greater than that in production ponds, but pond depth was nearly always between 1.0 and 1.5 m, indicating that normal seine harvesting would not be compromised in P+S ponds. Groundwater use in a one production pond linked to one P+S pond module was reduced by 47.0% and effluent volume was reduced by 82.7% compared to ponds that were filled when the water level dropped to 7.5 cm below the top of the drainage structure (Cathcart et al. 1999). Comparable reductions in groundwater use and effluent volume were obtained in the configuration with one P+S pond linked to three production ponds. Seepage in these simulations was 1 mm/d. A study to validate model results is currently being conducted at the National Warmwater Aquaculture Center in Stoneville, Mississippi. Preliminary data indicate that linked pond configurations conform to model expectations. In 2000, a particularly dry year, effluent volume was reduced by 65 to 75% and groundwater consumption was reduced by 25 to 38% compared to conventionally managed ponds (SRAC 2000).

*Best Management Practices to
Reduce Effluent Volume*

The practices described above are most effective when combined into an overall water management plan for the farm. Boyd and Queiroz (2001) suggested the following practices to conserve water and reduce effluent volume.

For levee and watershed ponds:

- Harvest fish without draining ponds unless necessary to repair levees or adjust fish inventory.
- Maintain at least 20 cm of storage capacity in watershed ponds during the summer and fall. Paint the upper 20 cm of the overflow pipe with a bright color to guide decisions to initiate groundwater addition.
- Flushing ponds with water does not improve water quality and should be avoided.

For watershed ponds:

- Construct new ponds with a watershed to pond area ratio of ≤10.
- Construct terraces in watersheds to divert excess runoff around ponds.
- Maintain good vegetative cover throughout the watershed. Replace short forage grasses with tall forage grasses or plant trees.

Trout Culture in Raceways

Production of rainbow trout in raceways is the second largest aquaculture enterprise by value in the United States. In 2000, the value of production was $75.8 million from sales of fish (95%) and eggs (5%) (USDA/NASS 2001a). Production of food fish was 26,800 mt in 2000, valued at $63.7 million (USDA/NASS 2001a). Commercial trout production was reported to occur in 20 states in 2000, although production is concentrated in Idaho (53% of total by value), North Carolina, California, and Pennsylvania. The major trout-producing area in Idaho is located in the Hagerman Valley along a 40-km length of the Snake River. The trout industry in Idaho is dominated by three large producers. One company is responsible for 50% of the trout produced in Idaho (Brannon and Klontz 1989). In 2000, there were 447 operations selling trout for food and an additional 256 operations that distributed trout for stocking natural waters for restoration, conservation, or recreation (USDA/NASS 2001a).

Facilities and Water Supplies

Rainbow trout are cultured in linear or circular tanks. Linear tanks are commonly called raceways. Construction specifications for raceway tanks include dimensional ratios of length, width and depth. Typically, raceways are constructed of reinforced concrete with a length-to-width-to-depth ratio of approximately 33:3:1. Many raceway tanks are 32.8-m long, 3.28-m wide and 1-m deep (100 ft x 10 ft x 3 ft). Raceways can be linked in series of four to 10 sections. The number of sections is dictated by source water quality, topography, and the extent to which water can be reaerated between raceway sections.

Water sources for raceways include surface

water and ground water. The use of surface water is more prevalent on trout farms in the Appalachian Mountains of the eastern United States. Groundwater is used almost exclusively on trout farms in Idaho. The groundwater used in Idaho emerges from springs in the canyon walls of the Hagerman Valley that provide a constant flow of 14.8 C water. Some Idaho trout farms are supplied with as much as 8.5 m^3/sec of flow (Brannon and Klontz 1989).

Raceway Water Budgets

Raceways are configured as open systems, with water flowing in one end of the production facility and out the other. If the water source is an existing stream or water from a natural spring, trout aquaculture is not a net consumer of water. Water is simply diverted for temporary use in aquaculture and there is no net loss or gain of water. Although there is no consumption of water in trout raceways, water quality will be impacted as it passes through production facilities.

In comparison to ponds, water budgets for raceways are simple. The contribution of rainfall and water losses from evaporation are negligible. Raceways do not have watersheds that contribute water directly from overland flow. Raceways are usually constructed of impermeable concrete, so seepage does not occur.

Raceways are designed to operate in flow conditions that approximate plug flow, and flow rates are generally not adjusted as fish grow. Water flow rates range from 57 to 142 L/sec (2–5 cfs) per raceway, with greater flow rates used for raceways with a higher number of raceway sections. The culture of rainbow trout from stocking as a 5-cm fish to harvest as a 30.5-cm fish requires approximately 300 d. Raceway carrying capacity ranges from 48 to 160 kg/m^3 (3–10 lb/ft^3) depending on fish size, elevation, and number of sections in a raceway unit (Willoughby 1968; Piper 1975). However, loading (mass/flow) is a better measure of

raceway carrying capacity than density (mass/volume).

Water requirements for single-pass raceways range from 250 to 350 m^3/kg production (1.5–2.0 kg/L per min), but can be reduced to 55 m^3/kg production by configuring raceways for serial water reuse (Table 1). The total water requirement for the production of trout as food fish can be estimated by multiplying the total production of food fish by the ranges for water use per unit production. Estimates ranging from 1.43 to 9.10 x 10^9 m^3 are conservative because water use for facilities producing trout for distribution are not considered.

Opportunities for Water Conservation in Trout Raceways

Given the inherent characteristics of the raceway production system, opportunities for cost-effective water conservation in raceways are extremely limited. There are no terms in the standard hydrological mass balance equation that can be manipulated to the same extent as in pond aquaculture to achieve water conservation. Water budgets in raceways are dominated by regulated water inflows.

Considerable research has been dedicated towards determination of maximum loading rates and fish densities in raceways. The most efficient use of water in raceways will be achieved when production units are operated near carrying capacity. Carrying capacity in raceways is more easily defined than in ponds. Carrying capacity in raceways can be defined exclusively as a function of trout metabolic characteristics, which exert an overwhelming influence on changes in water quality. In ponds, assessment of carrying capacity is complicated by the effects of phytoplankton metabolism, which exerts a dominating influence on pond water quality.

Water recirculation has been attempted to address water quality changes that occur upon passage through fish production raceways, not as a means of water conservation. However, effluent from raceways is extremely dilute with respect to nutrient and metabolite concentration. Low nutrient concentration means that an extremely large biofilter surface area would be required to fully treat effluent because microbial process rates will be substrate-limited.

Water conservation through recirculation or other water treatment means may be more justified from the perspective of reducing solid and nutrient mass discharge, a topic that is discussed elsewhere in this volume, rather than on the basis of water conservation. Sedimentation of settleable solids is the only practical water treatment method suitable for the treatment of large volumes of dilute effluent.

Summary

Globally, competition for water resources is becoming increasingly intense. Demands for water for potable and household use, industrial uses, and for agriculture are all increasing, yet the availability of water resources has not kept pace. Irrigated agriculture is assuming increasing importance in food production as approximately 40% of global food is derived from the 17% of the world's cropland that is irrigated (Postel 1989).

In the context of irrigated agriculture in the United States, catfish aquaculture is a greater consumer of water per unit area than irrigated peanuts, cotton, corn, soybeans, and wheat, but a lesser or comparable consumer of water as rice and alfalfa (Table 6). If ponds are operated to maintain water storage capacity, unit area water consumption by irrigated row crops and catfish aquaculture is similar. However, the value of various crops per unit volume water used is much greater for catfish than other crops (Table 6). If this value is taken

TABLE 6. *Annual crop yield, water use, gross returns and value of water use for selected crops in the United States.*

Crop	Yield (kg/ha)[a]	Water use (cm)[b]	Gross returns ($/ha)[c]	Value of water ($/m^3)[d]
Channel catfish	6,163	50	10,600	2.12
		75		1.41
		100		1.06
Peanuts	3,715	34	2,326	0.69
Cotton	870	46	1,183	0.26
Corn	10,229	37	781	0.21
Soybeans	2,757	24	499	0.20
Rice	6,612	73	1,296	0.18
Alfalfa	11,430	73	1,110	0.15
Wheat	5,043	43	491	0.12

[a] Yield for channel catfish from Heikes and Killian (1996); yield for irrigated field crops from USDA/NASS (1998).

[b] Water use for channel catfish based on three scenarios that bracket the range of expected groundwater use for levee ponds under various water management approaches; groundwater use for irrigated field crops from USDA/NASS (1998).

[c] Gross returns for channel catfish based on $1.72/kg; gross returns for field crops based on 1998 prices (USDA/NASS 2001b).

[d] Value of water is calculated by dividing gross returns by water use.

as an index of the economic efficiency of water use, then catfish aquaculture is a much more efficient user of water than irrigated row crops.

Water is a fundamental requirement for aquaculture. However, unlike other production inputs, such as feed, fingerlings and labor, the value of water is often underrated. Water is often considered a "free" and inexhaustible resource that is available simply for the cost of pumping. A true cost accounting of water use in aquaculture would consider the economic benefits, relative to costs, of alternative uses of water. To be comprehensive, such an accounting would consider effects (externalities) beyond farm boundaries.

Until recently, aquaculture has been a minor component of local economies. However, the two most successful aquaculture industries in the United States are notable for the concentration of development in specific locations, i.e., the lower Mississippi River valley in the case of catfish culture and Hagerman Valley, Idaho, in the case of rainbow trout culture. The continued development of these aquaculture industries relative to the availability of water supplies in those geographic areas is an open question. Fish producers and aquaculture research scientists have begun to recognize that the availability and utilization of water resources may impose limitations on the conduct of existing and future operations. Concerted efforts are now being directed toward improving the efficiency of water use in aquaculture to ensure continued development of sustainable aquaculture industries.

Acknowledgments

The authors acknowledge the assis-tance of G. Fornshell, who provided information on water use in trout raceways. Approved for publication as Article No. J-9914 of the Mississippi Agricultural and Forestry Experiment Station, Mississippi State University.

Literature Cited

Boyd, C. E. 1982. Hydrology of small experimental fish ponds at Auburn, Alabama. Transactions of the American Fisheries Society 111:638–644.

Boyd, C. E. 1985a. Pond evaporation. Transactions of the American Fisheries Society 114:299–303.

Boyd, C. E. 1985b. Hydrology and pond construction. Pages 107–133 *In* C. S. Tucker, editor. Channel catfish culture. Elsevier, Amsterdam, the Netherlands.

Boyd, C. E. 1986. Influence of evaporation excess on water requirements for fish farming. Pages 62–64 *In* Proceedings of the Conference on Climate and Water Management—A Critical Era. American Meteorological Society, Boston, Massachusetts, USA.

Boyd, C. E. and A. Gross. 2000. Water use and conservation for inland aquaculture ponds. Fisheries Management and Ecology 7:55–63.

Boyd, C. E. and J. F. Queiroz. 2001. Feasibility of retention structures, settling basins, and best management practices in effluent regulation for Alabama channel catfish farming. Reviews in Fisheries Science 9:43–67.

Boyd, C. E. and C. S. Tucker. 1998. Pond aquaculture water quality management. Kluwer Academic Publishers, Boston, Massachusetts, USA.

Boyd, C. E., J. Queiroz, J. Lee, M. Rowan, G. N. Whitis, and A. Gross. 2000. Environmental assessment of channel catfish *Ictalurus punctatus* farming in Alabama. Journal of the World Aquaculture Society 31:511–544.

Brannon, E. and G. Klontz. 1989. The Idaho aquaculture industry. The Northwest Environmental Journal 5:23–35.

Cathcart, T. P., J. W. Pote, and D. W. Rutherford. 1999. Reduction of effluent discharge and groundwater use in catfish ponds. Aquacultural Engineering 20:163–174.

EarthInfo, Inc. 1989. National Climate Data Center—Summary of the Day (CD-ROM). EarthInfo, Inc., Boulder, Colorado, USA.

Hargreaves, J. A., D. W. Rutherford, T. P. Cathcart, and C. S. Tucker. 2001. Drop-fill water management schemes for catfish ponds. The Catfish Journal 15(10):8–9.

Heikes, D. L and H. S. Killian. 1996. Catfish yield verification trials: final results. University of Arkansas at Pine Bluff, Pine Bluff, Arkansas, USA.

McGee, M. V. and C. E. Boyd. 1983. Evaluation of the influence of water exchange in channel catfish ponds. Transactions of the American Fisheries Society 112:557–560.

Parsons, D. A. 1949. The hydrology of a small area near Auburn, Alabama. Publication SCS-TP-85. U.S. Department of Agriculture/U.S. Soil Conservation Service, Washington, D.C., USA.

Phillips, M. J., M. C. M. Beveridge, and R. M. Clark. 1991. Impact of aquaculture on water resources. Pages 568–606 In D. E. Brune and J. R. Tomasso, editors. Aquaculture and water quality. World Aquaculture Society, Baton Rouge, Louisiana, USA.

Piper, R. G. 1975. A review of carrying capacity calculations for fish hatchery rearing units. Information leaflet no. 1. U.S. Fish and Wildlife Service—Fish Culture Development Center, Bozeman, Montana, USA.

Postel, S. 1989. Water for agriculture: facing the limits. Worldwatch paper 93. Worldwatch Institute, Washington, D.C., USA.

Pote, J. W., C. L. Wax, and C. S. Tucker. 1988. Water use in catfish production: sources, uses, and conservation. Special bulletin 88-3. Mississippi Agricultural and Forestry Experiment Station, Mississippi State, Mississippi, USA.

Pringle, H. C. and D. A. Pennington. 1991. Rice evapotranspiration rates and deep percolation losses—1990 preliminary results. Information sheet 1329. Mississippi Agricultural and Forestry Experiment Station, Mississippi State, Mississippi, USA.

Shelton, J. L., Jr. and C. E. Boyd. 1993. Water budgets for aquaculture ponds supplied by runoff with reference to effluent volume. Journal of Applied Aquaculture 2(1):1–27.

SRAC (Southern Regional Aquaculture Center). 2000. Thirteenth annual progress report. Southern Regional Aquaculture Center, Stoneville, Mississippi, USA.

Tucker, C. S. 1996. The ecology of channel catfish culture ponds in northwest Mississippi. Reviews in Fisheries Science 4:1–55.

Tucker, C. S. 2000. Channel catfish culture. Pages 153–170 In R. R. Stickney, editor. The Encyclopedia of Aquaculture. John Wiley and Sons, New York, New York, USA.

Tucker, C. S., S. K. Kingsbury, J. W. Pote, and C. L. Wax. 1996. Effects of water management practices on discharge of nutrients and organic matter from channel catfish (*Ictalurus punctatus*) ponds. Aquaculture 147:57–69.

USDA/APHIS (United States Department of Agriculture/Animal and Plant Health Inspection Service). 1997. Catfish NAHMS '97, part II: Reference of 1996 U.S. catfish management practices.

Centers for Epidemiology and Animal Health, USDA/APHIS, Fort Collins, Colorado, USA.

USDA/NASS (United States Department of Agriculture/National Agricultural Statistics Service). 1998. 1998 farm and ranch irrigation survey—Census of agriculture. United States Department of Agriculture, National Agricultural Statistics Service, Washington, D.C., USA.

USDA/NASS (United States Department of Agriculture/National Agricultural Statistics Service). 2000. Census of Aquaculture (1998). United States Department of Agriculture, National Agricultural Statistics Service, Washington, D.C., USA.

USDA/NASS (United States Department of Agriculture/National Agricultural Statistics Service). 2001a. Catfish and trout production. United States Department of Agriculture, National Agricultural Statistics Service, Washington, D.C., USA.

USDA/NASS (United States Department of Agriculture/National Agricultural Statistics Service). 2001b. Crop values—2000 summary. United States Department of Agriculture, National Agricultural Statistics Service, Washington, D.C., USA.

USDA/SCS (United States Department of Agriculture/Soil Conservation Service). 1982. Ponds—Planning, design and construction. Agriculture handbook no. 590. U.S. Department of Agriculture, Soil Conservation Service, Washington, D.C., USA.

Willoughby, H. 1968. A method for calculating carrying capacities of hatchery troughs and ponds. Progressive Fish-Culturist 30:173–174.

Yoo, K. H. and C. E. Boyd. 1994. Hydrology and water supply for aquaculture. Chapman and Hall, New York, New York, USA.

Characterization and Management of Effluents from Warmwater Aquaculture Ponds [1]

CRAIG S. TUCKER

Thad Cochran National Warmwater Aquaculture Center, Mississippi State University, P. O. Box 197, Stoneville, Mississippi 38776 USA

CLAUDE E. BOYD

Department of Fisheries and Allied Aquacultures, Auburn University Auburn, Alabama 36849-5419 USA

JOHN A. HARGREAVES

Department of Wildlife and Fisheries, Mississippi State University, Box 9690, Mississippi State, Mississippi 39762 USA

ABSTRACT

This chapter reviews the characteristics of effluents from warmwater aquaculture ponds, the potential environmental impacts of pond effluents, and management practices to reduce discharge from ponds. The discussion is based on studies of channel catfish *Ictalurus punctatus* ponds in Mississippi and Alabama, which are the most thoroughly studied aquaculture systems in the United States. The two major sources of effluent from catfish ponds are overflow when rainfall exceeds pond storage capacity and water discharged intentionally when ponds are drained. Nutrient and organic matter concentrations in overflow from catfish ponds never approach levels calculated from the quantities of waste produced by fish because natural processes remove much of the waste from the water before it is discharged. In general, pond overflow is comparable in quality to streams draining mixed cropland-forest watersheds in the southeastern United States. Concentrations of solids, nutrients, and organic matter in effluents can, however, be relatively high for brief periods when ponds are drained because poorly consolidated sediment near the drain pipe is discharged along with the pond water. Total mass discharge (the product of concentration and volume) from ponds varies with phase of fish production and pond hydrology. Fingerling ponds are drained each year and discharge a disproportionate share of total nutrients, solids, and organic matter compared to food-fish ponds, which represent a much larger total pond area but are drained less frequently. Watershed ponds have greater discharge volumes than levee ponds because much of the effluent volume is runoff from the watershed above the pond. The unique nature of pond aquaculture poses challenges for effluent management. First, there is scant evidence that pond effluents cause negative environmental impacts, so there is little incentive for farmers to change production practices or implement costly treatment practices. Second, the nature and timing of discharge make treatment a difficult engineering problem. Discharge is infrequent, which makes monitoring difficult, and discharge—when it occurs—may be large in volume but dilute in concentration, which makes treatment expensive. Accordingly, treating water to improve quality after it is discharged from ponds does not appear to be technically or economically feasible on most farms. The best approach to reducing mass discharge of nutrients, solids, and organic matter is to implement a system of best management practices, whose centerpiece is maintaining pond water storage capacity so that rainfall is captured rather than allowed to overflow. Maintaining storage capacity and reusing water for multiple fish crops can decrease nitrogen and phosphorus discharge from food-fish ponds by over 60% compared to ponds drained annually and not managed to capture rainfall.

---

[1] This article is approved for publication as Article BC-9902 of the Mississippi Agricultural and Experiment Station, Mississippi State University.

Pond aquaculture can be a profitable way to raise fish because nature provides many of the resources needed to grow a crop. When pond aquaculture is conducted properly, the aquaculturist manages the pond within limits imposed by the rates of natural processes, and high yields of food are possible with little technological intervention. The same processes that make it possible to grow fish in ponds with low management inputs also make it possible to operate ponds with minimal impact on the environment.

In this chapter, we review the characteristics of effluents from aquaculture ponds, the potential environmental impacts of pond effluents, and approaches to managing ponds to reduce those impacts. We base our discussion on studies of channel catfish *Ictalurus punctatus* ponds in Alabama and Mississippi. Focus on that segment of commercial aquaculture is justified because catfish farming dominates domestic aquaculture production, and catfish ponds are the most thoroughly studied aquaculture systems in the United States. Furthermore, the ecological processes that affect effluent volume and quality are the same in all warmwater aquaculture ponds, whether they are used to grow baitfish in Arkansas or hybrid striped bass in North Carolina, so the principles derived from studies of catfish ponds should apply to all ponds used in large-scale warmwater aquaculture. With some modification to suit specific cultural practices, the same approaches to managing effluents from ponds should be effective across species.

Pond Effluents, Production Facilities, and Management Practices

Characteristics of the catfish industry and descriptions of facilities, production practices, and water use are presented in Chapter 2. This information will not be repeated in detail here, but the following points are emphasized because they affect pond effluent characteristics.

Sources of Pond Effluent

Water may be discharged from catfish culture ponds for three reasons. First, water is unintentionally discharged when rainfall exceeds the water storage capacity of the pond. Rainfall is, of course, beyond the farmer's control, but overflow can be reduced by manipulating pond storage volume. Second, discharge occurs when water is intentionally pumped into the pond for water exchange or "flushing." Water exchange may be practiced in an attempt to improve pond water quality or as a desperation measure when a fish disease does not respond to normal therapies. Finally, water is discharged intentionally when ponds are drained.

Types of Ponds

Three hydrological types of ponds are used in catfish framing. Levee ponds (also called embankment ponds) are built by scraping soil from the area that will be the pond bottom and using that soil to form levees or embankments. Watersheds of levee ponds are small, consisting only of the pond area and the inside levee slopes, so there must be a source of pumped water to fill ponds and maintain water levels during droughts. Water can be pumped from underground aquifers or surface waters. Levee ponds are common in regions with flat topography and can be built in large contiguous tracts, which aids in pond management.

Watershed ponds are built in hilly areas by damming a watercourse. The major source of water is runoff from the drainage basin above the dam, although a source of pumped water is desirable to help offset evaporation and seepage during extreme drought conditions. It may be possible to build a closely spaced series of several watershed ponds in the same catchment basin, with one pond discharging into the next pond downstream, but large groups of watershed ponds generally cannot be built in close proximity to one another.

The third pond type is a hybrid between

levee and watershed ponds. These ponds may have two or three sides consisting of levees (actually low dams) across a relatively small, shallow drainage basin. A significant amount of water may be obtained from runoff, but because the catchment area above the pond is relatively small, a source of pumped water also must be available. Hybrid watershed-levee ponds are common in regions with gently rolling topography that is not ideally suited for levee ponds or watershed ponds.

The distinction between types of ponds is important because pond hydrology affects the quantity and, to a lesser extent, the quality of effluent. Discharge volumes for levee ponds and watershed ponds are quantified in Chapter 2: in general, levee ponds have much less overflow than watershed ponds, with the overflow volume from hybrid ponds being intermediate. Also, the quality of effluents from watershed ponds may be affected (either positively or negatively) by upstream water quality, which is, in turn, affected by land-use practices in the catchment area. This can be especially important during periods of high runoff.

Phase of Production

Over the long term, effluent volume depends on whether ponds are used as brood ponds, fingerling nursery ponds, or for food-fish production because the interval between pond drainings varies with use. Ponds used to hold brood fish are drained every 1 to 5 years; fry nursery ponds are drained annually; and food-fish production ponds are drained every 1 to 20 years. However, given equal pond size and (in the case of watershed ponds) equal catchment areas, the volume of water discharged as overflow and during each pond draining will be the same, regardless of use. The effect of pond draining frequency on discharge volume is quantified in Chapter 2.

The Fate of Nutrients and Organic Matter Added to Ponds

Generation of waste nitrogen and phosphorus by fish can be estimated by calculating quantities of nutrients added to the pond in feed and subtracting the amount recovered in fish that are harvested. For example, if annual fish production is 5,000 kg/ha from 10,000 kg of feed, approximately 400 kg of waste nitrogen and 80 kg of phosphorus are generated (Tucker 1991). At a fixed feed conversion efficiency, these quantities of waste will be produced by any fish in any culture system. For fish grown in net pens or flow-through raceways, those wastes would be released into the environment.

The amount of organic matter added to ponds is more difficult to calculate than for nitrogen and phosphorus. Carbon in the feed consumed by fish may be retained by fish, respired as carbon dioxide, or lost as organic matter in fish waste products. Accordingly, a simple accounting of carbon in feed and fish cannot be used to estimate organic loading to the pond. Empirical values for organic matter generation in fecal solids range from 0.3 to 0.7 kg/kg of feed (Henderson and Bromage 1988). Using a median value of 0.5 kg of fecal solids per kg of feed provides an estimate of 5,000 kg of waste organic matter when 10,000 kg of feed is consumed.

Further complications arise because organic matter added directly to the water in fish fecal matter greatly underestimates actual organic matter production in ponds. Inorganic nutrients in fish wastes stimulate phytoplankton growth that produces two to four times more organic matter through photosynthesis than added directly in fish waste (Boyd 1985). Thus autotrophic production may add at least an additional 10,000 kg of organic matter during culture of 5,000 kg of fish in a 1-ha pond, resulting in total organic matter production exceeding 15,000 kg/ha per year.

Most catfish ponds are managed with little or no water exchange, so wastes are not continuously released from the culture system. Assuming no loss of substances from the pond, nutrient and organic matter concentrations should increase to extraordinary levels over the production period. For example, assuming that 400 kg/ha of nitrogen, 80 kg/ha of phosphorus, and 15,000 kg/ha of organic matter are generated annually in a catfish pond that is 1.5 m deep, total nitrogen concentrations should reach approximately 25 mg/L, total phosphorus concentrations should be about 5 mg/L, and organic matter concentrations should exceed 900 mg/L at the end of 1 year of production. However, even after several years of production, total nitrogen concentrations seldom exceed 10 mg/L in commercial ponds, and they average about 6 mg/L, which is roughly 25% of that expected from simple mass balance for a single year of production. Similarly, actual total phosphorus concentrations average less than 0.5 mg/L. This is less than 10% of what is expected if no phosphorus was lost over 1 year. Organic matter concentrations average about 75 mg/L, or less than 10% of the expected value.

Nutrients and organic matter concentrations in catfish ponds never attain the levels predicted by simple mass balance because natural biological, chemical, and physical processes remove much of the wastes from the water. Total nitrogen concentrations are lower than expected because combined nitrogen is constantly lost from the water column when nitrogen-containing particulate organic matter settles and decomposes on the pond bottom. Nitrogen is also lost from the water as a gas through denitrification and ammonia volatilization. Total phosphorus concentrations in the water column remain at low levels because phosphorus is continuously lost to pond bottom soils as particulate organic phosphorus and precipitates of calcium phosphates. Organic matter does not accumulate at the predicted rate because it is rapidly and continuously oxidized in microbial decomposition processes. These processes are well known in limnology and waste treatment engineering, and are not unique to catfish ponds.

Loss processes for nutrients and organic matter have important implications for fish culture in ponds. In the absence of removal processes, concentrations of certain waste products, such as ammonia, would rapidly accumulate to levels that could kill fish. Accumulation of other substances would have a less direct, but important, impact on fish. For example, if organic matter accumulated to high levels, the oxygen demand expressed by the respiration and decomposition of that material would make it difficult to maintain adequate dissolved oxygen concentrations, even with abundant aeration. In fact, a central tenet of pond aquaculture is that production intensity must be maintained within the limits imposed by the capacity of natural processes to remove or assimilate wastes. These same processes are also important in reducing the quantities of waste discharged from ponds.

Quality of Effluents from Catfish Ponds

Four studies have been conducted to describe the general characteristics of overflow from catfish ponds. Two studies (summarized below) were conducted to describe long-term changes in the quality of effluents from typical commercial catfish ponds in northwest Mississippi (Tucker et al. 1996) and west-central Alabama (Schwartz and Boyd 1994b). In both studies, samples were collected from inside the pond at sites adjacent to the overflow device. These data therefore represent the quality of potential effluents that would result from overflow. A third data set (Boyd et al. 2000) was derived from samples collected from 55 pond outfalls in west-central Alabama. These samples depict the character of actual effluents, although samples were collected over a relatively short time-frame in a single year. The mean concentrations and ranges of variables from that study did not differ from those of the surface waters reported in the study of Schwartz and Boyd (1994b), so only the larger data set from Schwartz and Boyd will be summarized here. The fourth data set (Shireman and Cichra 1994) was collected from six small experimental ponds (0.06 ha) and eight commercial ponds (0.1 to 0.4 ha) in northwest Florida. Fish stocking densities and feed inputs were lower, and the ponds much smaller, than those used in modern commercial catfish production, so those results will not be summarized here.

Two studies have been conducted to characterize specific components of catfish pond water. Boyd and Tanner (1998) quantified coliform bacteria in Alabama catfish pond water and compared the coliform load in catfish pond water to that in sportfish ponds. Boyd and Gross (1999) then characterized the biochemical oxygen demand of catfish pond waters in Alabama.

The data sets described above are useful in characterizing the quality of effluents resulting from overflow, but do not necessarily reflect the quality of water released when ponds are drained. For example, some watershed ponds must be partially drained to facilitate fish harvest. Fish are concentrated into a small water volume during drawdown, and the activities of the fish and seining crew resuspend pond sediments. Water discharged during harvest contains solids and other substances derived from the disturbed sediments, and is therefore unlike typical pond water. The quality of water discharged when ponds are drained is described in the next section.

Quality of Overflow from Levee Ponds in Northwest Mississippi

Tucker et al. (1996) sampled 20 levee ponds in northwest Mississippi in summer, autumn, winter, and spring over 2 years. On each sampling date, water samples were collected from the surface 30 cm and the bottom 30 cm of each pond at a site adjacent to the discharge pipe. Samples were collected at two depths because waters may be discharged from ponds from either the surface or the bottom of the pond depending on the structure of the drain device. However, samples collected from near the surface and bottom of each pond on a given date did not differ for any variable measured, indicating that water columns were well-mixed when samples were collected. Analytical results for surface and bottom samples were therefore averaged to obtain a single value for each pond and sampling date. Means and ranges for selected water quality variables for each sampling period are presented in Table 1.

Quality of potential effluents varied considerably from pond to pond and from season to season. Generally, potential effluent quality was poorest (highest concentrations of solids, organic matter, total phosphorus, and total nitrogen) in the summer, a trend that has been confirmed in other studies of catfish pond water quality (Tucker and van der Ploeg 1993).

Seasonal variation in pond water quality (and therefore effluent quality) is associated, in large part, with seasonal variation in phytoplankton biomass and metabolism. Phytoplankton biomass in catfish ponds is greatest from late spring through early autumn because warm water temperatures, seasonally high solar radiation, and large inputs of plant nutrients (derived from feed via fish metabolic waste) support rapid phytoplankton growth. Most organic matter in channel catfish pond water consists of living phytoplankton cells and phytoplankton-derived detritus. Accordingly, biochemical oxygen demand, which is a measure of the organic matter content of water, varies with phytoplankton biomass and is generally highest in summer. Concentrations of suspended solids also correspond to phytoplankton biomass because phytoplankton and phytoplankton-derived detritus constitute most of the particulate material in catfish pond waters, except in ponds with high levels of suspended mineral particles.

TABLE 1. *Concentrations of selected water quality variables (means and ranges) in potential effluents from 20 commercial channel catfish ponds in northwest Mississippi from summer 1991 through spring 1993. Adapted from Tucker et al. (1996).*

Season	Settleable solids (mL/L)	Suspended solids (mg/L)	Total nitrogen (mg N/L)	Total ammonia (mg N/L)	Total phosphorus (mg P/L)	Biochemical oxygen demand (mg O_2 /L)
Summer 1991	0.20 (0-0.90)	127 (40-225)	6.1 (2.1-14.1)	1.22 (0.01-3.19)	0.54 (0.23-1.24)	26.1 (14.6-41.2)
Autumn	0.02 (0-0.25)	80 (40-225)	6.1 (2.9-10.8)	2.63 (0.05-6.35)	0.26 (0.14-0.58)	9.7 (1.9-26.4)
Winter 1992	0.06 (0-0.70)	109 (51-194)	5.1 (2.1-8.8)	0.86 (0.04-3.85)	0.34 (0.13-0.62)	13.7 (5.7-29.7)
Spring	0.11 (0-1.35)	123 (72-204)	4.5 (1.8-6.7)	1.06 (0.04-3.04)	0.31 (0.15-0.56)	14.8 (8.2-27.1)
Summer	0.09 (0-0.58)	117 (47-197)	7.0 (2.6-10.9)	0.71 (0.03-2.02)	0.51 (0.26-0.87)	21.2 (10.5-36.4)
Autumn	0.02 (0-0.15)	93 (41-175)	6.9 (3.8-10.4)	2.76 (0.07-8.10)	0.35 (0.15-1.03)	12.3 (5.4-34.0)
Winter 1993	0.01 (0-0.03)	93 (39-165)	5.5 (0.6-8.8)	1.48 (0.02-5.14)	0.34 (0.14-0.62)	11.9 (4.8-22.9)
Spring	0.12 (0-0.70)	135 (46-289)	5.2 (1.5-7.9)	2.21 (0.03-4.44)	0.37 (0.24-0.58)	14.9 (8.5-25.5)

Similarly, most of the combined nitrogen and phosphorus in catfish pond waters are present as particulate organic matter, primarily within phytoplankton cells, so total nitrogen and total phosphorus concentrations also tend to vary with phytoplankton biomass and usually are highest in the warmer months.

Quality of Overflow from Watershed Ponds in West-Central Alabama

Schwartz and Boyd (1994b) collected water samples from 25 commercial channel catfish ponds in central and west-central Alabama. Samples were collected as described above for ponds in Mississippi, and analytical results obtained from surface and bottom samples are averaged here to obtain one value for each pond and sampling date. Means and ranges for selected water quality variables for each sampling period are presented in Table 2.

Data were highly variable among ponds and seasonal trends in water quality were not consistent. Settleable solids concentrations were highest during summer, and concentrations were generally greater in surface waters than in bottom waters. Suspended solids concentrations peaked during summer and remained high into autumn.

Biochemical oxygen demand was also lowest in spring, with winter and summer levels being similar, and highest levels occurring during autumn. Total phosphorus concentrations were highest during summer and autumn. Total ammonia-nitrogen concentrations were highest in autumn and lowest in winter. Total Kjeldahl-nitrogen concentrations were high throughout the growing season and peaked in autumn.

Coliform Bacteria in Catfish Pond Waters

Boyd and Tanner (1998) studied coliform abundance in water samples collected in the

TABLE 2. *Concentrations of selected water quality variables (means and ranges) in potential effluents from 25 commercial channel catfish ponds in central and west-central Alabama from winter 1991 through autumn 1992. Adapted from Schwartz and Boyd (1994b).*

Season	Settleable solids (mL/L)	Suspended solids (mg/L)	Kjeldahl nitrogen (mg N/L)	Total ammonia (mg N/L)	Total phosphorus (mg P/L)	Biochemical oxygen demand (mg O_2/L)
Winter 1991	0.06	81	3.7	0.7	0.25	9.0
	(0-0.33)	(22-202)	(0.9-9.2)	(0.07-2.47)	(0.04-0.57)	(1.9-21.9)
Spring	0.05	52	4.4	1.07	0.21	6.5
	(0-0.40)	(5-134)	(1.8-10.6)	(0.02-3.45)	(0.07-0.37)	(2.4-21.4)
Summer	0.19	96	5.0	0.85	0.36	10.7
	(0-1.80)	(14-240)	(1.7-11.3)	(0.05-4.71)	(0.12-0.75)	(4.3-20.3)
Autumn	0.03	103	6.1	1.86	0.46	18.1
	(0-0.54)	(18-232)	(2.2-11.5)	(0.10-8.07)	(0.12-1.85)	(6.1-35.6)
Winter 1992	0.01	29	1.9	0.27	0.09	9.2
	(0-0.10)	(1-100)	(0.6-3.7)	(0.03-1.08)	(0-0.31)	(5.5-17.5)
Spring	0.06	45	3.8	0.91	0.18	5.3
	(0-0.35)	(1-68)	(1.5-6.8)	(0.01-4.08)	(0-0.39)	(2.1-7.6)
Summer	0.15	102	3.9	1.89	0.19	8.0
	(0-0.28)	(10-308)	(1.6-8.4)	(0.06-3.30)	(0-0.47)	(1.4-15.9)
Autumn	0.03	73	6.0	1.91	0.27	7.6
	(0-0.25)	(14-337)	(2.2-14.0)	(0.09-5.26)	(0.05-0.83)	(1.2-23.4)

summer from 48 channel catfish ponds on commercial farms in Alabama. Water samples were also collected from four sportfish ponds stocked with bluegill *Lepomis macrochirus* and largemouth bass *Micropterus salmoides.*

Catfish pond water contained 0 to 4,220 total coliforms/100 mL; 0 to 736 fecal coliforms/100 mL; and 0 to 905 fecal streptococci/100 mL. Most samples contained fewer than 1,250 total coliforms/100 mL, 200 fecal coliforms/100 mL, and 500 fecal streptococci/100 mL. Seven ponds received runoff from cattle pastures, but total and fecal coliforms were no more abundant in these ponds than in other ponds not located on watersheds with cattle pastures. Also, there were no differences in average abundance of total and fecal coliform bacteria between catfish ponds and sportfish ponds on any sampling date, suggesting that the microbial quality of catfish pond waters reflects the general characteristics of other impounded surface waters in the region. The fecal coliform:fecal streptococci ratio (which can be used to infer the source of fecal contamination) and the absence of any source of human fecal contamination in any of the ponds strongly suggest that fecal coliforms in aquaculture ponds originate from warm-blooded animals such as livestock, domestic pets, other small mammals, and birds.

Biochemical Oxygen Demand of Catfish Pond Waters

Biochemical oxygen demand (BOD) is a measure of biodegradable organic matter in water and provides a rough indication of the "pollutional strength" of a wastewater. The standard BOD test (the 5-d BOD, or BOD_5) measures the amount of oxygen consumed during sample incubation for 5 d at 20 C. Organic matter is incompletely oxidized within 5 d, so the standard test determines only one point in a curve relating oxygen concentration over time. Sometimes it is useful to know the amount of oxygen consumed during complete biochemical oxidation of organic matter. This is called the ultimate BOD (BOD_u), and can be determined from data obtained in longer-term incubation of samples. Oxygen consumed in the BOD test is used primarily by heterotrophic bacteria in the decomposition of organic matter. This fraction of the total BOD is called the carbonaceous biochemical oxygen demand, or CBOD. Oxygen may also be used by nitrifying bacteria to oxidize ammonia, and this fraction is called the nitrogenous oxygen demand, or NOD. The CBOD and NOD can be determined for the 5-d period ($CBOD_5$ and NOD_5) or as ultimate oxygen demands ($CBOD_u$ and NOD_u)

Boyd and Gross (1999) studied the variation and components of BOD in waters

TABLE 3. *Fractions of the biochemical oxygen demand (averages and standards errors of the mean, expressed as mg O_2/L) for waters from ten catfish ponds in east Alabama. Adapted from Boyd and Gross (1999).*

Variable	9 Jun	11 Jun	26 Jul	19 Aug	4 Sep	27 Sep	Overall mean
BOD_5	7.9±1.3	5.1±0.6	2.4±0.3	3.9±0.6	6.7±1.5	3.1±0.5	4.8±0.8
$CBOD_5$	3.8±0.9	3.4±0.5	1.9±0.2	3.2±0.6	3.8±1.2	2.1±0.4	3.0±0.6
NOD_5	4.1±0.7	1.7±0.5	0.5±0.3	0.7±0.4	2.9±0.1	1.0±0.2	1.8±1.4
BOD_u	22.2±3.9	14.0±2.4	9.6±2.1	13.9±2.9	26.2±8.2	5.2±1.7	15.2±3.5
$CBOD_u$	12.6±2.9	9.8±1.9	4.6±0.7	7.1±1.6	15.4±5.2	2.1±0.6	8.6±2.2
NOD_u	9.6±2.8	4.2±2.0	5.0±1.4	6.8±1.9	10.4±7.1	3.1±1.5	6.6±2.6

collected from 10 catfish ponds in east Alabama over the summer growing season. Data for various BOD fractions were averaged among ponds for each date and resulting means and standard errors provided in Table 3. Oxygen demand varied greatly among ponds on a given date; for example, on one date, $CBOD_5$ ranged from 0.6 to 14.2 mg/L and BOD_5 ranged from 1.4 to 17.2 mg/L. Overall, values for BOD_5 were considerably larger than $CBOD_5$, and the NOD_5 averaged 37.5% of the BOD_5. The BOD_u values also were larger than the $CBOD_u$, and NOD_u was about 43% of the BOD_u. Thus, oxygen used in nitrification contributed heavily to BOD_5 and BOD_u.

This study was conducted in ponds with stocking and feeding rates similar to those encountered on commercial farms, but BOD_5 concentrations were not high compared to wastewater. For example, in one survey (Hunter and Heukelekian 1965), the BOD_5 of municipal wastewater ranged from 75 to 275 mg/L, with a mean of 147 mg/L. The maximum value for BOD_5 in pond waters was 17.2 mg/L, and 90% of BOD_5 values were less than 8 mg/L. None of the BOD_5 values exceeded 30 mg/L and only six of the BOD_u values (10%) exceeded 30 mg/L—a value often used as a BOD_5 limitation in discharge permits. These results are comparable to those of the 2-year studies of commercial catfish ponds summarized in Tables 1 and 2, in which only three of 200 samples (1.5%) collected from Alabama catfish ponds (Schwartz and Boyd 1994a) and six of 160 samples (3.8%) collected from Mississippi catfish ponds (Tucker et al. 1996) had BOD_5 values above 30 mg/L.

The average $CBOD_5$:$CBOD_u$ ratio was 0.35 in catfish pond waters, compared to a typical $CBOD_5$:$CBOD_u$ of 0.67 for wastewater. The lower ratio of short-term to long-term oxygen demand is not surprising, because organic matter in catfish pond water is primarily living plankton, as compared to non-living organic matter of wastewater. Plankton apparently does not die quickly when confined in BOD bottles, and the oxygen demand of plankton in samples from catfish ponds is expressed more slowly than that of heterotrophic bacteria associated with non-living organic matter in wastewater. Because channel catfish pond waters do not have extremely high BOD and the oxygen demand is not expressed rapidly, a rapid depletion of dissolved oxygen in the mixing zone when catfish pond effluents are released into natural waters is not expected. Thus, relatively large BOD_u values can be tolerated in effluents, and the primary concern would be the maximum BOD load that could be assimilated by the receiving water body.

Effluent Quality When Ponds Are Drained

The quality of water discharged when catfish ponds are drained is different from the quality of water discharged as overflow. When ponds are drained, there are periods—often quite brief—when solids from the pond bottom are discharged along with pond water. Accordingly, these effluents will have higher concentrations of solids, nutrients, and organic matter than overflow waters.

Effluent Quality When Levee Ponds Are Drained

Many watershed ponds and all levee ponds are constructed to allow fish harvest without draining the pond. Ponds of this type represent well over 90% of the ponds used for catfish culture (USDA-APHIS 1997). When fish are harvested, a seine is pulled through the pond one or more times to capture the fish. Fingerlings in nursery ponds are completely harvested each year so that the pond can be drained and prepared for the next crop of fry. Food-fish production ponds are, however, operated without draining for several years. Market-sized fish are harvested with a large-mesh seine that allows smaller fish to escape and continue growing. Fingerlings are added

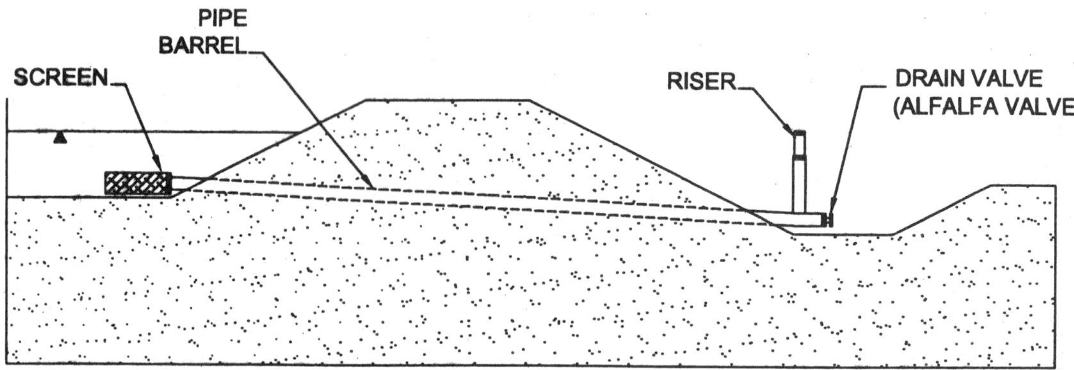

FIGURE 1. *The drainpipe structure most commonly used for levee ponds in Mississippi.*

to replace the market-sized fish that are removed. Final fish harvest occurs only when ponds need to be drained for renovation. Regardless of whether the pond is used for fingerling or food-fish production, water is discharged to drain the pond only after the harvest operation is complete. This practice can be contrasted with water management in deeper watershed ponds described in the following section. In watershed ponds, some water is discharged *before* harvesting begins.

Most levee ponds are equipped with fixed drains that extend from the deepest part of the pond through the levee. A screened extension is fitted to the end of the pipe (usually 25 to 30 cm in diameter) inside the pond, with a riser and an "alfalfa" valve on the outside. The height of the riser determines pond depth (Fig. 1). The riser may either be permanent or removable.

If the riser is permanent, the pond is drained by fully opening the alfalfa valve. If the riser has a removable extension, draining is initiated by removing the riser extension, then, after the water level in the pond has dropped at least half way, the valve at the bottom of the drain is opened and the remaining water in the pond is drained. When draining is initiated, shear forces generated by water moving into the drain pipe scour the pond bottom around the entrance to the drain pipe. The initial flush of water discharged therefore

consists of pond water and a slurry of sediment that has accumulated over the screen inside the pond. Depending on the depth of sediment around the drain, the effluent clears in 5 to 30 min and all water subsequently discharged is simply pond water.

Craig Tucker and John Hargreaves (unpublished data) measured changes in the quality of effluent from levee ponds in northwest Mississippi that were drained after fish harvest in early spring. Trends were similar for all ponds, so data from one representative pond is presented here. Pond area was 3.72 ha and water depth at the drain was 0.96 m. The discharge pipe was 25 cm in diameter. Pond draining commenced immediately after the last attempt to harvest fish by seining, and draining was completed within 5 d.

When the riser was removed, the initial discharge was high in total suspended solids (43,860 mg/L), volatile suspended solids (3,770 mg/L), BOD_5 (118 mg/L), and nutrients. Within 2.5 min after the initial discharge, total suspended solids had decreased to 1,790 mg/L, volatile suspended solids to 205 mg/L, and BOD_5 to 29 mg/L. By 30 min, the poorly consolidated sediments from around the drain structure inside the pond had been discharged, and effluent quality was identical to the bulk pond water for the remainder of the draining period (Fig. 2a, b).

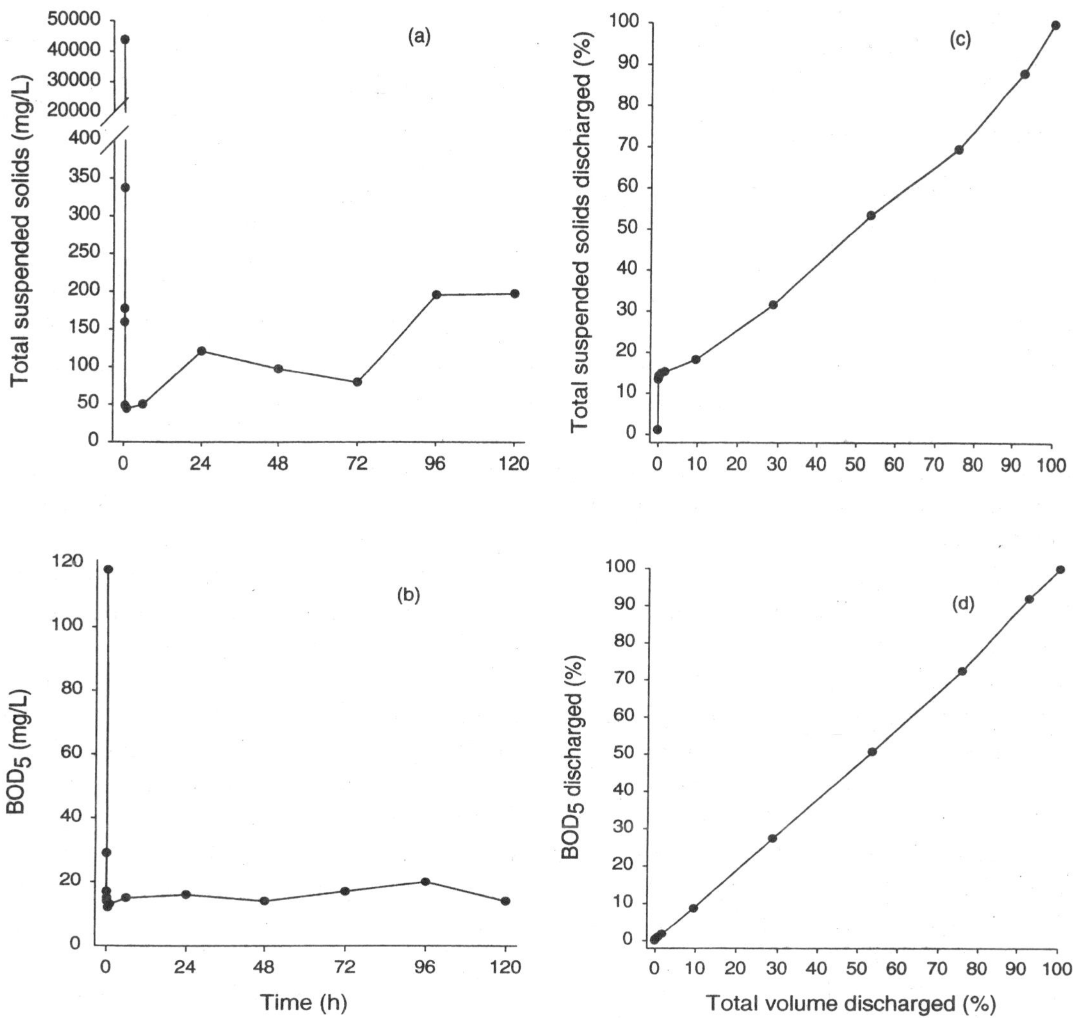

FIGURE 2. *Quality of effluents discharged after harvest of a levee pond in northwest Mississippi: temporal changes in concentrations of (a) total suspended solids and (b) 5-d biochemical oxygen demand (BOD$_5$); and percentage of total water volume discharged and cumulative discharge of (c) total suspended solids and (d) BOD$_5$ (Craig Tucker and John Hargreaves, unpublished data).*

For the effluent depicted in Fig. 2, 15.2% of the total solids were discharged with the first 1.7% of the total effluent volume (Fig. 2c). The initial "spike" in the cumulative discharge of solids was due to the high solids concentration in the initial flush of water, which was 900 times greater than the solids concentration in the bulk pond water. The BOD$_5$ of the initial flush of water discharged was only about nine times greater than the BOD$_5$ of the bulk pond water, and concentrations decreased more rapidly than total suspended solids. Therefore, cumulative discharge of BOD$_5$ as a function of water volume discharged (Fig. 2d) did not show the initial spike seen for total suspended solids; rather, cumulative discharge of BOD$_5$

was roughly proportional to the volume of water discharged. These results suggest that the ratio of mineral solids to organic matter in the initial flush of water was relatively higher than in water subsequently discharged.

Effluent Quality During Fish Harvest in Watershed Ponds

Some watershed ponds have deep areas near the dam that prevent normal fish harvest by seining. These ponds must be partially drained to concentrate fish into relatively shallow water near the dam. Fish are then harvested by seining and the remaining water is discharged to drain the pond. Ponds that must be drained to facilitate fish harvest constitute a small proportion of the ponds used in catfish farming, although they are not uncommon in west-central Alabama and east-central Mississippi.

Two studies have been published to describe the quality of effluents discharged from watershed ponds during fish harvest. In one of the first studies of catfish pond effluents, Boyd (1978) clearly showed that effluent quality is affected by activities associated with fish harvest. However, the ponds sampled in that study were stocked at much lower fish densities than those currently used in the industry, and those data will not be discussed. A more recent study (Schwartz and Boyd 1994a) was conducted under commercial conditions and will be summarized below.

Schwartz and Boyd (1994a) collected samples from three watershed ponds in east Alabama that were drained in the winter to facilitate fish harvest. Drains were opened in early morning and closed in the evening. This procedure was repeated daily until water levels were low enough to permit seining. Drains were then closed and ponds seined. After the first seining, drains were reopened to allow further drawdown so remaining fish could be easily captured. After as many fish as possible had been removed by seining, ponds were allowed to drain completely so fish that escaped seining could be removed by hand. Water samples were collected twice daily, 1–2 h after drains were opened in the morning, and 6 h after morning samples were collected.

Daily mean concentrations of each variable were multiplied by daily effluent volumes to calculate total masses released in effluents. There was considerable variation among ponds in amounts of each water quality variable released. This difference was thought to be related to the fact that fish were seined from one pond in a single day, whereas the other two ponds were seined on more than one day. Each time ponds were seined, sediments were stirred into the water and concentrations of various substances increased. Most of the nitrogen and phosphorus in effluent were contained in dissolved organic matter or particulate matter, for there was little total ammonia-nitrogen, nitrite-nitrogen, and nitrate-nitrogen in relation to total Kjeldahl-nitrogen, and only a small proportion of the total phosphorus could be accounted for in soluble reactive phosphorus. Ponds did not contain heavy phytoplankton blooms at harvest because of low winter temperatures. The BOD_5 was expressed by soluble organic matter and particulate organic matter stirred into the water from bottom sediments during seining. The major component of total settleable solids was suspended soil particles.

Changes in effluent quality during draining and harvest followed similar trends in all three ponds. Data from one pond are summarized for illustration. Seining commenced after approximately 96 h of draining. Total ammonia-nitrogen concentrations remained below 0.1 mg/L during the first 48 h of draining and did not exceed 0.5 mg/L until the drain had been open for over 120 h. Concentrations in surface and effluent samples were roughly equivalent throughout the draining event. Kjeldahl-nitrogen concentrations remained below 5.0 mg/L for the first 72 h of draining in surface and effluent samples. After that time, the surface concentration increased slightly to peak at 10 mg/L while the effluent concentration increased sharply to 137 mg/L. Concentrations of nitrite-nitrogen fluctuated with no apparent trend during the first 120 h of draining. After this time, the surface concentration increased from 0.002 mg/L to 0.006 mg/L while the effluent concentration increased from 0.002 mg/L to a peak of 0.007 mg/L at 144 h. Nitrate-nitrogen concentrations fluctuated greatly throughout the draining event with surface concentrations showing almost no increase over time. Effluent concentrations of nitrate-nitrogen increased from 0.36 mg/L at the beginning of draining to 0.64 mg/L at the end. Soluble reactive phosphorus concentrations remained stable at 0.001 mg/L for the first 48 h of draining, increased gradually to 0.02 mg/L after 96 h and then increased sharply to peak at 0.12 mg/L at the surface and 0.16 mg/L in the effluent. Total phosphorus concentrations were about 0.15 mg/L in the surface water and effluent for the first 72 h of draining. They then increased markedly to 1.0 mg/L at the surface and 1.3 mg/L in the effluent. The BOD_5 was around 10 mg/L in surface water for the first 48 h and then increased gradually to peak at 18 mg/L. Effluent values for BOD_5 remained about the same as surface values for the first 120 h and then increased sharply to peak at 296 mg/L. Concentrations of total settleable solids remained negligible in the surface water throughout the draining event. Effluent concentrations of total settleable solids paralleled those at the surface for the first 120 h of draining, after which they increased sharply to peak at 60 mL/L. Temporal changes in BOD_5 and settleable solids are shown in Fig. 3a, b.

In general, concentrations of total Kjeldahl-nitrogen, BOD_5, and settleable solids were fairly constant throughout the draining phase. It was not until the seining phase that these variables increased in concentration. Total ammonia-nitrogen, soluble reactive phosphorus, and total phosphorus steadily increased during the draining phase and sharply increased during the seining phase. Increases in phosphorus were most likely a result of sediments being resuspended, while the elevated total ammonia-nitrogen concentrations were probably a result of fish metabolic wastes becoming more concentrated in a decreasing volume of water. Values of pH steadily decreased throughout pond draining. As the fish in a pond became more concentrated in a decreasing volume of water, carbon dioxide from their respiration was a major factor causing pH to decline. Both nitrite-nitrogen and nitrate-nitrogen concentrations fluctuated widely in all three ponds during draining and seining.

Cumulative percentages of total nitrogen, total phosphorus, settleable solids (Fig. 3c), and BOD_5 (Fig. 3d) discharged relative to the cumulative percentages of effluent volume were calculated for the pond described above. Approximately 50% of the nitrogen, phosphorus, and BOD_5 were discharged in the last 15–20% of effluent. For settleable solids, 50% of the total was discharged in the last 5% of effluent.

The pattern of cumulative discharge of substances from watershed ponds drained *during* fish harvest (Fig. 3c, d) differs from the pattern seen for effluents from levee ponds drained *after* harvest (Fig. 2c, d). For levee ponds, cumulative discharge of material is either proportional to discharge volumes or shows an initial spike in the first few percent of water volume discharged (essentially the reverse of the pattern seen for watershed ponds drained during harvest).

The best way to minimize the pollution potential of effluents from ponds drained for harvest is to harvest fish as quickly as possible, and either not discharge water during the

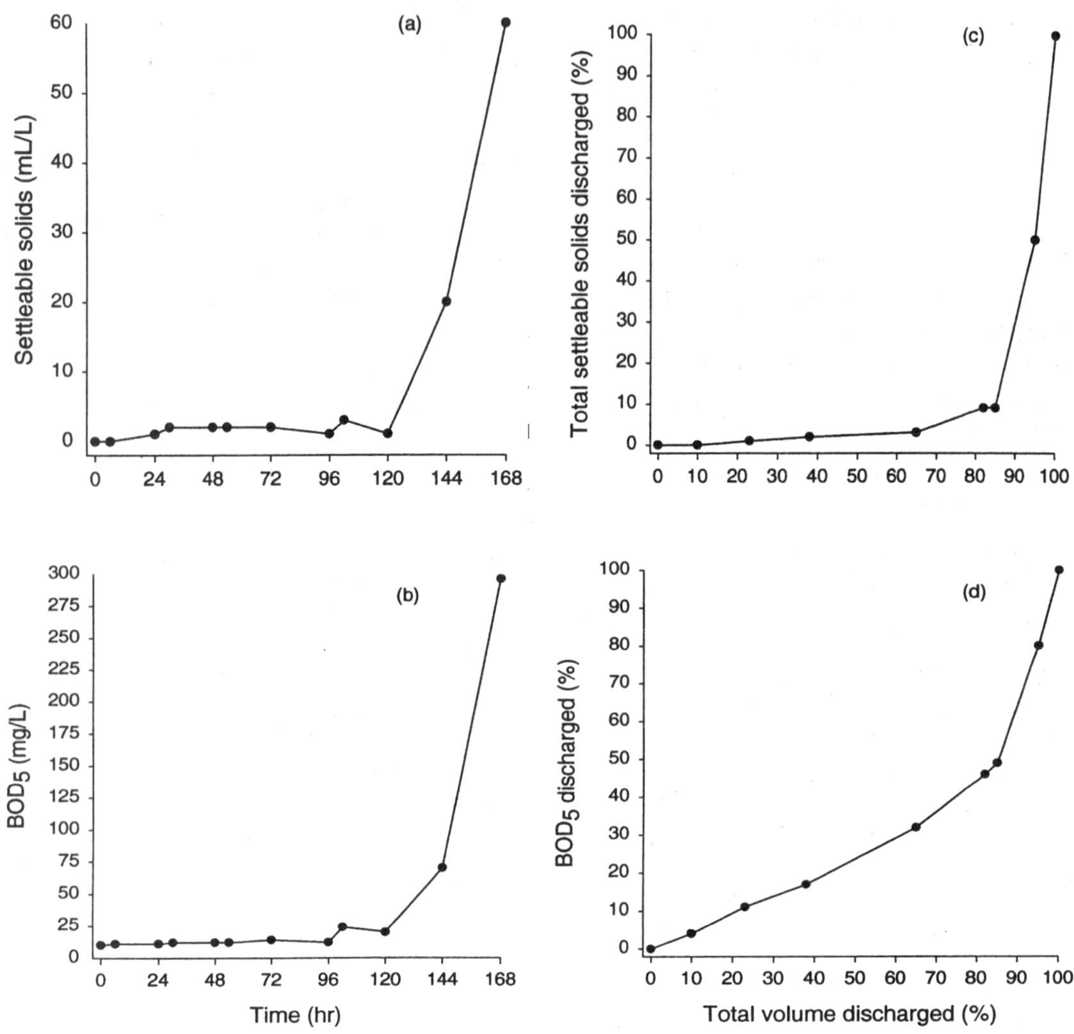

FIGURE 3. *Quality of effluents discharged during harvest of a watershed pond in east Alabama: temporal changes in concentrations of (a) settleable solids and (b) 5-d biochemical oxygen demand (BOD₅); and percentage of total water volume discharged and cumulative discharge of (c) settleable solids and (d) BOD₅. Redrawn from Schwartz and Boyd (1994a).*

seining phase or to discharge the contaminated water into a settling basin or retention pond. It also may be feasible to allow effluents to flow untreated into the environment during the pre-seining phase of draining, because concentrations of potential pollutants are low during this phase of draining.

Potential Environmental Impacts of Pond Effluent Constituents

Similar to nearly all forms of agriculture, catfish farms release solids, organic matter, and nutrients into receiving streams. These substances may affect receiving water bodies in the following ways: 1) particulate matter in the effluent is deposited in the receiving body of water, altering substrate composition, reducing water depth, and causing potential alterations in benthic invertebrate and fish community composition; 2) organic matter loading results in a greater oxygen demand, leading to possible limitations on the distribution of biota sensitive to low dissolved oxygen concentration; and 3) nitrogen, phosphorus, and other nutrients can stimulate excessive growth of aquatic plants. The major constituents of concern in catfish pond effluents are solids, organic matter, phosphorus, and nitrogen. Pesticides and other toxic materials are not present in effluents from aquaculture ponds, as they may be in effluents from other industries.

Suspended Solids

The substance in catfish pond effluents that most frequently exceeds existing water quality standards is total suspended solids (Schwartz and Boyd 1994b). Suspended solids in pond overflow are primarily organic, and consist mainly of phytoplankton cells and detritus of phytoplankton origin. The quantity of inorganic suspended solids discharged from catfish ponds is usually low, although concentrations may be quite high when sediment particles on the pond bottom are resuspended and discharged when ponds are drained. Additional inorganic solids may be generated when effluents erode levees and ditch banks.

The main concern with solids discharged from aquaculture ponds is not the effects of sedimentation in the receiving stream, but rather the oxygen demand expressed by decomposition of organic solids. However, the BOD of effluents from catfish ponds is relatively low (usually less than 15 mg/L) and is expressed over a longer time period than most wastewaters (Boyd and Gross 1999). Further, concentrations of suspended solids of algal origin are highest during summer and autumn, but discharge volume from ponds is low or nonexistent. When pond discharge volumes are greatest in winter and spring, streams in the southeast already carry large loads of inorganic suspended solids derived from erosion of fallow fields and other areas. It is therefore unlikely that catfish ponds effluents affect suspended solids levels in receiving streams at that time.

Although catfish farming is an important industry in certain areas of the southeast, aquaculture development accounts for a small proportion of land area devoted to intensive agriculture. Under such conditions, the contribution of catfish farming to overall solids or nutrient load of watersheds in the region is low. Row crop farming yields about 6,500 kg/ha per year of total suspended solids in runoff (USDA/USEPA 1975), primarily inorganic solids caused by soil erosion. Estimates of total suspended solids from catfish farming (Boyd et al. 2000) ranged from 432 kg/ha per year for food-fish grown in levee ponds to 4,600 kg/ha for fry and fingerlings grown in watershed ponds. When weighted for the total pond area devoted to the various pond types and uses, the average is about 1,400 kg/ha per year, or less than a quarter of that produced by row crop farming. This should not be surprising because construction of ponds is an established method of reducing solids loads into streams (Swingle and Smith 1939).

The yield of suspended solids from catfish ponds calculated by Boyd et al. (2000) is based on conditions in Alabama where a significant portion of catfish production is derived from watershed ponds. Industry-wide yield of suspended solids from ponds is undoubtedly lower than that estimate because most catfish are grown in levee-type ponds that discharge less water and are not subject to inputs of inorganic suspended solids derived from erosion on the watershed upstream of the pond. Furthermore, when soil loss from land converted to ponds is compared to that lost from row crop acreage, the difference is even greater because most of the suspended solids discharged from catfish ponds are particulate organic matter and not soil.

Sediment that accumulates in catfish ponds is removed every 15 to 20 years, but this renovation involves moving sediment from the bottoms of ponds and using it to repair erosion damage to levees or dams. The sediment is not disposed of outside the ponds because it is needed to replace the soil eroded from the inside of the levee. The work is usually conducted during the dry season, so if drains are closed during the work, little turbid runoff will result when a rainstorm occurs.

Nutrients and Oxygen Demand

Other than fry and fingerling ponds, which comprise about 10% of the production area, catfish ponds are seldom drained and most ponds are managed (intentionally or not) to provide water storage capacity. Thus, most of the nutrients and organic wastes from feeding fish are assimilated or dissipated by naturally occurring processes in ponds rather than discharged. The combination of low discharge volume and within-pond waste processing reduces mass discharge of nutrients and organic matter into receiving streams. Further, climatic conditions are such that nearly all overflow occurs during the wet winter and spring season when streams are flowing high with runoff from non-point sources in the watershed. Thus, the localized impact of potential pollutants in pond effluents is small because nutrients are quickly diluted and transported to larger streams.

Water quality sampling in streams upstream and downstream of outfalls from catfish farms did not reveal significant water quality deterioration related to pond effluents (Boyd et al. 2000). This supports the conclusion, based on timing, volume, and composition of effluent, that catfish farming is not a major source of nutrients or oxygen demand to streams in the southeastern United States.

Salinization

Common salt, NaCl, is often added to catfish ponds to prevent toxicosis from nitrite, which occurs naturally in fish ponds as a result of bacterial nitrification of the fish metabolite ammonia (Boyd and Tucker 1998). Salt is added to achieve chloride concentrations of 25–100 mg/L. The impact of adding salt to ponds on stream water chloride concentrations is not significant because the amounts of salt added to ponds are low and do not raise chloride concentrations or specific conductance above levels readily tolerated by freshwater species endemic to the region. The only possible problem would be large, accidental spills of sodium chloride onto the land or into water bodies that would temporarily raise the salinity of small streams to levels that could endanger certain organisms.

Dissolved Oxygen

Most overflow from catfish ponds occurs in the cooler months of the year, when pond waters have high concentrations of dissolved oxygen (Tucker 1996). In summer, dissolved oxygen concentrations in ponds may be temporarily low in the early morning hours when nighttime respiratory losses of dissolved oxygen exceed oxygen production in photosynthesis during daylight hours. However, pond oxygen concentrations are near or above saturation for most of the day. If water is discharged from a pond at night when dissolved oxygen concentrations are low, the water will be rapidly reaerated in the mixing zone of the effluent-receiving stream. Therefore, pond effluents should not cause dilution of stream dissolved oxygen concentrations in the outfall area. Results of the stream survey conducted by Boyd et al. (2000) verified this conclusion, and in several cases, dissolved oxygen concentrations were higher downstream from catfish farms than upstream.

pH

The pH of catfish pond waters at equilibrium with atmospheric carbon dioxide lies between 6.5 and 8.5, depending on total alkalinity and the acid-base status of pond bottom soils (Boyd and Tucker 1998). Pond water pH can temporarily deviate outside that range due to net carbon dioxide uptake by phytoplankton during photosynthesis (causing pH to rise) or net carbon dioxide release from respiration (causing pH to fall). In winter, when overflow from ponds is greatest, biological activity in ponds is slow because water temperatures are low, and pond water pH fluctuates little around the equilibrium pH. In summer, intense biological activity in ponds causes pH to fluctuate over a much wider range, changing by as much as 2 or 3 pH units over a 24-h period. In particular, rapid photosynthesis by phytoplankton during daylight hours may cause pH to rise 8.5 to 9.5. However, pH excursions outside the 6.0 to 8.5 range, which is often set as an effluent discharge limitation, are not due to the presence of strong mineral acids or bases, as in some industrial effluents, but rather to the effects of biological activity on the carbon dioxide-bicarbonate equilibrium system. Discharge of pond water during periods when pH is high or low will not have a significant effect on the pH of the receiving stream because the effluent will rapidly reestablish equilibrium pH through dilution, reactions with the buffer system of the receiving water, or reaeration, which will move dissolved carbon dioxide concentrations towards equilibrium with the atmosphere and moderate the pH. Also, the alkalinity naturally present in most catfish pond waters buffers against wide changes in pH.

Toxic Chemicals

Catfish pond effluents contain environmentally insignificant quantities of pesticides and therapeutants because: 1) only EPA-approved chemicals are used in ponds, 2) those chemicals are used infrequently because it is expensive to treat large volumes of water, and 3) the long hydraulic residence time and intense microbiological activity in catfish ponds provides ample time and opportunity for dissipation of the chemical before discharge. Other than salt, lime, and fertilizer, the most commonly used chemical is copper sulfate for algal control. Copper sulfate can be toxic to fish at high concentrations but algicidal concentrations are usually less than 0.5 mg Cu/L—well below levels that kill fish. Copper from copper sulfate has a short residual life (less than 1 d) in pond water because it quickly precipitates to pond soils as copper oxide (Masuda and Boyd 1993). Following precipitation, copper will redistribute to soil fractions with progressively lower bioavailability (Han et al. 2001). Thus, effluents from ponds will not have significant residues of copper, and any copper in the effluent will be in the form of sparingly soluble copper oxides. In addition, Han et al. (2001) measured no effect on sensitive toxicity test organisms exposed to sediment from ponds with a 3-year history of copper sulfate application. Therefore, there will likely be no impact of copper on receiving-stream biota.

Fuels and Oils

Most mechanical aerators and pumps are operated by electricity, so fuel and oils are infrequently used and stored around ponds. There is always the possibility of accidental spill when fueling or lubricating tractors and other equipment, but fish farming presents no greater risk of fuel and oil spills than does any other type of mechanized agriculture. Further, petroleum products at low concentrations can impart an undesirable off-flavor to fish, so fish farmers are particularly careful to avoid contamination of the culture water.

Human Pathogens

Most catfish are raised in levee ponds filled with ground water. Levee ponds have no true watershed and there is no source of human fecal contamination. Catfish ponds, like all impounded surface waters, provide habitat for other warm-blooded animals, and coliforms are, therefore, present. Coliform levels are low, even in watershed-type ponds (usually < 200 organisms/100 mL; Boyd and Tanner 1998), and it appears that coliform abundance is no greater than other impounded surface waters. As such, catfish pond effluents pose little threat of contributing to coliform abundance in receiving streams.

Assessments of the Effects of Pond Effluents on Stream Water Quality

Assessing the effect of aquaculture effluents on a receiving stream is difficult because the cause-and-effect relationship between discharge and ecological impacts is not necessarily clear and direct. Nutrients, solids, and organic matter in pond effluents have the potential to cause pollution if they are discharged at a rate that exceeds the capacity of the receiving waters to assimilate or otherwise process the discharged materials. The effect on receiving waters will therefore depend on the volume and concentration of substances in effluents in relation to the discharge rate of the receiving water body and the timing of discharge. Further complications arise because pond effluents may modify stream water quality, particularly if discharge occurs during periods of low flow in receiving streams, but this may not result in a negative ecological impact.

Although the following discussion addresses the impact of pond effluents on stream water quality, water discharged from catfish ponds also affects stream hydrology. For example, water discharged from catfish ponds will make a variable contribution to total stream flow, depending on the relative proportion of the watershed that is developed into ponds and water management practices on individual farms. During droughts, water discharged from catfish ponds may contribute a large share of total stream flow in small, intermittent streams. In such streams, water may be absent if effluent is not discharged from catfish ponds. Therefore, catfish pond effluents may sustain aquatic communities in intermittent streams in which other sources of flow are absent. Ponds also affect steam hydrology by storing precipitation. Storage will reduce peak stream discharges and dampen and delay storm hydrographs downstream from ponds. In areas with significant watershed development into ponds, the risk of damage caused by downstream flooding may be reduced.

Two studies have been conducted to assess the effect of catfish pond effluents on water quality of receiving streams. In one study, Tucker and Lloyd (1985a) compared concentrations of potential pollutants in effluents from levee ponds in northwest Mississippi to concentrations in streams in the same area. Such information is of limited usefulness because it does not account for the relative volumetric flows of effluent and receiving stream; however, some insight is provided on potential impacts of pond effluents on receiving bodies of water. In the second study, Boyd et al. (2000) compared water quality in Alabama streams upstream and downstream from catfish farms. These data allow direct assessment of the impact of pond effluents on receiving streams.

Comparison of Catfish Pond and Stream Water Quality in Mississippi

Water samples were collected from 25 commercial catfish ponds and 24 small, permanent streams in Northwest Mississippi. Samples were collected during the "wet season" when stream flows were high (March)

TABLE 4. *Average concentrations of eight water quality variables in catfish ponds and streams in northwest Mississippi. Within each sampling period (spring or summer), an asterisk indicates that the higher value differs from the lower value at the 0.05 level of probability (Tucker and Lloyd 1985a).*

Variable (units)	Spring		Summer	
	Ponds	Streams	Ponds	Streams
Total solids (mg/L)	481	499	500*	344
Total volatile solids (mg/L)	149	123	162*	66
Chemical oxygen demand (mg/L)	61*	32	97*	33
Soluble reactive phosphorus (mg P/L)	0.07	0.18*	0.16	0.24
Total phosphorus (mg P/L)	0.56	0.63	0.85*	0.58
Total ammonia (mg N/L)	0.93	0.62	0.42*	0.13
Nitrate + nitrite (mg N/L)	0.05	0.26*	0.24	0.19
Total nitrogen (mg N/L)	4.41*	2.07	5.55*	1.47

and during the "dry season" when stream flows were low (August and September). A summary of the data collected in that study is presented in Table 4.

During the spring sampling period, concentrations of chemical oxygen demand and total nitrogen were higher in catfish pond water than in streams. However, on average, concentrations of nitrate plus nitrite and soluble reactive phosphorus were higher in streams than in pond waters. Concentrations of other variables did not differ. The relatively high concentrations of oxidized inorganic nitrogen and soluble phosphorus were attributed to fertilizer nutrients entering streams in runoff from row-crop fields during this period of seasonally high precipitation. In the summer sampling period, means of all variables except nitrate plus nitrite and soluble reactive phosphorus were higher in pond waters than in streams because nutrient and organic matter concentration in catfish ponds increase during the summer growing season as described above. Thus, it is possible that pond effluents may impact receiving stream water quality in late summer. However, there is generally little or no water discharged from catfish ponds in summer and autumn because water losses from pond evaporation plus seepage exceed water inputs from precipitation during that period.

So, in winter and spring, when discharge volume from ponds is highest, the proportional contribution of nutrients derived from runoff from adjacent row crop fields is high relative to the contribution from pond overflow. In addition, high stream flows will dilute any pond discharge. Further, aquaculture ponds represent less than 1% of the land in the area studied, which makes it unlikely that pond effluents add significantly to the total nutrient load of streams during periods of high runoff. Conversely, at the time of greatest potential impact of pond effluents on receiving stream water quality, climatic conditions are such that little, if any, water is discharged from ponds.

Impact of Catfish Farms on Water Quality in Alabama Streams

Boyd et al. (2000) collected water samples upstream and downstream from catfish farms on eight streams in west-central Alabama. With one exception, there were no catfish farms above the upstream sampling points. The streams drained mostly cropland, pastures, and

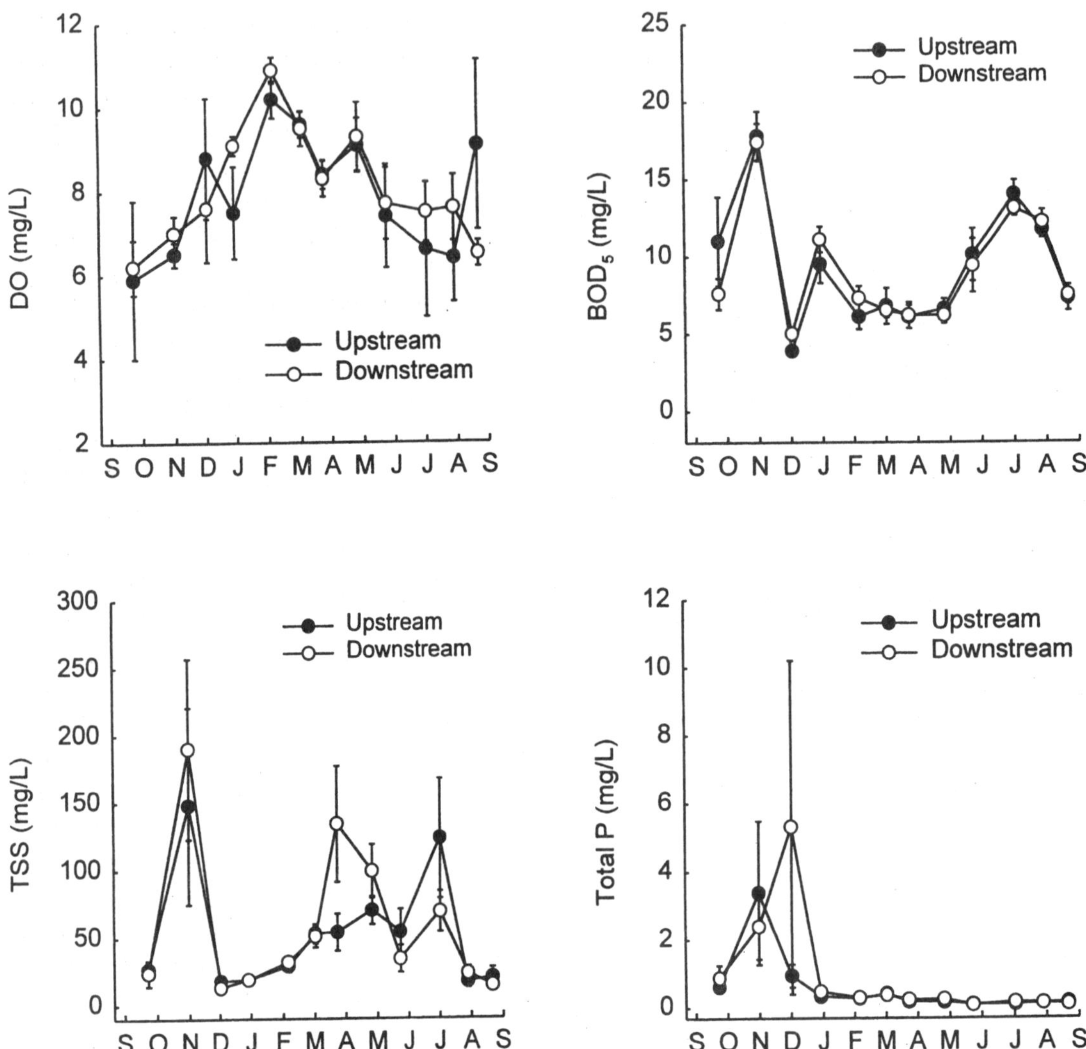

FIGURE 4. *Average concentrations of dissolved oxygen, 5-d biochemical oxygen demand, total phosphorus, and suspended solids in eight east-central Alabama streams at sampling locations upstream and downstream of catfish farms (redrawn from Boyd et al. 2000).*

woods, and were considered typical for the rural southeast. Water samples were collected monthly from streams between July 1997 and August 1998, except on a few occasions when some streams dried up in summer and autumn. Samples were analyzed for turbidity, total suspended solids, soluble reactive phosphorus, total phosphorus, total ammonia nitrogen, nitrate-nitrogen, total nitrogen, BOD$_5$,

dissolved oxygen, pH, specific conductance, and temperature. Average concentrations of dissolved oxygen, BOD$_5$, total phosphorus, and suspended solids at upstream and downstream sites are provided in Fig. 4.

Average water temperature was similar between sites upstream and downstream from catfish farms. The pH was slightly greater downstream of catfish farms than above, but the average pH was between 7.1 and 8.2 at all sites. Dissolved oxygen concentrations were similar downstream and upstream of catfish farms (Fig. 4), and all average concentrations exceeded 5 mg/L. Specific conductance tended to be elevated downstream of catfish farms during the drier months; however, the highest average value was less than 600 µS/cm. Average BOD_5 was similar between upstream and downstream sites (Fig. 4), except in June and July when BOD_5 was greater upstream. In autumn, total ammonia-nitrogen tended to be higher downstream, but the opposite was true on some dates in summer. Except for a single sampling date in autumn, there was little difference in average nitrate-nitrogen and total nitrogen concentrations upstream or downstream. There were no clear differences in either soluble reactive phosphorus or total phosphorus between sampling sites. Total suspended solids tended to be greater downstream of catfish farms in April and May, but higher upstream in June and July (Fig. 4), and similar trends were observed in turbidity values.

Data for individual streams showed that there were times when certain water quality variables were degraded downstream of catfish farms. However, there were almost an equal number of cases where water quality was improved below catfish farms. Furthermore, there were no cases where extremely high concentrations of variables (or very low dissolved oxygen concentrations) were noted downstream of farms. These findings strongly suggest that catfish farm effluents were not having adverse impacts on stream water quality in the study area.

Estimation of Mass Discharge of Nutrients and Organic Matter

Most of the preceding discussion has focused on the concentration of substances in pond effluents. Environmental impact is, however, related to the amount of material discharged (mass discharge or loading) rather than concentration alone. Mass discharge is the product of concentration and volume, so effluent volume is an important factor affecting potential environmental impacts.

Boyd et al. (2000) used data on annual discharge and the composition of overflow, partial drawdown, and final drawdown effluent to compute mass discharge of substances from Alabama catfish farms. Mass discharge was calculated for levee ponds and watershed ponds, and for fingerling and food-fish ponds. Intervals between pond drainings and practices used for pond drawdown were obtained from farm surveys, so the results are specific to conditions in Alabama. However, the general approach used to calculate mass discharge can be applied to any set of pond management conditions.

Overflow occurred mostly in winter and early spring when little or no feed was applied, phytoplankton was not abundant, and BOD_5 was low in ponds. Overflow from watershed ponds was caused by rain falling directly into ponds and by runoff. Thus, water discharged from ponds was a mixture of pond water, rainfall, and runoff, and substances in the water came from the ponds and the other sources. Thus, for watershed ponds, the load should be attributed to the pond plus its watershed. The average ratio of watershed area to pond area was 6.32, so a 1-ha pond represented 7.32 ha of water and land surface. Mass discharge attributed to the pond was therefore calculated as:

$$\text{Mass discharge} = (\text{overflow volume} \times \text{concentration}) \div 7.32$$

Partial drawdown effluent was considered to be any effluent released while pond volumes exceeded 20% of total pond volume. Fry and fingerling ponds are drained completely in one event. Based on the farm survey, food-fish pond water levels are reduced by about 50% at about 6-year intervals for complete harvests, and ponds are completely emptied about every 15 years. Thus, partial drawdown effluent in food fish ponds is 1.8 pond volumes in 15 years and final drawdown effluent is 0.2 pond volumes in 15 years. Thus, only 10% of the total drawdown effluent is final drawdown effluent and 90% partial drawdown effluent. Mass discharge of substances in partial drawdown effluent from fry and fingerling ponds was calculated as 0.8 x drawdown volume x concentration. For food-fish ponds, mass discharge of substances in partial drawdown effluent was calculated as 0.9 x drawdown volume x concentration. Drawdown volumes are provided in Chapter 2 (Table 4) and concentrations were obtained from Schwartz and Boyd (1994b).

In fry and fingerling ponds, 20% of the drawdown volume is represented by final drawdown. In food-fish ponds, 10% of the drawdown volume can be considered final drawdown. Mass discharge of substances in final drawdown effluents from fry and fingerling ponds was calculated as 0.2 x

TABLE 5. *Mass discharge (kg/ha per yr) of total suspended solids (TSS), 5-d biochemical oxygen demand (BOD$_5$), total nitrogen (TN), and total phosphorus (TP) from channel catfish ponds in Alabama (Boyd et al. 2000).*

Pond type and source of effluent	TSS	BOD$_5$	TN	TP
Fry and Fingerling Ponds				
Levee ponds				
Overflow	65	8.8	5.0	0.54
Partial drawdown	922	125.8	84.3	3.34
Final drawdown	3,430	106.2	2.0	5.31
Total	4,417	240.8	121.3	9.19
Watershed ponds				
Overflow	260	35.3	11.0	2.18
Partial drawdown	921	125.8	84.3	3.34
Final drawdown	3,430	106.2	32.0	5.31
Total	4,611	267.3	127.3	10.83
Food-fish Production Ponds				
Levee ponds				
Overflow	65	8.8	5.0	0.54
Partial drawdown	138	18.9	6.8	0.50
Final drawdown	229	7.1	21.3	0.35
Total	432	34.8	33.1	1.39
Watershed ponds				
Overflow	827	57.1	17.7	3.53
Partial drawdown	138	18.9	6.8	0.50
Final drawdown	229	7.1	21.3	0.35
Total	1,194	83.1	45.8	4.38

drawdown volume x concentration, and from food-fish ponds as 0.1 x drawdown volume x concentration.

Mass discharges in Table 5 represent averages for a large farm or an area with several farms. An individual food fish pond would have greater mass discharge on a year when it was drawn down for partial harvest or drained than in other years. Also, fingerling ponds harvested without draining should have mass discharges similar to food-fish ponds. Levee ponds discharge lower quantities than watershed ponds because they have smaller effluent volumes than watershed ponds.

Total phosphorus loads were high, but the majority of phosphorus was in particulate form (83% in overflow, 96% in drawdown effluent and final draining effluent). Overflow occurs mostly in the winter and spring during heavy rains, and ponds are turbid with suspended soil particles at this time. Thus, most phosphorus is apparently associated with soil particles. During drawdown and final draining, most phosphorus is probably associated with plankton and dead organic matter. Most of the organic nitrogen and BOD_5 is associated with plankton and dead organic matter.

Because they are drained annually, fry and fingerling ponds release proportionately greater nutrients and organic matter loads than food-fish ponds. Using the data in Table 5 and the relative area represented by fingerling and food-fish ponds in Alabama, Boyd et al. (2000) showed that fry and fingerling production release about 35% of the total suspended solids, 30% of BOD_5, 35% of total nitrogen, and 18% of total phosphorus, but they represent only 10.6% of total pond area.

Effects of Water Management on Mass Discharge

Because mass discharge is the product of concentration and volume, reducing effluent volume can be an effective approach to reducing potential environmental impacts.

Effluent volume is affected by how often the pond is drained and how much water storage capacity is maintained in the pond. Manipulating these two factors can significantly reduce the amount of water discharged from ponds.

Reusing Water for Multiple Fish Crops

The most obvious procedure for reducing effluent volume is to harvest the fish without draining ponds. This practice is usable only if year-to-year water reuse does not cause reduced fish yields due to deterioration of water quality over time. Seok et al. (1995) conducted a study to compare water quality and fish production between annually drained ponds and undrained ponds over a 3-year period and found no differences in net fish production, average fish weight at harvest, and feed conversion ratio attributable to draining ponds between crops. Phytoplankton blooms, as estimated from chlorophyll *a* concentrations, increased each summer growing season in response to greater nutrient inputs from fish feed. However, averaged over the entire study period, chlorophyll *a* concentrations did not differ between treatments. There were also no differences in dissolved oxygen concentration between treatments or among years. However, total ammonia-nitrogen concentrations were higher in undrained ponds during the second and third year of the study, although on the last sampling dates in the third year, total ammonia-nitrogen concentrations were similar in both treatments. Concentrations of suspended solids, BOD_5, Kjeldahl-nitrogen, and total phosphorus increased as the growing seasons progressed each year, but there were no treatment-related differences for any variable.

In the study conducted by Seok et al. (1995), few differences in water quality and no differences in fish production were observed between annually drained and undrained catfish ponds over a 3-year period. By extrapolation, these findings suggest that overflow from catfish ponds that are not drained for 3 years will be similar in quality to those from catfish ponds drained each year. Using ponds for several years without draining and refilling is possible because natural microbiological and physicochemical processes continually remove nutrients and organic matter from pond water. The rate at which these processes act is such that channel catfish ponds in the southeastern United States can be used for many years without significant long-term accumulation of nutrients and organic matter in the water column, despite large inputs of metabolic waste resulting from fish feeding practices and settled phytoplankton. When ponds are drained, the amount of nutrients and organic matter discharged is therefore much lower than the total waste loading to the pond.

Operation of ponds without draining makes better use of the waste assimilation capacity of the pond ecosystem; it is not merely

TABLE 6. *Predicted discharge (kg/ha of pond surface per yr) of total nitrogen, total phosphorus, and biochemical oxygen demand from levee-type channel catfish ponds under two water-management scenarios (No S = pond water level managed with no water storage potential; S = pond water level managed to maintain a minimum 7.5 cm of water storage potential). Values were calculated as the product of measured concentrations of each variable (Table 1, this chapter) and pond overflow volumes (Table 4, Chapter 2). Adapted from Tucker et al. (1996).*

Season	Average yr		Wet yr		Dry yr		Wet quarter		Dry quarter	
	No S	*S*	*No S*	*S*	*No S*	*S*	*No S*	*S*	*No S*	*S*
Total Nitrogen (as N)										
Spring	14.7	4.2	24.6	11.3	9.7	0.0	28.0	15.7	5.1	0.0
Summer	12.4	1.0	16.9	0.0	8.6	0.0	19.9	6.8	5.2	0.0
Autumn	15.2	2.0	23.2	1.9	12.9	0.0	28.2	13.1	6.5	0.0
Winter	17.2	10.1	22.8	18.1	6.8	0.0	33.9	24.7	6.2	0.0
Total	59.5	17.3	87.5	31.3	38.8	0.0	109.0	60.6	23.0	0.0
Total Phosphorus (as P)										
Spring	1.0	0.3	1.7	0.8	0.6	0.0	2.0	1.1	0.4	0.0
Summer	0.9	0.1	1.2	0.0	0.6	0.0	1.5	0.6	0.4	0.0
Autumn	0.7	0.2	1.1	0.1	0.6	0.0	1.3	0.6	0.3	0.0
Winter	1.1	0.7	1.5	1.2	0.4	0.0	2.2	1.6	0.4	0.0
Total	3.7	1.3	5.5	2.1	2.2	0.0	7.0	3.9	1.5	0.0
Biochemical oxygen demand (as O_2)										
Spring	45	13	75	35	28	0	86	48	16	0
Summer	41	3	56	0	28	0	66	22	17	0
Autumn	25	3	39	3	22	0	48	22	11	0
Winter	42	25	56	45	17	0	84	61	15	0
Total	153	44	226	83	95	0	284	153	59	0

a way to delay discharge of wastes. Increasing the interval between pond drainings allows natural processes to remove more wastes before discharging the water. Water reuse for multiple crops will result in significant savings in water use and will also reduce overall effluent volume.

Maintaining Water Storage Capacity

Another practice that can be used to reduce mass discharge is to maintain some water storage capacity in the pond so that rainfall is captured rather than allowed to overflow. Maintaining storage capacity is easy: when water is added to ponds to replace water lost to evaporation and seepage, the pond is not refilled completely. This leaves several inches of storage capacity so that rainfall is captured rather than allowed to overflow. The effect of this practice on effluent volume was discussed in Chapter 2. Tucker et al. (1996) combined the modeled overflow volumes and actual concentrations of pond water quality variables to assess the impact of storage volume on mass discharge from levee ponds. That exercise revealed that seasonal changes in overflow volume are much more important than seasonal changes in pond effluent composition in determining mass discharge of nutrients and organic matter. Specifically, predicted mass discharge was greatest in the winter when overflow volume was maximum and not in the summer when concentrations of nutrients and organic matter in the pond were highest (Table 6).

The data in Table 6 show that nitrogen, phosphorus, and organic matter discharged in an average year from ponds managed to capture rainfall is only about 30% of that discharged from ponds not managed to maintain water storage capacity. Even in wet years, discharge of nutrients and organic matter was predicted to be considerably lower from ponds managed to capture rainfall. In dry years, no water overflowed from ponds at all. The greatest relative reduction in predicted nutrient discharge occurred in the summer months. In an average summer, managing pond water levels to capture rainfall reduced the predicted discharge of nitrogen and phosphorus by about 90% compared to summertime discharge from ponds not managed to capture rainfall. Reducing waste discharge in summer is particularly important because the quality of potential pond effluents is poorest and stream flows are at their annual minimum, resulting in low rates of dilution of any water discharged from ponds. Most of the water lost from catfish ponds in the southeastern United States occurs in the winter and early spring when rainfall is greatest. During those periods of high rainfall, pond effluents will have little impact on the water quality of effluent receiving streams because stream flows are high, which greatly dilutes any material discharged in pond effluents, and stream water quality is already heavily impacted by erosion from fallow row-crop lands.

The effect of combining water reuse for multiple crops and maintenance of water storage volume was modeled for levee ponds operated with three intervals (1, 3, and 5 years) between total pond drainings (Tucker et al. 1996). Harvesting fish without draining ponds between crops substantially reduced the average volume of water discharged each year, and the reduction was greatest when ponds were also managed to maintain water storage potential (Table 7). For ponds not managed to maintain surplus water storage, the model showed that using ponds for 5 year before draining reduced annual average waste discharge by approximately 45% compared to annually drained ponds. When pond water levels were managed for water storage potential, discharge of nutrients and organic matter was reduced relative to annually drained ponds by more than 60% when ponds were used for 5 years between drainings.

The budget in Table 7 illustrates how

TABLE 7. *Annual quantities (kg/ha of pond surface) of nitrogen and phosphorus added to catfish pond in feed, produced by catfish as metabolic waste, and discharged from ponds. Discharge was calculated for two water management scenarios (with and without water storage capacity) and two intervals between pond drainings. Adapted from Tucker (1991) and Tucker et al. (1996).*

	Nitrogen	Phosphorus
Amount added in feed	500	100
Waste produced	400	80
Waste discharged		
Ponds drained annually		
Rainfall not captured	110	7
Rainfall captured	75	5
Ponds drained every 5 yr		
Rainfall not captured	70	4
Rainfall captured	30	2

natural waste-reduction processes within the pond interact with simple water management practices to reduce the amounts of nitrogen and phosphorus eventually discharged from catfish ponds. This budget was produced by assuming a net annual production of 5,000 kg of fish from 10,000 kg of feed. The amounts of waste nitrogen and phosphorus produced within the pond (400 kg/ha for nitrogen and 80 kg/ha for phosphorus) are the amounts that would reach the environment if wastes were immediately discharged to the environment. However, if catfish are grown in ponds managed to capture rainfall and drained every 5 years, 30 kg/ha of nitrogen and 2 kg/ha of phosphorus are discharged annually. This is only about 7% of the waste nitrogen and 3% of the waste phosphorus produced in growing 5,000 kg of fish/ha. So, without resorting to any formal waste treatment technology, and using only the two simple water management practices already in common use on most catfish farms, waste treatment efficiencies exceeding 90% can be attained for nitrogen and phosphorus.

Managing Ponds to Reduce Mass Discharge

In this section, three broad approaches to reducing quantities of substances discharged from catfish ponds will be discussed. These include: 1) reducing discharge volume; 2) decreasing waste production within the pond by reducing feeding rates, by increasing feed nutrient retention by fish, or by optimizing the amount of nitrogen and phosphorus in the feed; and 3) increasing rates of in-pond biological and physicochemical loss processes for nitrogen, phosphorus, and organic matter to reduce concentrations of those substances in the pond before water is discharged. Managing pond effluents by treating the water after it has been discharged from ponds is discussed in the next section

Reduce Effluent Volume

Reducing effluent volume appears to be the most practical way to reduce nutrient and organic matter discharge from catfish ponds. This can be accomplished by reusing water in food-fish production ponds for multiple fish crops, maintaining water storage capacity in ponds so rainfall is captured rather than allowed to overflow, and reducing or eliminating water exchange using pumped water during the production cycle. This topic is discussed in further detail in Chapter 2.

The effects of water management on effluent volume and mass discharge were discussed in the previous section. Beyond the simple practices of harvesting fish without draining the pond and maintaining water storage capacity in ponds, there are other possibilities for reducing effluent volume in catfish farming. For example, when it is necessary to completely drain ponds, water can be pumped or drained into adjacent ponds and stored. Water stored in this manner could then be drained or pumped back into an empty pond for reuse. On large farms, it may be possible to transfer water to a storage reservoir. Water quality would improve over time in the storage reservoir through natural water purification processes and the water could then be reused.

Another approach to reducing effluent volume is based on using some ponds on a farm for both fish production and water storage (Cathcart et al. 1999). The production/storage ponds are 0.3 to 1 m deeper than typical production ponds to provide additional volume for storage of rainfall. The production/storage ponds are linked via culverts to 1 to 3 production ponds so that overflow from all ponds in the linked system drains into the production/storage pond. Overflow occurs only when storage capacity of the linked system is exceeded. Mathematical modeling using a 26-year climatological record showed that effluent discharge from linked ponds can be reduced by 40–90% (depending upon rainfall) relative to single ponds refilled to the top of the overflow device every time the water level drops 7.6 cm. A field study to validate the model and identify any practical limitations to this approach is currently being conducted at the National Warmwater Aquaculture Center in Stoneville, Mississippi.

Prior to about 1985, water exchange was used liberally in catfish ponds in attempts to correct water quality and fish health problems. Catfish farmers believed that flushing the pond with pumped water would substantially improve environmental conditions and benefit the fish population. However, the displaced pond water represents a pollution load in receiving waters, and no benefit of water exchange as a water quality management procedure in large commercial aquaculture ponds has been shown. Because incoming water is greatly diluted when added to large ponds, it is unlikely that sufficient water can be exchanged quickly enough to have a beneficial effect during acute water quality crises. Also, research has shown that routine water exchange at low rates (several pond water exchanges over the growing season) has little or no effect on pond water quality and should not be used (McGee and Boyd 1983). When catfish ponds are properly managed within their assimilative capacity by using conservative fish stocking and feeding rates, natural biological activity and mechanical aeration maintain adequate environmental conditions for culture, and pumped water is needed only to fill ponds and replace evaporation and seepage losses.

Reduce Waste Production within Ponds

Despite the use of high quality manufactured feeds, relatively little of the feed nutrient content is ultimately converted to catfish flesh and removed from ponds at harvest. Under commercial conditions, only about 10–20% of the carbon, 15–25% of the nitrogen, and 20–30% of the phosphorus in feed is removed in fish at harvest (Boyd 1985). The remainder represents the nutrient load to the ponds and, when feeding rates are high during the summer months, plant nutrients derived from fish wastes stimulate luxuriant phytoplankton growth. Phytoplankton then contributes to the organic load in ponds by fixing two to four times as much carbon in net photosynthesis as that added to ponds in feed. Reducing the waste load in ponds can be accomplished by decreasing the amount of feed added to the pond or by maintaining high feeding rates but increasing the efficiency of food utilization.

Water quality in ponds is directly related to the amount of feed added to ponds, which is a function of fish biomass and water temperature. In static pond aquaculture, phytoplankton abundance is regulated by plant nutrient availability (derived from fish wastes) up to daily feeding rates of about 30–50 kg/ha per d. At feeding rates below that level, average phytoplankton abundance is moderate and concentrations of soluble reactive phosphorus and dissolved inorganic nitrogen are low. At those feeding rates, maximum fish production can be no more than 2,000–3,000 kg/ha per year assuming a 150- to 180-d growing season and a feed conversion ratio of about 2.0. Annual fish yields this low are generally considered too low for profitable large-scale commercial aquaculture.

To achieve greater production and, presumably, profitability, it is necessary to stock more fish and provide more feed. However, when feeding rates exceed 30–50 kg/ha per d, water quality rapidly deteriorates as a result of excessive phytoplankton production, which leads to increased oxygen demand. Supplemental aeration allows feeding rates in excess of 50 kg/ha per d to be used profitably without high risk of fish loss. However, supplemental aeration does not allow unlimited feeding rates because variables other than dissolved oxygen availability (such as accumulation of ammonia, carbon dioxide, or other unidentified substances) begin to affect feed consumption (appetite) and ultimately limit fish production in ponds operated without water exchange. The maximum daily feeding rate at which net fish production does not increase in proportion to increases in feeding rate appears to be in the range of 100 to 150 kg/ha (Cole and Boyd 1986). These feeding rates will limit net annual production to 8,000 to 10,000 kg/ha per year, a level seldom achieved under commercial conditions.

Most catfish farmers manage ponds at a level between the two critical feeding rates described above. Pond water quality (and, therefore, effluent quality) could be improved by reducing feeding rates, but based on the few data available (primarily Cole and Boyd 1986), truly significant improvement in water quality appears possible only by reducing average daily feeding rates to values less than about 50 kg/ha per d. It is doubtful if catfish farming would be profitable on a large scale at the production levels possible with that level of feed input.

Rather than reducing the total amount of feed offered to fish, another approach is to improve nutrient utilization, thereby reducing waste production per unit of feed consumed by fish. Most efforts have concentrated on improving utilization of phosphorus, which is considered to be the most important potential nutrient pollutant released from fish culture facilities. There are obvious benefits to reducing waste phosphorus generation from fish cultured in raceways, net pens, and other facilities that discharge directly to receiving streams. However, the benefits of improved phosphorus utilization in static-water ponds with high fish densities are less clear. The reduction in phosphorus loading to the water possible by diet modification is overwhelmed by the complex fates of phosphorus within the pond ecosystem. As such, modest improvements in feed phosphorus utilization have not resulted in improved quality of potential pond effluents.

A series of unpublished studies have been conducted by Edwin Robinson and Craig Tucker at the National Warmwater Aquaculture Center, Stoneville Mississippi, to evaluate the effect of improving phosphorus utilization by catfish on water quality and phytoplankton abundance in ponds. Three approaches were evaluated: 1) use feeds with the lowest possible phosphorus supplementation to meet the dietary requirements; 2) use a water-insoluble phosphorus supplementation (deflourinated rock phosphate); and 3) use microbial phytase to improve the bioavailability of plant phytate phosphorus. Regardless of the approach used to reduce waste phosphorus loading in ponds by modifying practical diets, no differences were seen in concentrations of soluble phosphorus, total phosphorus, or phytoplankton abundance.

In another study, Gross et al. (1997) measured phosphorus concentrations in channel catfish ponds receiving equal amounts of feed with phosphorus concentrations of 0.60, 0.68, 0.75, 0.81, and 1.03%. At the end of the grow-out period when feeding rates were highest, similar amounts of phosphorus were present in all pond waters regardless of diet. Phosphorus did not accumulate in waters of ponds with high phosphorus diets because more phosphorus was adsorbed by bottom soils in those ponds than in ponds with low phosphorus diets. Bottom soil adsorption and sediment oxidation-reduction potential largely control phosphorus concentrations in pond water, so lowering phosphorus concentrations in diets is beneficial, even if it does not result in direct reduction of water-borne phosphorus levels. Long-term reduction in phosphorus loading reduces the input of phosphorus to bottom soils and conserves the capacity of soils to adsorb phosphorus.

Feed utilization efficiency can be improved through careful feeding. Feeding fish more than they can consume in a relatively short period (20–30 min) is wasteful and leads to poor feed conversion. Because feed typically represents about 50% of production costs, inefficient feeding practices can have a large effect on profitability. In addition, decomposition of uneaten feed exerts an unnecessary oxygen demand and releases nutrients that contribute to overabundance of phytoplankton.

Enhance Within-Pond Removal of Nutrients and Organic Matter

Natural biological and physicochemical processes within ponds act to reduce nutrient and organic matter levels in potential effluents far below the levels expected from simple mass balance of inputs and outputs. Effluent quality could therefore be improved if rates of these natural processes could be enhanced. Approaches that have been examined include the following: 1) aeration or circulation to enhance the dissolved oxygen supply; 2) precipitating inorganic phosphorus from water with soluble calcium or aluminum salts; 3) using bacterial or enzyme amendments (bioaugmentation); and 4) assimilation of nutrients by aquatic plants.

Aeration. Mechanical aeration is the most common procedure for improving water quality in catfish ponds. Aeration provides a zone of elevated dissolved oxygen sufficiently large to maintain the cultured fish biomass. It also prevents thermal stratification and reduces the development of anaerobic conditions in deeper water and at the soil-water interface. By enhancing dissolved oxygen concentrations, aeration increases the capacity of ponds to assimilate organic matter by aerobic processes. Higher dissolved oxygen concentrations also increase the nitrification rate of ammonia to nitrate, which is then lost from the pond through denitrification in the large volume of anoxic sediment.

Aeration and water circulation also affect rates of phosphorus loss from pond water. Formation of oxidized ferric phosphates and phosphorus occluded in ferric oxyhydroxide coatings on soil particles in the surface layers of sediment are important sinks for orthophosphate from the overlying water. The thin layer of oxidized surface sediment also functions as a barrier preventing the release of orthophosphate from deeper, anaerobic layers of mud into the overlying water. In the absence of an oxidized surface layer, soluble reactive phosphorus will be released from the sediment in response to a concentration gradient between sediment porewater and the overlying water column (Masuda and Boyd 1994b; Hargreaves and Tucker 1996). This was demonstrated under field conditions by Masuda and Boyd (1994a), who showed that soluble reactive phosphorus concentrations were higher in unaerated ponds than in aerated ponds. Apparently, the sediment-water interface in aerated ponds was sufficiently oxidized to function as an effective barrier to the movement of soluble reactive phosphorus from sediment porewaters. So, adequate aeration and circulation can enhance rates of inorganic phosphorus removal from pond water, with the overall effect of reducing phosphorus availability to phytoplankton as well as reducing concentrations of soluble reactive phosphorus.

Despite a clear need for aeration in aquaculture ponds, excessive aeration and circulation can have deleterious effects. Erosion of pond bottoms and levees by strong water currents produced by certain types of aerators can suspend large amounts of particles and potentially increase the suspended solid concentration in pond effluents.

Chemical removal. Alum (aluminum sulfate) has a long history of use in potable water supplies to reduce particulate turbidity. Alum treatment also quickly reduces the amount of phosphorus in water by co-precipitating phosphorus with flocs of aluminum hydroxide formed after alum is added to water. Alum dosages of 0.5–25 mg Al/L have been used to remove phosphorus from water in lake restoration programs (Cooke and Kennedy 1981). Smaller doses are commonly used to precipitate phosphorus from the water column while larger doses are used to tie up phosphorus in the sediment to reduce phosphorus release from bottom soils.

Masuda and Boyd (1994a) found that treatment of channel catfish culture ponds with 20 mg/L alum (about 1.8 mg Al/L) reduced soluble reactive phosphorus concentrations by about 50% and total phosphorus concentrations by about 80%. Much of the phosphorus

removal from treated water was attributable to precipitation of phosphorus-containing suspended matter. Alum treatment does not cause a long-term increase in Al^{3+} concentrations because it is quickly converted to insoluble aluminum hydroxides that precipitate to the bottom soil. Thus, alum has little residual effect, and phosphorus concentrations quickly increase in aquaculture ponds in response to continuing inputs in feed. Frequent treatment would be needed for long-term control of phosphorus levels.

In soft-water ponds, phosphate can be removed by increasing the concentrations of calcium, which forms poorly soluble calcium phosphates at pH values above neutrality. Gypsum ($CaSO_4 \cdot 2H_2O$) is an inexpensive and highly soluble source of calcium. Gypsum has an advantage over alum as a phosphorus-precipitating agent in that calcium is only slowly lost from pond waters (unless the water is rapidly diluted with low-calcium water), so treatment of pond waters with gypsum should influence phosphorus levels for a longer period of time than alum treatment. Wu and Boyd (1990) used gypsum to increase calcium concentrations of fertilized fish ponds from 2–3 mg/L to about 50–60 mg/L. Soluble reactive phosphorus concentrations in gypsum-treated ponds were reduced by about 95% relative to those in control ponds. It should be noted, however, that many catfish ponds are supplied with ground water that is naturally high in calcium, and addition of gypsum will have little effect on phosphorus concentrations in those ponds. Also, most of the phosphorus in catfish pond waters is present in the particulate organic fraction, principally in living phytoplankton or phytoplankton-derived detritus. Increasing water hardness using gypsum will have little effect on phytoplankton abundance (and, by extension, total phosphorus concentrations) as evidenced by abundant phytoplankton in many nutrient-enriched waters with extremely high calcium concentrations.

Bioaugmentation. Static pond aquaculture is possible because natural pond microbial communities perform many of the functions required to maintain adequate environmental conditions for fish growth. These functions include organic matter decomposition and the transformation and eventual loss of waste nitrogen from the pond. The central role of the microbiological community in pond ecology has led to the notion that water quality can be improved by augmenting native microbial communities with microorganisms produced in culture. This approach to ecosystem manipulation is called "bioaugmentation" and is based on the belief that the inoculum increases total microbial biomass or abundance of certain microbes in the pond (either directly or through subsequent growth of the inoculum), thereby increasing rates of certain microbial processes. Another approach involves bypassing the role of microorganisms by adding enzymes directly to the water.

Numerous studies of catfish pond bioaugmentation have been conducted at Auburn University, in Alabama, and at the National Warmwater Aquaculture Center, in Stoneville, Mississippi. However, only three studies have been published (Tucker and Lloyd 1985b; Queiroz and Boyd 1998; Queiroz et al. 1998) because much of the data is proprietary or the results were equivocal. In one study (Queiroz and Boyd 1998) application of a bacterial inoculum improved catfish survival but the difference in survival could not be directly linked to the inoculum. In no study conducted at either location (published or unpublished) has bioaugmentation significantly improved water quality.

The lack of response to pond bioaugmentation is not surprising given the current level of understanding of catfish pond ecology. Large organic matter additions to ponds (either as fish waste or through algal production) provide ample substrate for bacterial growth, sustaining abundant and diverse microbial communities. The activity of the microbial community, and the resulting rate of organic matter decomposition, in catfish ponds is more likely limited by the oxygen availability than by the abundance of a particular group of microorganisms.

Nutrient removal by aquatic plants. Considerable work has been conducted on the use of floating aquatic plants that are intentionally cultured in fish ponds to remove nutrients and improve water quality. Most research has focused on the use of water hyacinths, a tropical to subtropical floating plant that is usually considered to be a highly undesirable weed. Water hyacinths remove large quantities of nutrients from the water when they are actively growing and, when present over at least 10% of the pond surface, they reduce phytoplankton abundance through competition for nutrients and light (McVea and Boyd 1975; Boyd 1976). However, dissolved oxygen concentrations are reduced in ponds with hyacinths, and plants must be harvested frequently to maintain rapid nutrient removal rates. The labor involved in harvesting hyacinths and the resulting disposal problem (hyacinths are 90% water and have no commercial value) are significant obstacles to the widespread use of this practice. Large mats of water hyacinth may also interfere with normal culture operations, especially fish harvest, and may serve as a refuge for fish predators. Evapotranspiration rates from ponds with water hyacinths exceed evaporation rates from open ponds, thereby increasing the need for pumped groundwater. Also, hyacinths thrive only in warm water and will have no effect on water quality during the winter when most water is discharged from catfish ponds. On the contrary, if hyacinths are not harvested in late autumn, they will release nutrients and organic matter back into the pond when they die and decompose. Under those conditions, effluent quality will be worse in the winter than if plants were not present at all.

Treating Effluents to Remove Potential Pollutants

Many schemes have been proposed for treating pond effluents or using water discharged from ponds for a beneficial purpose. Proposed practices include: using retention ponds to prevent discharge, removing nutrients and organic matter with traditional wastewater treatment processes; using effluents for hydroponics or to grow another crop of aquatic animal; and using effluents for irrigation of row crops or rice, treatment of effluents in wetlands or settling basins, and water reuse. It is important to emphasize that the intermittent nature of discharge and the winter-spring peak in discharge volume impact the cost and potential effectiveness of nearly all treatment options

Concentrated animal feeding operations (CAFOs), such as cattle feedlots, are not allowed to discharge wastewater. Discharge from CAFOs is prevented by constructing retention ponds to retain all wastewater plus storm overflow from a 25-year rain. Using retention ponds to prevent discharge from catfish ponds does not, however, appear feasible because the amount of water discharged from ponds during storm overflow is much greater than that discharged from CAFOs. Boyd and Queiroz (2001) calculated that for every ha of levee pond, 1.53 ha of retention pond would be needed to retain the overflow from a 25-year rainfall event in Alabama. The situation is even worse for watershed ponds because of additional runoff from the catchment area above the ponds. Retention ponds would need a surface area of 11 ha for every 1 ha of watershed production pond. Obviously, committing more land area to retention ponds than for fish production is not economical. The most significant implication of the work by Boyd and Queiroz (2001) is that catfish ponds in the southeast and other areas where rainfall is relatively abundant cannot be operated profitably without effluent.

Wastewater treatment procedures such as mechanical or biological filtration and activated-sludge processes are not economically feasible because nutrient and organic matter concentrations in pond effluents are too dilute to make these treatment procedures effective. Another complicating factor is that average annual hydraulic loading to treatment systems will be low, but when discharge occurs, the volume will be large for a brief period. This is a difficult engineering problem because systems must be designed to rapidly treat a large volume of dilute wastewater. The intermittent nature of pond discharges will also affect economic performance of treatment systems because they will be idle for many more days than they are used. In short, conventional wastewater treatment technologies are too expensive to use with pond effluents.

Nutrients in catfish pond effluents are not concentrated enough for use in hydroponics unless nutrient concentrations are supplemented, which defeats the purpose of using this procedure to "treat" effluents. Also, it is difficult to visualize hydroponics being developed to the scale where any significant proportion of the catfish industry's discharge can be treated in that manner. Filter-feeding fish and mollusks and certain plants have been successfully cultured in effluents, but this practice has seldom been economical (in part because of limited markets for the "second" crop), and it does not greatly improve effluent quality. Again, a significant limitation to these uses of pond effluent is the intermittent and seasonal nature of discharge.

The three processes that appear to have some application in treatment of pond effluents are wetlands, settling basins, and crop irrigation. Even these three approaches have serious drawbacks that make their use problematic.

Wetlands

Using wetlands to treat wastewaters is based on removal of nutrients and solids as the water is slowly passed through a shallow, vegetated impoundment. Nutrients are assimilated by wetland plants, removed by physicochemical processes, such as precipitation and adsorption reactions in the soil, and transformed and removed by biological reactions associated with the vast surface area provided by plant roots and above-ground plant biomass. Solids are removed by filtration and settling as water slowly passes through the system.

Schwartz and Boyd (1995) studied constructed wetlands for treating catfish pond effluents in west-central Alabama. Water from a single pond was passed through a constructed wetland consisting of two cells, one planted with California bulrush *Scirpus californicus* and giant cutgrass *Zizaniopsis miliacea* and one planted with Halifax maidencane *Panicum hemitomon*. Removal of potential pollutants from water flowing through the wetland was determined for 1-, 2-, 3-, and 4-d hydraulic residence times (HRTs). Concentrations of potential pollutants were much lower in effluent from the wetland than in influent from the catfish pond. The following reductions in concentrations were recorded: total ammonia-nitrogen, 1–81%; nitrite-nitrogen, 43–98%; nitrate-nitrogen, 51–75%; total Kjeldahl-nitrogen, 45–61%; total phosphorus, 59–84%; BOD_5, 37–67%; suspended solids, 75–87%; volatile suspended solids, 68–91%; and settleable solids, 57–100%. Overall performance of the wetland was best when operated with a 4-d HRT in the vegetative season, but good removal of potential pollutants was achieved for shorter HRTs. The wetland was relatively efficient in improving water quality even in late fall and winter when vegetation was dormant.

The disadvantage of constructed wetlands for treating wastes from channel catfish ponds is the large area necessary to provide adequate hydraulic residence time when the process is used on large farms. In particular, the relationship between pond area and wetland area needed for effective effluent treatment during the high-discharge winter periods is not known.

An alternative to using wetlands designed to treat all effluents from a farm would be to construct a small wetland to treat only the most concentrated effluents released when ponds are drained in the drier seasons—the time of greatest potential environmental impact because of low rates of dilution in effluent-receiving streams. In other words, effluents released during peak discharge would not be treated because of the large wetland area needed to provide an effective HRT and the great dilution provided by high receiving stream flows. This approach would minimize the land needed for constructed wetlands and significantly improve effluent quality during dry periods. However, the overall reduction in mass discharge of nutrients and organic matter, and the costs associated with this scaled-down approach are unknown.

Settling Basins

Settling basins are easier to construct and operate than wetlands because they do not have to be seeded with plants. Findings summarized in Table 8 suggest that settling basins can be just as effective as wetlands in improving the quality of catfish pond effluents. Laboratory-scale studies of sedimentation (Boyd et al. 1998) also revealed that large reductions in concentrations of key water quality variables could be effected. Removal of total phosphorus, total suspended solids, turbidity, and BOD from water discharged during the final stages of pond draining was usually greater than 50% after 8 h. This results because much of the phosphorus and organic matter in pond effluents are associated with inorganic suspended solids, primarily suspended soil

TABLE 8. *Effectiveness of settling basins and constructed wetlands for treating catfish pond effluent. Values represent percentage removal of the variables. Adapted from Boyd et al. (1998).*

Variable	Settling basin, 2-d retention	Constructed wetland, 2-d retention
Total suspended solids	96%	89%
Kjeldahl nitrogen	74%	65%
Total phosphorus	69%	64%
Biochemical oxygen demand	59%	60%

particles, which settle quickly if effluent is held in settling basins. Boyd (1995) discusses the sedimentation process and methods for computing the hydraulic residence time for removing solids of different sizes. Unfortunately, most organic suspended solids in aquaculture ponds are associated with phytoplankton (most are less than 30 µm in diameter) that does not settle readily in sedimentation basins.

Boyd and Queiroz (2001) studied the feasibility of using settling ponds to treat effluent from pond draining alone or effluent from pond draining and storm overflow. Their calculations used data on Alabama climatology, land-use, and soil types, and were based on using a settling pond with mean depth of 1.5 m and an 8-h hydraulic residence time. Average settling basin areas to treat effluents from pond draining alone are 0.25 ha per hectare of levee pond and 0.28 ha per hectare of watershed pond. Watershed ponds tend to be slightly deeper than levee ponds, which accounts for the need for a larger settling pond on a per-hectare of production pond basis.

Because of the effect of watershed area on runoff volume, settling ponds to treat overflow from storms must be much larger for watershed ponds than for levee ponds. For levee ponds, settling pond areas (per hectare of production pond area) are 0.07 ha for rainfall from a 25-year storm, 0.08 ha for a 50-year storm, and 0.09 for a 100-year storm. Settling pond areas (per hectare of production pond area) for watershed ponds are 0.43 ha for rainfall from a 25-year storm, 0.50 ha for a 50-year storm, and 0.57 for a 100-year storm.

Estimates of settling pond size were used by Boyd and Queiroz (2001) to calculate settling pond areas needed to treat draining effluent or draining effluent plus storm overflow for catfish farms of various sizes. Percentages of the farm area devoted to settling ponds for draining effluent from levee ponds ranged from 0.7% for a 200-ha farm to 14% for a 10-ha farm. The corresponding range for watershed ponds is 0.8% to 15%. To treat overflow from a 25-year storm on farms greater than 25 ha, settling ponds would constitute 7% of the farm area for levee ponds and 43% of the farm area for watershed ponds.

Boyd et al. (2000) observed that most catfish farms in Alabama extend downslope on watersheds to streams or property lines, which precludes installation of settling ponds unless existing ponds are taken out of production and reconfigured as settling ponds. Even where space is available, Boyd and Queiroz (2001) believe that land and construction costs for settling ponds large enough to treat draining and overflow effluents would be prohibitive.

The obvious disadvantage of settling basins is that a substantial amount of land must be available for construction. However, an empty production pond can be used as a temporary settling basin, although on large farms, the logistics of pumping water over great distances may make this practice impractical. The economics of these practices are unknown and, once again, the fact that most discharge occurs after heavy rain events in the wet season will probably introduce scaling problems that make it impractical to use settling basins to treat all the effluent from ponds.

For watershed ponds that must be partially drained to facilitate fish harvest, another possible approach, and one that does not require additional investment, is simply using the production pond as its own settling basin. When ponds are drained during fish harvest, most of the solids, organic matter, and nutrients are released in the last 20% of water discharged from ponds. So, when ponds are drained, the final volume of water may be held in the pond for 2–3 d to allow solids to settle before draining completely. Holding this last portion of water without discharge is even more desirable and can further minimize the potential environmental impacts of catfish pond effluents. Alternatively, the last 20% of water discharged from ponds can be held in farm drainage ditches for settling prior to final discharge.

Crop Irrigation

Most catfish farming is practiced in areas already used for intensive row-crop or rice agriculture. As such, it appears logical that pond effluents could be used to irrigate terrestrial crops. The primary goal of integrating catfish aquaculture and terrestrial agriculture would be to make productive use of pond effluents by supplementing the water supply for irrigation. Nutrients in the pond effluent might also be beneficial to the crop.

Problems with using pond waters for irrigation are as follows: 1) peak water demand for irrigation occurs at the time when there is little or no overflow from fish ponds, 2) catfish producers object to draining ponds during the summer growing season, and 3) evaporation and seepage losses from ponds would be greater than that if the water were directly applied to fields. Also, in some types of irrigation (such as flooding of rice fields) water is needed quickly, in relatively large volumes, and at a specific time—which may or may not correspond to the availability of water from a catfish pond. For example, a crop of rice requires about 0.75 to 1.5 m of water, which corresponds to nearly the entire volume of most catfish ponds for an equivalent area of rice. Previously, we mentioned that low water exchange did not improve water quality in catfish ponds, so use of catfish pond water for irrigation would not benefit catfish production. Further, the need for large water volumes delivered quickly means that gravity flow from pond discharge devices may not be sufficient to provide the flow rates that are needed. Rapid delivery of large water volumes would require installation of large pumps and water distribution systems to convey water where it is to be applied in the required amounts.

The benefit of the "nutrient load" in pond effluents to the irrigated crop is largely an illusion because effluents from catfish ponds are, in fact, quite dilute with respect to major plant nutrients and most nutrients are bound in relatively unavailable, organic forms. For example, to supply 1 kg of ammonia-nitrogen to 1 ha of land would require the application of 0.2 m of irrigation water, assuming an ammonia-nitrogen concentration of 0.5 mg/L in the irrigation water. Nitrogen is usually applied to rice as a top dressing at 35–50 kg/ha, an amount contained in 7 to 10 m of catfish pond water. Thus, the contribution of nutrients in catfish pond water to the irrigated crop is small. The lack of benefit of the nutrients in catfish pond water to the irrigated crop was verified in an unpublished study conducted at Stoneville, Mississippi, where rice fields were irrigated over a 3-year period using water pumped from a catfish pond. Rice yield across several nitrogen fertilization practices did not differ when using catfish pond water or ground water as a source of irrigation water.

Management Practices to Reduce Environmental Impacts

As indicated by the discussion in the previous two sections, the unique nature of pond aquaculture poses challenges for effluent management, particularly when traditional "monitor and treat" approaches to pollution abatement are considered. First, there is little indication that discharges from aquaculture pond pose a significant environmental threat. Accordingly, there is limited incentive for farmers to change production practices or implement costly treatment practices, particularly in times of thin profit margins. Second, the nature and timing of discharge makes treatment a difficult engineering problem. Discharges are infrequent, which makes monitoring difficult, and when discharge occurs, it may be large in volume but dilute, which makes treatment difficult.

Overall, adoption of management practices that minimize environmental impacts may be a more effective means of implementing environmental management for the pond-raised catfish industry than monitoring and post-discharge treatment. These practices, taken as a whole, will optimize mass discharge relative to fish yield by reducing effluent volume or by improving nutrient utilization within ponds.

Certain environmental management practices may also have collateral economic benefits by improving operational and production efficiency. For example, good feeding practices will improve feed conversion efficiency. Improving feed conversion efficiency in turn equates to proportionally fewer nutrients released to the pond environment and therefore improved water quality. In another example, managing ponds to maintain some capacity to store rainfall reduces the requirement for groundwater, thereby reducing costs associated with operation of well pumps, simultaneously reducing the volume of effluent discharged.

Below is a list of recommended management practices that will make farm operations more efficient and provide environmental protection. The list is based on practices described by Schwartz and Boyd (1996), Brunson (1997), Boyd and Tucker (1998), Tucker (1998), Boyd (1999), and Boyd et al. (2000). Practices are divided into sections describing techniques to minimize environmental impacts of pond aquaculture with reference to pond operation and management, fish harvest and draining, and pond construction. A wide variety of culture techniques and culture facilities are used in catfish farming, so some practices may not be applicable to all situations.

Pond Operation and Management

Operate food-fish production ponds for several years without draining. Research has shown that it is possible to maintain adequate water quality for good fish production for at least 3 years in undrained ponds harvested each year by seining. Practical experience indicates that multiple water reuse is possible for much longer. For example, many catfish producers operate food-fish production ponds for 10 years, or longer (average = 6.5 years; USDA/APHIS 1997) before draining. Water reuse for multiple crops not only reduces effluent volume, but also reduces the need for pumped groundwater to refill ponds.

Capture rainfall to reduce effluent volume. Maintaining storage volume by keeping the pond water level 10 to 15 cm (or more) below the drain overflow level greatly reduces discharge volume during rainfall events. Capturing rainfall in this manner also reduces the need for pumped groundwater to maintain pond water levels in levee ponds.

Use high quality feeds and efficient feeding practices. Feeds are the origin of most potential pollutants in catfish pond effluents. Using high quality feeds improves feed conversion efficiency and reduces amounts of metabolic waste and uneaten feed. Efficient feeding practices will improve feed conversion ratios, reduce fish production costs, and decrease waste generation per unit of fish produced. Fish should be fed no more than they will consume in 20–30 min.

Manage within the pond assimilative capacity. Water quality deteriorates when the nutrient load from feeding exceeds the capacity of the pond to assimilate those nutrients. Impaired water quality associated with high fish stocking rates and high feeding rates stresses fish and reduces the efficiency of feed conversion and fish production. High stocking and feeding rates also lead to effluents with a greater pollution potential. Thus, by stocking and feeding at moderate rates, production is more efficient, and the likelihood of water pollution by effluents is less. Maximum daily feeding rates with supplemental aeration should not exceed 150 kg/ha.

Provide adequate aeration and circulation of pond water. Maintaining dissolved oxygen concentrations above 4 or 5 mg/L enhances fish appetite and encourages good feed conversion ratios. Circulation prevents stratification and enhances organic matter decomposition in pond soils. An evenly oxygenated pond will oxidize organic matter rapidly. Oxidation of organic matter within ponds diminishes the amount of organic matter in effluents. Adequate aeration and circulation also enhances nitrogen loss through enhanced nitrification coupled with denitrification, and reduces the solubility of phosphorus in pond water.

Position mechanical aerators to minimize erosion. Aerators produce strong water currents that can erode pond bottoms and levees. Aerators should not be positioned so that strong currents impinge on levees. Areas immediately in front of aerators could be hardened by compaction or rip-rap installed to reduce erosion. Erosion causes sediment deposition on pond bottoms, reduces pond depth, and can result in high concentrations of inorganic suspended solids in effluents.

Eliminate water exchange. Water exchange can reduce high concentrations of ammonia or other toxic substances only if large volumes are exchanged quickly, which is difficult to accomplish in practice because commercial aquaculture ponds contain large water volumes relative to well pumping capacities. Because incoming water is greatly diluted when added to large ponds, it is unlikely that enough water can be exchanged over a short period of time to have a beneficial effect during acute water quality crises. For example, pumping 2,000 L/min of water into a 4-ha, 1.5-m deep catfish pond for 3 d will reduce ammonia concentrations by only about 10%. Research has also shown that routine water exchange at rates possible on most farms (less than 5% of the pond volume exchanged daily) has little or no effect on pond water quality over the long term and should not be used. Furthermore, the displaced pond water represents a pollution load in receiving waters, and high rates of water exchange should not be used unless absolutely necessary.

Harvest and Draining Practices

Allow solids to settle before discharging water. In ponds that are partially drained to facilitate fish harvest, minimize pond draining during seining to avoid the discharge of suspended sediment. After seining, hold water in the pond for 2–3 d to allow solids to settle before draining completely. Alternatively, this last portion of water can be held without discharge. Holding water for 2 d after seining can greatly reduce discharge of solids, organic matter, and nutrients.

Reuse water that is drained from ponds. Rather than draining ponds for fish harvest, water can be discharged or pumped to adjacent ponds and reused in the same or other ponds. Current pond construction practices may not allow water reuse in this way because adequate freeboard may not be available in adjacent ponds to store water pumped or discharged from the pond that is being drawn down. However, production ponds can be built with higher levees, or water levels can be maintained with more freeboard to provide greater storage volume. Water from one pond can be transferred to another with a low-head lift pump and then transferred back by siphon. If topography of the landscape allows, ponds can be constructed to allow draining one pond into adjacent, lower ponds.

Treat pond effluents in constructed wetlands prior to discharge. Although the economics of using wetlands to treat all water discharged from a catfish farm are not encouraging and may be prohibitive for smaller operations, constructed wetlands are very efficient at removing potential pollutants from pond water provided that the wetlands are constructed and managed with at least a 2-d hydraulic retention time. Because large land areas are required for constructed wetlands, treating only the most concentrated effluents in the final stages of draining would minimize the land needed for constructed wetlands and significantly improve effluent quality.

Where possible, release pond effluents into natural wetlands. Wetlands are effective waste treatment systems. Of course, care must be exercised to prevent overloading wetlands with effluents. If hydraulic loadings and waste inputs are too high, ecological perturbations will occur. As above, effective effluent treatment will require a 2-d hydraulic retention time.

Use effluents to irrigate terrestrial crops. Under certain conditions, water discharged from ponds may have value as irrigation water for agronomic crops. However, routine overflow from ponds cannot be relied upon for irrigation water because crop water requirements seldom correspond to the availability of pond effluents. Also, pumping water through a pond solely to provide irrigation water should not be practiced because water is lost to evaporation while the water is in the pond and the nutrient content of water from aquaculture ponds is too low to significantly reduce the crop's fertilizer requirements. In addition, dilution of pond water can reduce the capacity of natural processes to maintain water quality.

Pond Construction and Renovation Practices

Optimize the ratio of watershed to pond area. Watershed ponds should not have watershed areas larger than necessary to keep ponds full because excessively large watersheds increase runoff into ponds and result in high discharge. Runoff from watersheds may be partially diverted from ponds by terracing, supply stream diversion, or other means. The Natural Resources Conservation Service recommends an approximate watershed-to-pond area ratio of 10.

Divert excess runoff from large watersheds away from ponds. This practice can divert turbid runoff where it is not possible to provide erosion control on watersheds. Diversion can extend the production interval between draining and renovation in watershed ponds by minimizing reduction in pond depth caused by deposition of sediment derived from turbid runoff. Less water passing through ponds will reduce erosion of embankments and farm ditches, and the output of suspended solids will be less.

Construct ditches to minimize erosion and

establish plant cover on banks. Ditches should be sufficiently large and of proper cross section to convey farm discharges without excessive currents that can cause bottom scouring and erosion of banks. Plant cover can protect against bank erosion from both in-stream flows and rainfall. This practice can reduce the input of suspended solids to streams and minimize ditch maintenance.

Protect embankments in drainage ditches from erosion. Extending drain pipes beyond the toes of the levees at the point of discharge protects streams from inputs of suspended solids derived from levee erosion. In addition, rip-rap can be placed on the ditch bank opposite and around the point of discharge. Maintaining plant cover on the exterior of pond levees will prevent erosion and reduce the amount of solids entering drainage ditches. Less maintenance will be needed to repair levees and remove accumulated sediment in ditches.

Maintain plant cover on pond watersheds. Vegetative cover will reduce erosion of watershed soils and the loading of suspended solids to ponds in runoff. In watersheds used for grazing, maintenance of appropriate livestock stocking rates will allow maintenance of sufficient plant cover. In degraded watersheds or those subject to erosion, stream-side management zones or buffer strips adjacent to water supply streams should be maintained.

Avoid leaving ponds drained in winter, and close valves once ponds are drained. Rain falling on drained ponds erodes inside slopes of levees and pond bottoms, causing sedimentation in deeper areas of ponds and discharge of suspended solids through the open drain. Refilling ponds promptly and keeping valves closed in empty drained ponds can protect the pond infrastructure and reduce suspended solids loads to farm ditches and finally to streams.

Close drain valves when renovating ponds. Even in the dry season when empty ponds are being repaired, heavy rains may occur. If valves are open, large amounts of suspended solids may be discharged from ponds into drainage ditches and streams.

Use sediment from within the pond to repair levees rather than disposing it outside of ponds. Improperly disposed sediment can result in erosion and suspended solid contributions to streams.

During pond renovation, excavate to increase operational depth. Increasing depth during renovation can increase water storage and permit greater fluctuation in water level without compromising fish production. Increased water storage will reduce the volume of effluent.

Literature Cited

Boyd, C. E. 1976. Accumulation of dry matter, nitrogen and phosphorus by cultivated water hyacinths. Economic Botany 30:51–56.

Boyd, C. E. 1978. Effluents from catfish ponds during fish harvest. Journal of Environmental Quality 7:59–62.

Boyd, C. E. 1985. Chemical budgets for channel catfish ponds. Transactions of the American Fisheries Society 114:291–298.

Boyd, C. E. 1995. Bottom soils, sediment, and pond aquaculture. Chapman and Hall, New York, USA.

Boyd, C. E. 1999. Codes of practice for responsible shrimp farming. Global Aquaculture Alliance, St. Louis, Missouri, USA.

Boyd, C. E. and A. Gross. 1999. Biochemical oxygen demand in channel catfish *Ictalurus punctatus* pond waters. Journal of the World Aquaculture Society 30:349–356.

Boyd, C. E. and J. F. Queiroz. 2001. Feasibility of retention structures, settling basins, and best management practices for Alabama channel catfish farming.

Reviews in Fisheries Science 9(2):43–67.

Boyd, C. E. and M. Tanner. 1998. Coliform organisms in waters of channel catfish ponds. Journal of the World Aquaculture Society 29:74–78.

Boyd, C. E. and C. S. Tucker. 1998. Pond aquaculture water quality management. Kluwer Academic Publishers, Boston, Massachusetts, USA.

Boyd, C. E., A. Gross, and M. Rowan. 1998. Laboratory study of sedimentation for improving quality of pond effluents. Journal of Applied Aquaculture 8:39–48.

Boyd, C. E., J. Queiroz, J.-Y. Lee, M. Rowan, G. Whitis, and A. Gross. 2000. Environmental assessment of channel catfish *Ictalurus punctatus* farming in Alabama. Journal of the World Aquaculture Society 31:511–544.

Brunson, M. W. 1997. Catfish quality assurance. Mississippi Cooperative Extension Service, Publication 1873, Mississippi State University, Mississippi, USA.

Cathcart, T. P., J. W. Pote, and D. W. Rutherford. 1999. Reduction of effluent discharge and groundwater use in catfish ponds. Aquacultural Engineering 20:163–174.

Cole, B. A. and C. E. Boyd. 1986. Feeding rate, water quality, and channel catfish production in ponds. Progressive Fish-Culturist 48:25-29.

Cooke, G. D. and R. H. Kennedy. 1981. Precipitation and inactivation of phosphorus as a lake restoration technique. Report 600/3-81-011. United States Environmental Protection Agency, Washington, D.C., USA.

Gross, A., C. E. Boyd, and R. T. Lovell. 1997. Phosphorus budgets for channel catfish ponds receiving diets with different phosphorus concentrations. Journal of the World Aquaculture Society 29:31–39.

Han, F. X., J. A. Hargreaves, W. L. Kingery, D. B. Huggett, and D. K. Schlenk. 2001. Accumulation, distribution, and toxicity of copper in sediments of catfish ponds receiving periodic copper sulfate applications. Journal of Environmental Quality 30:912–919.

Hargreaves, J. A. and C. S. Tucker. 1996. Evidence for control of water quality in channel catfish *Ictalurus punctatus* ponds by phytoplankton biomass and sediment oxygenation. Journal of the World Aquaculture Society 27:21–29.

Henderson, J. P. and N. R. Bromage. 1988. Optimising the removal of suspended solids in aquaculture effluents in settlement lakes. Aquacultural Engineering 7:167–181.

Hunter, J. V. and H. Heukelekian. 1965. Composition of domestic sewage fractions. Journal of the Water Pollution Control Federation 37:1142–1147.

Masuda, K. and C. E. Boyd. 1993. Comparative evaluation of the solubility and algal toxicity of copper sulfate and chelated copper. Aquaculture 117:287–302.

Masuda, K. and C. E. Boyd. 1994a. Effects of aeration, alum treatment, liming, and organic matter application on phosphorus exchange between pond soil and water in aquaculture ponds. Journal of the World Aquaculture Society 25:405–416.

Masuda, K. and C. E. Boyd. 1994b. Phosphorus fractions in soil and water of aquaculture ponds built on clayey, Ultisols at Auburn, Alabama. Journal of the World Aquaculture Society 25:379–395.

McGee, M. V. and C. E. Boyd. 1983. Evaluation of the influence of water exchange in channel catfish ponds. Transactions of the American Fisheries Society 112:557–560.

McVea, C. and C. E. Boyd. 1975. Effects of

water hyacinth cover on water chemistry, phytoplankton, and fish in ponds. Journal of Environmental Quality 4:375–378.

Quieroz, J. and C. E. Boyd. 1998. Effects of a bacterial inoculum in channel catfish ponds. Journal of the World Aquaculture Society 29:67–73.

Quieroz, J., C. E. Boyd, and A. Gross. 1998. Evaluation of a bio-organic catalyst in channel catfish, *Ictalurus punctatus,* ponds. Journal of Applied Aquaculture 8(2):49–61.

Schwartz, M. F. and C. E. Boyd. 1994a. Effluent quality during harvest of channel catfish from watershed ponds. Progressive Fish-Culturist 56:25–32.

Schwartz, M. F. and C. E. Boyd. 1994b. Channel catfish pond effluents. Progressive Fish-Culturist 56:273–281.

Schwartz, M. F. and C. E. Boyd. 1995. Constructed wetlands for treatment of channel catfish pond effluents. Progressive Fish-Culturist 57:255–266.

Schwartz, M. F. and C. E. Boyd. 1996. Suggested management to improve quality and reduce quantity of channel catfish pond effluents. Leaflet 108, Alabama Agricultural Experiment Station, Auburn University, Alabama, USA.

Seok, K., S. Leonard, C. E. Boyd, and M. Schwartz. 1995. Water quality in annually drained and undrained channel catfish ponds over a three-year period. Progressive Fish-Culturist 57:52–58.

Shireman, J. V. and C. E. Cichra. 1994. Evaluation of aquaculture effluents. Aquaculture 123:55–68.

Swingle H. S. and E. V. Smith. 1939. Fish production in terrace-water ponds in Alabama. Transactions of the American Fisheries Society 69:101–105.

Tucker, C. S. 1991. Quality of potential effluents from channel catfish culture ponds. Pages 177–184 *in* J. Blake, J. Donald, and W. Magette, editors. National livestock, poultry, and aquaculture waste management. American Society of Agricultural Engineers, St. Joseph, Michigan, USA.

Tucker, C. S. 1996. The ecology of channel catfish culture ponds in northwest Mississippi. Reviews in Fisheries Science 4(1):1–55.

Tucker, C. S., editor. 1998. Characterization and management of effluents from aquaculture ponds in the southeastern United States. Final project report no. 600. Southern Regional Aquaculture Center, Stoneville, Mississippi, USA.

Tucker, C. S. and S. W. Lloyd. 1985a. Water quality in streams and channel catfish *Ictalurus punctatus* ponds in west-central Mississippi. Technical bulletin 129. Mississippi Agricultural and Forestry Experiment Station, Mississippi State, Mississippi, USA.

Tucker, C. S. and S. W. Lloyd. 1985b. Evaluation of a commercial bacterial amendment for improving water quality in channel catfish ponds. Research report 10. Mississippi Agricultural and Forestry Experiment Station, Mississippi State, Mississippi, USA.

Tucker, C. S. and M. van der Ploeg. 1993. Seasonal changes in water quality in commercial channel catfish culture ponds in Mississippi. Journal of the World Aquaculture Society 24:473–481.

Tucker, C. S., S. K. Kingsbury, J. W. Pote, and C. L. Wax. 1996. Effects of water management practices on discharge of nutrients and organic matter from channel catfish (*Ictalurus punctatus*) ponds. Aquaculture 147:57–69.

USDA/APHIS (United States Department of Agriculture/Animal and Plant Health Inspection Service). 1997. Catfish NAHMS '97, Part II: Reference of 1996 U. S. Catfish Management Practices.

Effluents from Raceways

JEFFREY M. HINSHAW

North Carolina State University, 455 Research Drive, Fletcher, North Carolina 28732 USA

GARY FORNSHELL

University of Idaho, Twin Falls County Extension Office, 246 Third Ave. E.,

Twin Falls, Idaho 83301 USA

ABSTRACT

The effluents from raceways, primarily used for salmonid culture, have been studied extensively over the past three decades. The stimulus for much of this work has been a growing perception by the public of potential environmental impact from these types of culture systems and by increasingly stringent water pollution control regulations. Although the impact of a particular culture system will depend upon the physical and chemical characteristics of the receiving water body and on the relative intensity of fish production, the two primary strategies available for reducing the impact of the systems are: 1) reduction of potential pollutants at the source through manipulating feeds and feeding strategies, and 2) use of effective and cost-efficient solids capture technologies. Estimates of waste reduction can be made using nutritional or bioenergetics models, but verification on an industry-wide basis is still needed. In spite of the potential for impact of raceway effluents on receiving waters, evidence of more than slight to moderate impacts in the U.S. is limited to a very few specific sites.

Production of fish in raceways in the U.S. is typically equated with commercial production of rainbow trout *Oncorhynchus mykiss*. Other coldwater fish species produced in raceway systems include brook trout *Salvelinus fontinalis* and brown trout *Salmo trutta*. A variety of additional trout species are produced in these systems by public hatcheries, but the quantity is small compared to commercial rainbow trout production. Raceway systems are also used on a limited scale for the production of warmwater fish, such as catfish *Ictalurus punctatus* or *I. furcatus*, and also for tilapia. Recently, raceway systems have been used to produce coolwater species such as yellow perch *Perca flavescens*, hybrids of *Morone spp.,* and several species of sturgeon. The focus of this chapter will be on effluents from salmonid culture in raceway systems. Other perceived impacts of aquaculture will be discussed elsewhere in this book.

A raceway in its simplest form is a flume for carrying water. Raceway production systems for coldwater fish in the U.S. aquaculture industry evolved from earthen pond systems, the most popular shape of which was long and narrow, with sufficient slope to allow for aeration by gravity between ponds. A typical raceway production system consists of a tank or series of tanks, usually rectangular with water flow along the long axis (Fig. 1). On farms, raceways may be divided into two or more tanks at each step in the series, but on smaller farms the tanks are usually in pairs for ease of access (Fig. 2). Larger farms may construct access on the tanks and have multiple raceways in parallel series (Fig. 3). The water in raceway production systems is rarely recirculated, but is 're-used' serially with aeration or oxygenation between tanks. Reviews of physical design, construction, and operational parameters of raceways can be found in Burroughs and Chenoweth (1955),

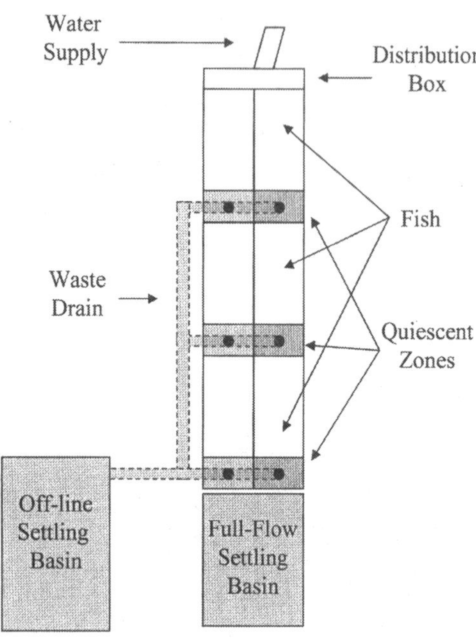

FIGURE 1. *Diagram of a raceway production system.*

FIGURE 2. *Paired concrete raceways for rainbow trout production.*

FIGURE 3. *Parallel series of raceways in a large trout farm.*

Wheaton (1977), Klapsis and Burley (1984), and Timmons and Youngs (1991).

Water flow in an ideal raceway will approximate plug flow with uniform velocity throughout the tank. However, due to friction losses at the boundary layer, flow may be reduced toward the bottom and the sides (Wheaton 1977) resulting in accumulation of metabolites from the fish in those areas. Individual tanks in a raceway system are approximately 7:1 length:width, with widths rarely more than six to seven meters to facilitate fish grading and harvesting. Tank depths are typically less than one meter to allow culturists to work within the tanks and to visually observe the fish. The shallow tanks can also facilitate efficient feeding. Most raceway systems are constructed of concrete.

Water Use and Fish Production

Design criteria for raceway systems are based first on satisfying metabolic requirements for the species of fish being produced, and only indirectly on waste collection or removal. For trout culture in the U.S., the synthesis of biological requirements of the organism with design and management of the facility began with a publication by Haskell (1955) elucidating the relationship between fish feeding, oxygen consumption, and the accumulation of metabolites. Haskell and others kept detailed hatchery records and initially correlated the carrying capacity of a particular tank or raceway with the amount of feed per unit of tank volume at a relatively constant flow rate. Maximum carrying capacity of fish was defined in terms of the amount of feed appropriate for the fish size. Water temperature was initially believed to have relatively little effect between 5 C and 15 C. Later, Buterbaugh and Willoughby (1967) incorporated site-specific influences into feeding estimates by developing a 'Hatchery Constant' that integrated feed conversion and growth rates typical for each facility. Water exchange was not ignored, but was incorporated into estimates of carrying capacity through the hatchery constant until knowledge of trout metabolism could be better defined. Consumption of oxygen, the first limiting factor encountered in most flow-through systems, per kg of feed eaten by hatchery trout was determined to be 0.22 kg (Willoughby 1968) and then 0.25 kg (Willoughby et al. 1972), figures still widely used in the U.S. trout industry to estimate carrying capacity relative to available oxygen. Subsequent empirical observations described the relationship between feeding and ammonia excretion in trout culture facilities to average 32 g ammonia/kg feed (Willoughby et al. 1972). Based on tolerance limits for available oxygen of 5 mg/L minimum and 0.0125 mg/L un-ionized ammonia (Smith and Piper 1975), Westers and Pratt (1977) presented a synthesis of flow-through system design criteria that reflects the existing U.S. trout industry, based on metabolic characteristics of trout. Much of the commercial industry still operates within the range of productivity outlined by Westers and Pratt (1977), normally carrying between 20 and 80 kg fish/m^3 of water volume with water exchange rates per tank of 3–6 times/h. The water is re-used serially 4–6 times or until ammonia becomes limiting (e.g., un-ionized ammonia approaches 0.0125 mg/L), then is discharged. In areas where the waters are

acidic, 10 or more uses are typical, and other factors such as carbon dioxide or suspended solids may become limiting before ammonia. Colt et al. (1991) describe methods for calculating carrying capacity in culture systems where oxygen is not a limitation. Most raceway facilities include a quiescent zone for primary settling and an off-line settling basin for collection of concentrated solids from the facility. Some facilities, particularly farms with earthen ponds and smaller farms with limited serial re-use, use full-flow settling basins for removing solids from the primary flow through the system.

Effluent Impacts

Even though as many as one-third of North American fisheries agencies had experienced pollution problems with hatcheries as early as 1970, published information on pollution from fish hatcheries was either absent or found to be unsatisfactory in an early survey of aquaculture, fisheries, water pollution, and public health sources (Liao 1970a). The potential for environmental impact of intensive commercial fish culture has also been recognized for an extended time, but implementation of economically feasible and effective technologies for effluent treatments by industry still remained minimal at least until 1990 (Stechey and Trudell 1990a). Since the addition of fish food and wastes to the water flow is analogous to an industry that takes water for use in processing and subsequently discharges the water containing wastes (Odum 1974), control authorization was established in 1972 by Federal Water Pollution Control Act Amendments and the creation of the National Pollutant Discharge Elimination System (NPDES) permit program. In the literature debate on environmental impacts of aquaculture, a fish farm with 50,000–100,000 salmon is estimated to produce more organic waste than 10,000 people (Barinaga 1990), or less organic waste than 320 people (Gagen

1990). But fish production wastes cannot be directly compared with human sewage (Pitts 1990) except perhaps in terms of individual components such as biochemical oxygen demand (BOD) or chemical oxygen demand (COD) (Bergheim and Selmer-Olsen 1978; Bergheim and Sivertsen 1981). In addition to proportional differences in BOD, nitrogen, and phosphorus, Krieger et al. (1987) found levels of nine potentially toxic elements in fish manure that were slightly to greatly reduced compared to municipal sludge and to cow manure.

The overall environmental impact of fish hatchery effluents in the U.S. has been suggested to be "less than significant" (Harris 1981), though potential environmental alterations can stimulate changes in the composition of plant and animal populations, changes in primary production, pH changes and lowered dissolved oxygen (DO) concentrations (Liao 1970a; Hinshaw 1973; Odum 1974; Alabaster 1982; Warrer-Hansen 1982a; Gowen et al. 1991). Reactions of receiving waters to effluents from raceways range from insignificant to 'heavily impacted,' but heavily impacted streams are few in the U.S., and tend to be very site specific (Sloan 2001). Taylor and Perrin (1989) summarize invertebrate reactions in freshwater sediments from aquaculture (flow-through and net-pen) ranging from mild enhancement to increased abundance of organic pollution indicators to reduction in species abundance and richness. In organic enrichment scenarios in freshwater, opportunists include tubificid oligochaetes, certain leeches, water louse, and chironomids. A general decline in the "clean-water" taxa such as plecoptera, ephemeroptera, and tricoptera is present downstream of Danish trout farms (Markmann 1981; Iwama 1991). Impacts on benthic invertebrate populations include decreases in taxa richness, increases in pollution tolerant species, and decreases in intolerant species for at least 1.5 km downstream from some eastern U.S. trout farms

(Loch et al. 1996; Selong and Helfrich 1998), although the effects vary with season and diminish significantly with distance from the farms. Using U.S. EPA metrics for rapid bioassessment, the streams immediately below four trout farms in Virginia were classified as moderately impaired (2), slightly impaired (1), or unimpaired (1) (Selong and Helfrich 1998). Boaventura et al. (1997) suggested that 3–5 km of stream length were needed before chemical parameters return to normal levels for streams below two trout farms in Portugal. They noted that a third stream had not completely recovered by 12 km downstream, but that there were other animal feedlots potentially discharging into the river system below the trout farm.

The qualities of potential pollutants from flow-through aquaculture facilities are highly variable, but can be considered in three primary groups: 1) pathogenic bacteria or parasites; 2) therapeutic chemicals and drugs; and 3) metabolic products and food wastes (Piper et al. 1982; Beveridge et al. 1991). The first two are sporadic and have only recently drawn attention as a major concern among the byproducts of aquaculture, as early evidence for significant impact from their release was limited (Odum 1974; Niemi and Taipalinen 1980; Solbé 1982; Sumari 1982; UMA Engineering Ltd. 1988). More recent studies have examined the persistence of antimicrobial compounds in sediment of aquacultural origin, development of antibiotic resistant microbes, and comparative differences in abundance of pathogens above and below flow-through systems. The potential for these impacts will be discussed elsewhere in this publication.

The third component discharged from flow-through systems, metabolic products and food wastes, have been characterized by a number of researchers. The following review summarizes published reports describing the characteristics of metabolic products and food wastes of raceway or flow-through culture system effluents.

Factors Impacting Waste Production

Feed

The single operating factor having the greatest impact upon raceway effluent is feeding. Feeding the proper amount at the proper times not only increases growth rates but also results in little or no residual food. Times and amounts of feeding vary with temperature, fish species and biomass, and type of feed, and must be optimized for each facility to reduce waste (Liao 1970a; Sparrow 1981; Merican and Phillips 1985; Seymour and Bergheim 1991). In addition to efficient feed utilization, the quantities of pollutants in the effluent are directly related to feeding rates rather than the size of the fish being fed (Liao and Mayo 1974; Querellou et al. 1982; Merican and Phillips 1985). Willoughby et al. (1972) even provided a simplified formula to estimate effluent pollutant levels relative to feed inputs, prior to effluent treatment:

$$ppm = \frac{(kg\ of\ pollutant/kg\ feed)\ (kg\ fed)}{.021\ (Lpm\ flow)}$$

Feed quality is also an important factor in the production of nutrient wastes in fish farm effluents. Reductions in phosphorus levels in fish farm effluents approaching 50% (below standard commercial fish diets) can be achieved without sacrificing growth rates by manipulating the bioavailability in the diets (Ketola et al. 1991). Nitrogen excretion rates in fish could also be reduced by optimizing dietary amino acid levels in feeds (Lovell 1989), although application of this practice to commercial feed production has been limited until very recently. Substitution of alternate sources of protein for fish meal in fish diets is feasible, but has the potential to exacerbate pollution problems if not properly formulated (Hardy 1996).

Feed production methods can also influence waste production. Extruded feeds are reported to be far more stable in water than milled pellets (Kazamzakeh 1990) and are more digestible, as extrusion gelatinizes the starch in the feed. Increased digestibility can reduce feces, and the extrusion process produces less dust (Mudrak 1981; Clarke 1990; Kazamzakeh 1990; Barrows and Hardy 2000). Ingredients in feed, such as indigestible binding agents, poor quality or indigestible carbohydrates, and excessive or poor quality proteins, can also affect amounts of BOD, ammonia, and levels of solids in effluent (Henderson and Bromage 1987; Clarke 1990).

Species

Most studies of fish farm effluents from raceway systems have examined discharges from salmonid culture, with only limited work reported on effluents from farming of other fish species (Ellis et al. 1978; Bohl 1988). Although theoretical estimates suggest that wastes from one kind of fish may be several times greater than another (Heinsbroek 1988), the range of values is so great that the role of species is unclear (Beveridge et al. 1991). One recent study comparing two species of trout frequently cultured in raceway systems, rainbow trout and brown trout, suggested that brown trout have significantly lower phosphorus retention (Dosdat et al. 1998).

Body Size and Temperature

Metabolic waste production is related to fish size and water temperature. Waste production per unit of body mass in individual fish decreases with increasing body weight (Bergheim et al. 1984; Beveridge et al. 1991). However, since larger fish are typically fed a smaller percentage of their body weight per day, the amount of waste generated per unit of feed used is not necessarily reduced. When expressed in terms of the amount of feed per day, Clark et al. (1985) found no difference in waste production between the size classes. The production of wastes is also proportional to metabolic rate, e.g., as temperatures rise through spring and summer, so does waste production (Beveridge et al. 1991). Again, the link between waste production and temperature is a function of feeding, which also increases with water temperatures. Since fish gain weight and are fed in proportion to their body size, nearly 90% of the feed is given and a proportionate amount of waste generated during the last half of the growing cycle within a given population of fish.

Daily Variations

The daily activities on farms will also influence pollution release through a variety of mechanisms, primarily related to feeding (Rychly and Marina 1977; Fivelstad et al. 1990). For example, Clark et al. (1985) observed that concentrations of ammonia, nitrite, phosphorus, and suspended solids remained low overnight and increased from midday to a peak in the early evening when feeding was started at 0830 h daily.

Cleaning versus Normal Operations

When measuring or characterizing fish farm wastes, an important consideration is whether a farm is in normal operation or if the tanks are being cleaned. Waste levels can be many times greater during cleaning operations (Liao 1970b; Sparrow 1981; Bergheim et al. 1984; Kendra 1991) if care is not exercised during the process. Mudrak (1981) recommended minimizing shock loading by staggering raceway cleaning periods and using flow-retaining shields to permit concentrating the flush of solids to a raceway clean-out area. Where possible, the measurements included in this review have been noted as to mode of farm operation. Unfortunately, many authors neglected to include this information.

Waste Characterization

Pollutants from metabolic products and fish feed wastes fall into three classes: organic, nutrient, and solids. A point of confusion exists between the terminology used to describe levels of pollutants in fish farm effluent. The use of mg/L gives no relation to the amount of production or weight of feed used and simply describes characteristics of the effluent without regard to changes due to the farms. More useful values are weight of pollutant per weight of fish held or weight of pollutant per weight of food fed. These measures more accurately reflect the change in water quality resulting from fish production. Regardless, the various units will all be given to allow maximum use of the information following in this review. Some of the publications cited are studies of other flow-through aquaculture systems such as circular tanks, or a mixture of tanks and raceways. When relevant, these differences are noted.

Organic Pollutants

Organic pollutants include those associated with biochemical oxygen demand (BOD) and the closely related measure of oxygen depletion. BOD is a measure of the organic matter content of the water, generally increases with increasing suspended solids (Alabaster 1982), and is a more long-term measure of the consumption of oxygen by the effluent from the farm. The oxygen depletion or oxygen consumption rate is a shorter-term measure integrating oxygen consumption primarily by the fish with the oxygen recharge capabilities built into the farm. Chemical oxygen demand (COD) is related to BOD and is a measure of the oxygen equivalent of organic matter susceptible to oxidation by a strong chemical oxidant. Liao and Mayo (1974) suggest using a relationship of BOD = 0.3 x COD as an estimate for fish farm effluents. COD is included in this section though rarely reported in literature describing raceway effluents.

Biochemical Oxygen Demand (Table 1). Solbé (1982) suggested a relationship between solids in the effluent and BOD which can be expressed by the formula: BOD = 0.132(SS) +

TABLE 1. *Biochemical oxygen demand (BOD) in fish farm effluent.*

Reference	Comments	BOD
	mg/L	
Liao 1970a	average increase, normal operation	5.36
Train et al. 1977	normal	4.0
	cleaning	21.2
UMA Engineering Ltd. 1979	normal	0
Solbé 1982	normal and cleaning average	2.8
	weighted arithmetic mean	1.5
Boaventura et al. 1997	average, 15 mt production	2.0
	average, 55 mt production	3.1
	average, 500 mt production	15.6
	g/kg fish per/d	
Liao 1970a		13.4
Liao 1970b		5–30
Brisbin 1971	average	8.3
	range	10–15
Knosche 1971	average	5.2
	range	2.3–8.1
Shanks 1971		12.0
EPA 1974		13.0
Walden and Birkbeck 1974		1.18–10.8
Train et al. 1977		13
Bergheim and Selmer-Olsen 1978		1.6–4.6
Caddy 1979		5.0
Harris 1981		3.5–8
Butz and Vens-Cappell 1982		1.7, 4.0
	g/kg fish	
UMA Engineering Ltd. 1979	normal	0
	g/kg feed per/d	
Willoughby et al. 1972		34
Liao and Mayo 1974		60
Fauvre 1977		52
Butz and Vens-Cappell 1982		117
Clark et al. 1985		140
Makinen 1988		100–300
	g/kg feed	
Knosche 1971	average	170
	range	80–270
Butz and Vens-Cappell 1982		165, 133
	kg/d per/mt annual production	
Alabaster 1982	range	1.4–2.7
	average	1.7
Solbé 1982		2.4
Boaventura et al. 1997	average, 15 mt production	0.35
	average, 55 mt production	0.69
	average, 500 mt production	1.51
	kg/mt fish per/d	
Warrer-Hansen 1978		0.9–1.5

TABLE 2. *Oxygen depletion (consumption) in fish farm effluent.*

Reference	Comments	Oxygen Depletion
	mg/L	
Solbé 1982	weighted arithmetic mean	1.63
Kendra 1991		0.4
Boaventura et al. 1997	average, 15 mt production	1.3
	average, 55 mt production	0.6
	average, 500 mt production	2.4
	g/kg fish per/d	
Liao 1970a	average	7.27
	range	2.46–17.54
Brisbin 1971		22
	kg/d per/mt annual production	
Boaventura et al. 1997	average, 15 mt production	0.55
	average, 55 mt production	0.26
	average, 500 mt production	0.26

1.47. In general, reported BOD levels averaged 2.0 mg/L during normal operations. BOD increased approximately 10-fold when raceways were being cleaned, reflecting disturbance of the settleable solids rather than typical farm effluents. Levels of BOD reported relative to fish or feeds ranged from 1.6–30 g/kg fish per/d, and 34–300 g/kg food per/d, respectively.

Oxygen Depletion (Table 2). Dissolved oxygen is consumed mainly by the fish in raceway facilities and may or may not be replenished prior to the water leaving the farm. Essentially, oxygen depletion is a comparative measure of the oxygen quantity entering the farm in the water relative to the levels leaving the farm. Important factors include stocking levels, proximity of fish to the downstream end of the farm, and the amount of fall from raceway to raceway, which impacts the amount of oxygen added through aeration during each step. Farms with particularly heavy stocking densities may also employ supplemental aeration with either atmospheric air or pure oxygen. The largest deficits of dissolved oxygen occur at lowest rates of flow, which reduces aeration within the farm. Norwegian data also shows larger deficits associated with long retention times and high ambient temperatures (Alabaster 1982). On average, values for oxygen depletion ranged from 0.4–2.4 mg/L or 7.2–22 g/kg fish per/d.

Chemical Oxygen Demand (Table 3). Chemical oxygen demand information was not presented in the literature as often as BOD, but can provide additional information about the oxygen depleting capabilities of fish farm effluent. Values for COD are generally much greater than for BOD, since BOD is measured over a 5-d period and organic matter decomposition, primarily by heterotrophic bacteria, will be incomplete. In contrast, chemical oxidation of organic matter in the test for COD is expected to be 95–100% of the theoretical value. For example, Train et al. (1977) reported respective BOD and COD levels of 4 versus 21.2 mg/L for normal operations and 25 versus 61 mg/L for

TABLE 3. *Chemical oxygen demand (COD) in fish farm effluent.*

Reference	Comments	COD
	mg/L	
Liao 1970b	average increase, normal operation	21.3
Train et al. 1977	normal	25
	cleaning	61
Kendra 1991		4
	g/kg fish per/d	
Bodien 1970		20
Bergheim et al. 1984		8.1, 10.4
	g/kg fish	
Train et al. 1977		60
	g/kg feed per/d	
Liao 1970b	calculated	189
	kg/d per/mt annual production	
Alabaster 1982	range	1.0–31
	average	25

measurements during cleaning operations. Alabaster (1982) also supports this with an average BOD of 1.7 kg/d per/t annual production as opposed to 25 kg/d per/t for COD.

Nutrient Pollutants

Ammonia (Table 4). Ammonia is the primary soluble metabolite resulting from protein metabolism in fish and is released mainly by excretion across the gills. To a lesser degree, ammonia can also come from the breakdown of protein in uneaten feed or fines (Querellou et al. 1982; Solbé 1982; Bergheim et al. 1984; Clark et al. 1985). Depending upon pH and temperature, ammonia exists as fractions of ammonium ion (NH_4^+) and un-ionized ammonia (NH_3). At the temperatures on most salmonid farms, unless the pH is greater than 7.3, the fraction of NH_3 is less than 1% of the total. The un-ionized molecule is generally considered to be the toxic form (Smart 1978; Solbé 1982), although some authors suggest that the ionized form (NH_4^+) does have some toxic action (Burrows 1968; Fromm and Gillette 1968; Liao et al. 1972; Liao and Mayo 1974). Ammonia is typically reported as the concentration of elemental nitrogen present as ammonia, regardless of ionization. Ammonia-nitrogen values can vary greatly depending upon the interval after feeding, water retention time, and stocking density. Reported values range from 0.01–1.52 mg/L, 2.9–37 g/kg food per/d and 0.03–1.6 g/kg fish per/d. In a study of one trout farm, Foy and Rosell (1991a) reported that ammonia represented 62.1% of the total nitrogen released

TABLE 4. *Total ammonia-nitrogen (TAN) in fish farm effluent.*

Reference	Comment	TAN
	mg/L	
Liao 1970a	average increase, normal	0.53
Train et al. 1977	normal	0.49
	cleaning	0.52
UMA Engineering Ltd. 1979	normal	0.01 –0.18
Solbé 1982	weighted arithmetic mean	0.172
Kendra 1991		0.19
Boaventura et al. 1997	average, 15 mt production	0.42
	average, 55 mt production	0.32
	average, 500 mt production	1.52
Cho and Bureau 1998	average, calculated from	total N 0.10
	g/kg fish per/d	
Bodien 1970		0.58
Liao 1970a		1.66
Brisbin 1971		0.5
EPA 1974		0.9
Walden and Birkbeck 1974		0.14–0.43
Train et al. 1977		0.9
Korzeniewski et al. 1982		0.03
Bergheim et al. 1984		0.53, 0.58
Clark et al. 1985		0.13–0.21, 0.3–0.8
	g/kg fish	
UMA Engineering Ltd. 1979	normal	0.5–2.0
	g/kg feed per/d	
Willoughby et al. 1972		32
Speece 1973	62 F	32
	49 F	26
Liao and Mayo 1974		29.9
Piper et al. 1982		32
Clark et al. 1985		9–15, 31–37
	g/kg feed	
Castledine 1986		38.3
	kg/d per/mt annual production	
Alabaster 1982	range	0.1–0.5
	average	0.3
Purdom 1982		0.1
Solbé 1982		0.1
Boaventura et al. 1997	average, 15 mt production	0.16
	average, 55 mt production	0.11
	average, 500 mt production	0.16

by the fish into the effluent.

Nitrate (Table 5). Nitrate is the end product of the decomposition of protein in fish food, in which ammonia is oxidized to nitrite and finally nitrate. Limited quantities of additional nitrate can come from direct production by the fish (Liao 1970a; Liao and Mayo 1974; Willoughby et al. 1972) and may vary considerably with protein content in the feed (Clark et al. 1985). Reported values for nitrate range from 0–2.1 mg/L, 0.05–0.9 g/kg fish per/ d and 0.5–20 g/kg food per/d. When compared to concentrations in the riverine water supply to the farms, the nitrate levels in the effluent from the farms were not significantly different from the inflow (Boaventura et al. 1997).

Phosphorus (Table 6). Phosphorus can be an important component in effluent because it is the main factor determining the degree of eutrophication in freshwater in many geographic areas (Sumari 1982). Soluble reactive phosphorus includes those forms that respond to analytical tests without preliminary hydrolysis or oxidative digestion, and is composed primarily of orthophosphate (APHA 1985). Many authors report phosphorus as

TABLE 5. *Nitrate nitrogen in fish farm effluent.*

Reference	Comment	Nitrate Nitrogen
	mg/L	
Liao 1970b	average increase	1.68
Train et al. 1977	normal operation	0–0.17
	cleaning	0.64
UMA Engineering Ltd. 1979	normal	0–0.017
Solbé 1982	weighted arithmetic mean	0.036
Boaventura et al. 1997	average, 15 mt production	*1.0
	average, 55 mt production	*2.1
	average, 500 mt production	*1.0
	*not significantly different from inflow	
	g/kg fish per/d	
Liao 1970a		4.0
Brisbin 1971		0.4
Liao et al. 1972		0.9
EPA 1974		0.6
Train et al. 1977		0.06
Korzeniewski et al. 1982		0.05
	g/kg fish	
UMA Engineering Ltd 1979		0.02–0.1
	g/kg feed per/d	
Liao 1970a		0.11
Willoughby et al. 1972		20, 8.7
Piper et al. 1982		0.5

TABLE 6. *Total phosphorus in fish farm effluent.*

Reference	Comment	Total Phosphorus
	mg/L	
Train et al. 1977	normal	0.09
	cleaning	0.38
UMA Engineering Ltd 1979	normal	0–0.13
Solbé 1982	weighted mean	0.038
Kendra 1991		0.09
Boaventura et al. 1997	average, 15 mt production	0.098
	average, 55 mt production	0.065
	average, 500 mt production	0.591
Cho and Bureau 1998	average	0.027
	g/kg fish per/d	
Bodien 1970		0.36
Liao 1970b		0.33
EPA 1974		0.3
Train et al. 1977		0.3
Bergheim and Selmer-Olsen 1978		0.05
Korzeniewski et al. 1982		0.10
Bergheim et al. 1984		0.18, 0.22
Clark et al. 1985		0.067–0.170
	kg/100 kg fish	
UMA Engineering Ltd. 1979	normal	0.02–0.06
	g/kg feed per/d	
Willoughby et al. 1972		5
Piper et al. 1982		5
Clark et al. 1985		5.2–5.9
	g/kg feed	
Ketola 1985		9
Castledine 1986		7.6
Makinen 1988		4.7–10.8
Cho et al. 1991		9
	kg/d per/mt annual production	
Alabaster 1982	range	0.06–0.31
	average	0.14
Purdom 1982		0.10
Solbé 1982		0.03
Boaventura et al. 1997	average, 15 mt production	0.036
	average, 55 mt production	0.024
	average, 500 mt production	0.063
	kg/mt fish per/d	
Warrer-Hansen 1978		0.03–0.05
Westers 1990		1.5–25

soluble reactive (in solution, requiring no hydrolysis or digestion), soluble unreactive (principally organic complexes requiring hydrolysis or oxidation for release), or particulate phosphorus (retained on a 0.45-micron filter). The total of all forms (organic plus inorganic) is reported as total phosphorus.

The phosphorus component of fish feed and feces is principally organic phosphorus in particulate form, and is unavailable in the water column until after chemical and biological reactions (Stechey and Trudell 1990a). Thus, the major source of dissolved phosphorus as the water leaves the fish farm is from direct excretion through the kidney (Forster and Goldstein 1969; Clark et al. 1985) with secondary production through leaching of feces and feed (Clark et al. 1985). Foy and Rosell (1991b) found the soluble reactive and unreactive forms of phosphorus excreted by fish on one trout farm to represent 60.0% and 10.0% of the excreted phosphorus, respectively. The farm they examined had a notably poor feed conversion rate (>1.8:1) and relatively high phosphorus content in the diet (average 1.655%). Samples in their study were collected prior to gravitational removal of solids, so their results are not included in the table describing effluent levels of phosphorus from raceway systems. Even so, their results still correspond to the conclusion by Persson (1990) that 60% of the phosphorus fraction in fish feces is ultimately released to the environment.

Dietary changes can markedly affect phosphorus content of effluent. The dietary salmonid requirement for phosphorus is between 0.4% and 0.6% feed by weight and although concentrations as low as 0.37% have shown no detrimental growth affects (Ketola 1985; Eskelinen 1986; Wiesmann et al. 1988; Rodehutscord 1996), slightly more may be required for maximum bone mineralization (Ketola and Richmond 1994). Commercial diets having 1.1-1.8% could be modified, resulting in less phosphorus in receiving waters (Stechey and Trudell 1990b). With one exception, phosphorus values reported for normal operations ranged from 0-0.13 mg/L with cleaning effluent approximately three times this level. Values in g/kg fish per/d and g/kg feed per/d varied considerably with levels of 0.05–0.3 g and 0.5–5.9 g, respectively.

Dissolved Phosphorus (Table 7). Phosphate rather than phosphorus is the biologically

TABLE 7. *Dissolved phosphorus in fish farm effluent.*

Reference	Comment	Soluble Phosphorus
	mg/L	
Liao 1970a	average increase	0.077
	g/kg fish per/d	
Bodien 1970		0.15
Brisbin 1971		0.06
Korzeniewski et al. 1982		0.033
Bergheim et al. 1984		0.10, 0.13
	g/kg feed per/d	
Liao 1970a		9.4
Liao and Mayo 1974		16.2

available end product of phosphorus from fish food decomposition and typically does not accumulate within fish farms due to the relatively short residence time of the water. Liao (1970a) reported an average effluent concentration of 0.077 mg/L, with other reported value ranges of 0.03–0.13 g/kg fish per/d and 1.6–16 g/kg food per/d, the latter of which exceeds the concentration in most commercial diets in current use. In more recent publications, authors may also include both soluble reactive and soluble unreactive (principally humic-associated) phosphorus as separate components of dissolved phosphorus.

Solids (Table 8). Solids in raceway effluents are a combination of dissolved and suspended solids. As reported by most authors, the suspended solids fraction typically is inclusive of both settleable and non-settleable materials. Settleable solids are defined as particulate material that settles within one hour to the bottom of a standardized cone (Imhoff cone). The particulate matter that remains in suspension is the non-settleable fraction of the suspended solids. When raceways are not being cleaned, about 92% of total solids in raceway effluents are dissolved solids, unless the water is exceptionally low in mineral content. The remaining 8% are suspended solids, about half of which are settleable. During tank cleaning, total solids concentration may increase to almost twice their former level, with only 45% made up of dissolved solids. Eighty-nine percent of the suspended solids during cleaning may be contained in the settleable fraction (Liao 1970a; Sparrow 1981). This example illustrates the importance of the mode of operation in reporting suspended solids, as well as the need to prevent re-suspension of solids during cleaning.

Feeding efficiency is also an important factor in solids production. Many authors (e.g., Willoughby et al. 1972; Mudrak and Stark 1981; Zeigler 1988) have suggested 300 g waste solids/kg feed as an acceptable industry average, but this figure is dropping rapidly to between 150 and 200 g/kg feed as feed qualities improve (e.g., Cho et al. 1994; Cripps and Bergheim 2000).

Water flow patterns in fish farms have a significant effect upon the solids levels leaving the farm. Raceway designs that minimize fragmentation of feces are extremely important as unbroken fecal matter has a rather high settling velocity (>1.5cm/sec; Warrer-Hansen 1978; Querellou et al. 1982). Properly designed settling basins can remove as much as 90% of the settleable solids in as little as 12 min (Jensen 1972). This can be critical since half or more of the suspended solids are settleable and removing 90% of the settleable solids removes about 85% of the BOD (Willoughby et al. 1972). Phosphorus, BOD, and COD are all closely tied to the solid fraction of wastes (Piper et al. 1982; Querellou et al. 1982), as well as a portion of the nitrogen.

As previously stated, concentrations of solids increase substantially when raceways are being cleaned. Reported levels for normal operations ranged from 0-35 mg/L while the range when cleaning was 61.9 to over 1,000 mg/L. No mode of operation was noted by authors reporting in g/kg fish per/d or g/kg feed per/d, but values ranged from 0.118–23.8 g and 30–90 g, respectively.

Summary and Recommendations

A tremendous amount of work has been done by scores of researchers on the characterization of effluent from fish culture in raceway and other flow-through systems, particularly those from salmonid production. The enormous variation seen in the reported values illustrates the importance of factors such as mode of operation during measurement, stocking density, composition of feed and feed conversion efficiency, and the intensity of water use. This variation in effluent quality and the unique characteristics of the receiving waters of each farm make most generalizations

TABLE 8. *Suspended solids in fish farm effluent.*

Reference	Comment	Suspended Solids
	mg/L	
Liao 1970a, 1970b	normal, range	0–55
	normal, average	7
	cleaning, average	96
Train et al. 1977	normal	3.7
	cleaning	61.9
UMA Engineering Ltd 1979	normal	0–2.8
Solbé 1982	weighted arithmetic mean	4.36
Bergheim et al. 1984	normal	1–100
	cleaning	30–5,800
Kendra 1991		1
Cripps 1995	average	6.9
	g/kg fish per/d	
Brisbin 1971		1.0–1.5
Shanks 1971		5.8
EPA 1974	range	19.8–23.8
Walden and Birkbeck 1974		0.118–0.941
Bergheim et al. 1984		4.8, 3.3
Clark et al. 1985		0.80–0.94
	g/kg fish	
UMA Engineering Ltd. 1979	normal	11.2–13.2
	g/kg feed per/d	
Liao 1970a		52
Willoughby et al. 1972		30
Liao and Mayo 1974		52
Fauvre 1977		45
Klontz et al. 1979		95* 1/feed conversion ratio
Clark et al. 1985		40–90
	g/kg feed	
Makinen 1988		80–280
	kg/d per/t annual production	
Alabaster 1982	range	1.3–11.0
	average	3.4
Purdom 1982		9.4
Solbé 1982		4.1
	kg/mt fish production	
Westers 1990		200–1000
Cripps and Bergheim 2000	calculated	150–200

regarding the impact of raceway culture systems of very limited value. However, two statements can be made that are consistent for raceway culture systems at all locations: 1) as in other intensive methods of fish culture, the source of nutrient pollutants is the feed for the fish, and 2) primary raceway effluents are characterized by high volumes of water with low concentrations of nutrients. Also of note is the relative paucity of published reports on raceway effluents in the U.S., particularly since 1990. One reason for this is the recent emphasis on modeling energetics, nutrition, and feeding efficiency in aquaculture as accurate methods for predicting waste output (Cho et al. 1991, 1994; Frier et al. 1995; Cho and Bureau 1998).

Management of Nutrient Discharges from Raceways

Feeds and Feeding

Since effluents from raceway culture systems in the U.S. discharge almost exclusively into freshwater, the primary nutrient of concern is phosphorus. Most producers, researchers, and regulators recognize that the most effective way to reduce the potential for impact from raceway systems is to increase nutrient retention efficiency and digestibility of the diets. Reported requirements for phosphorus in diets of rainbow trout range from 0.5 to 0.8% (Ogino and Takeda 1978; Ketola and Richmond 1994). Earlier studies often relied on growth and tissue content for such determinations, but recent studies using phosphorus excretion rates as well as growth and tissue content for a variety of trout sizes have suggested lower levels ranging from 0.322% to 0.524 % for rainbow trout (Sugiura et al. 2000). The requirement for larger trout (350 g) may also be somewhat lower than for smaller trout (<150 g), allowing the possibility of specific diet formulations for efficient production of different sizes of fish, and reduced waste. As mentioned earlier, nearly 90% of the feed is applied during the last half of the

production cycle for each group of fish in raceway systems. The amount of feed given to smaller fish on commercial facilities is relatively insignificant and the contribution to wastes in the effluent is proportionately small.

Techniques such as acidification of the diet could also enhance the availability of dietary minerals such as phosphorus, but research in this area is still needed (Sugiura and Hardy 2000). Ketola (1982) reported as late as 1982 that phosphorus retention in commercial trout facilities was only about 19%, with the remainder lost to the effluent. In the past decade, the reduction of phosphorus content of commercial diets from over 1.4%–1.8% to 1.1–1.3% and a concomitant increase in digestibility has resulted in increasing phosphorus retention per unit of feed from 32% or less (1.4% P-commercial diet, Ketola and Harland 1993), to over 46% retention (1.09% P-custom diet, Azevedo et al. 1998; 1.32% P-commercial diet, Sugiura et al. 1998, 2000). Commercial diets with less than 1% phosphorus are available, but are usually more expensive per unit of fish weight gain than feeds containing 1.1% or slightly higher.

Feeding rates (as % body weight/d) have often been suggested to affect the quantity of waste generated per unit of fish produced or per unit of feed used. Early studies of fish energetics suggested that for salmonids, optimal conversions of feed and therefore the lowest proportionate discharge of wastes could only be achieved at less than that required for maximum growth (Warren and Davis 1967; Elliot 1976; Windell et al. 1978; Brett and Groves 1979; Storebakken and Austreng 1987), although the distinction is not always made between maximum consumption and maximum growth. Recent research suggests that feeding rainbow trout at rates between 50% and 100% maximum growth ration does not result in a significant difference in feed conversion efficiency (Alanärä 1994; Einen et al. 1995; Azevedo et al. 1998), or waste generation per unit of fish growth (Azevedo et al. 1998). Increased temperature was noted to have a positive impact on feed digestibility, however.

The feeding method used can impact the percent of feed that is wasted, although limited documentation is available on this topic with modern extruded diets (floating or semi-floating) in raceways. If a properly formulated, highly digestible feed is consumed by fish, a relatively small percentage will end up in the water as waste solids, (approximately 15–25%). By comparison, 100% of feed not consumed will become waste solids. Warrer-Hansen (1982b) gave ranges for uneaten pelleted feed from 1% to 5% for rainbow trout cultured intensively in ponds, which reflects the maximum of what we have observed for estimates of losses in raceways. Newer extruded diets tend to approach 100% being eaten by fish, since the pellets are more water stable, typically float, and uneaten pellets simply pass to the next tank. Feeding method (automatic versus demand or self-feeding) does not significantly affect the digestibility of the feed (Yamamoto et al. 2001) nor the growth of trout when feed rations are adjusted to provide comparable amounts (Bailey and Alanärä 2001), although trout fed with self-feeders had slightly higher variation in growth among individuals. Acoustic and light or image-based technologies have been developed for limiting feed wastage in cages or circular tanks (Juell et al. 1993; Foster et al. 1995; Summerfelt et al. 1995), but these technologies have not been applied to raceways and may not be needed for these systems. Models incorporating energetics, nutrient availability and retention efficiencies, and fish growth potential can provide producers with accurate projections of feed requirements for specific diet formulations (e.g., Frier et al. 1995; Cho and Bureau 1998). These models can also assist in developing projected feed conversion

efficiencies and waste output allowing management intervention by the producer when necessary. Verification of this approach for predicting nutrient discharge from commercial raceway production systems would be valuable. Feeding by hand whenever practical is still recommended as the best way to observe fish appetite and behavior in raceways, and adjust feeding accordingly.

Solids Management

Once an efficient diet and feeding rate is determined for a raceway system, steps should be taken to ensure removal of a high percentage of the solid wastes excreted by the fish. The solids produced by salmonids typically represent about 9% of the nitrogen wastes and 30% of the phosphorus wastes released by the trout in a tank system prior to settling (Foy and Rosell 1991a). When collected directly from trout, fecal solids were 0.825% phosphorus by weight with 26.6% of the fecal phosphorus in a soluble fraction after 24 hr incubation in lakewater (Garcia-Ruiz and Hall 1996). Settled trout manure ranging from 1 mo to 9 mo old in sedimentation basins had an average of 0.69% phosphorus (Westerman et al. 1993). Total nitrogen content of trout manure typically ranged from 2% to just over 5% dry weight for settled material (Olson 1992; Westerman et al. 1993). Manure sludge captured from a trout and salmon farm using microseives, contained 8.1% nitrogen and 4.3% phosphorus (Bergheim et al. 1998), perhaps corresponding to rapidity of collection, but also to different feed formulations and much better feed conversion efficiency (1.07:1) than reported in Westerman et al. (1.5 to 1.6:1; 1993) and Garcia-Ruiz and Hall (1.5:1; 1996).

Stechey and Trudell (1990b) found sedimentation to be the most widely applicable and inexpensive method for removing solids from flow-through trout farms. Gravitational settling can provide a simple, low-maintenance, and only moderately expensive method of

FIGURE 4. *Quiescent zones at the end of a tank in a raceway system.*

FIGURE 6. *Paired off-line settling basins.*

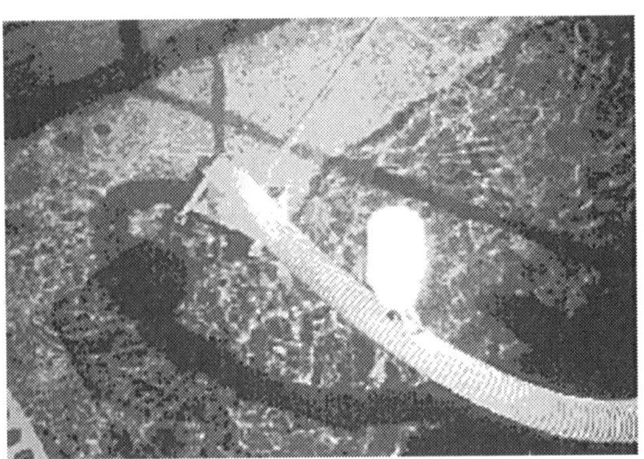

FIGURE 5. *Cleaning settled solids from a quiescent zone.*

FIGURE 7. *Full-flow settling basins.*

removing a high percentage of the solids and accompanying nutrients from raceway effluents, but may require considerable flat area (Summerfelt 1999) and must be properly designed. Earthen settling ponds were reportedly ineffective at reducing suspended solids concentrations below 6 mg/L, but this calculated minimum was attributed to resuspension of solids in inadequately designed basins (Henderson and Bromage 1988). Stechey and Trudell (1990b) estimate the cost of treatment of raceway effluent using settling basins at US$0.072 per kg fish produced ($0.09 Canadian per kg).

Within raceway production systems, three types of settling basins are used to settle solids: quiescent zones (Figs. 4, 5), off-line settling basins (Fig. 6), and full-flow settling basins (Fig. 7). A quiescent zone is an area downstream of the rearing area of each tank for initial separation of settleable solids from the water. A screen prohibits fish from entering the quiescent zone allowing the solids to settle undisturbed. Ideally, quiescent zones should be installed in every tank, but it is essential that the last tanks of each raceway series have quiescent zones to settle solids before the water is discharged into receiving waters. A common misconception is that the water velocity within raceway systems serves to 'flush' solids from the tanks. In fact, water velocities in virtually all commercial raceway systems are below 3.7 cm/sec suggested as the minimum flow required for self-cleaning (Youngs and Timmons 1991). Fecal solids and waste feeds will accumulate nearly anywhere in the system that fish are not present.

Quiescent zones serve as the pre-treatment system for solids, which are then either pumped or transported by gravity flow to off-line settling basins. Off-line settling basins receive the concentrated solids removed from quiescent zones, but not the full flow of the water from the facility. The combination of quiescent zones and off-line settling basins is the most commonly used system of treatment with concrete raceways in the U.S. to capture and remove solids. Full-flow settling basins receive the entire flow of a facility and are generally used where total facility flow is considered small, generally less than 283 L/sec, and level area is not limiting. Quiescent zones may or may not be included with full-flow basins depending upon the intensity of water use. Particle settling velocities vary with size and density and each type of settling basin is designed to retain a specific particle size fraction. Trout fecal casts are relatively large and heavy, and will have settling velocities ranging from 2 to 5 cm/sec, while fine, lighter particles settle at much slower velocities, ranging from 0.046 to 0.09 cm/sec (Warrer-Hansen 1982a; Stechey and Trudell 1990b). Settleable solids from a commercial trout raceway facility had a median settling velocity of 1.7 cm/sec in terms of total mass, but a median settling velocity of 1.15 cm/sec in terms of phosphorus mass (Wong and Piedrahita 2000), suggesting that a settling basin should be slightly more efficient at phosphorus removal compared to mass removal of solids. The Idaho Waste Management Guidelines for Aqua-culture Operations (IDEQ 1997) recommends for trout raceway production systems overflow rates of 0.94 cm/sec, 0.40 cm/sec, and 0.046 cm/sec for quiescent zones, full-flow settling basins, and off-line settling basins respectively, when operated under ideal conditions. For example, based on settling curves and a non-settleable faction of 27% as presented by Wong and Piedrahita (2000) for trout raceway system solids, a quiescent zone with an overflow rate of 0.94 cm/sec should capture approximately 51.2% of the total solids and 58.8% of the particulate phosphorus. The

full-flow settling basin with an overflow rate of 0.4 cm/sec operating under ideal conditions should capture 54.0% of the solids in the water, and 62% of the phosphorus associated with those solids. The off-line settling basin would have a calculated capture efficiency of the settleable solids of 84.1% and 96.6% of the settleable particulate phosphorus. The residence time of the off-line basins typically far exceeds 1 h, so we did not attempt to calculate the efficiencies based on total suspended and settleable solids. Basic design criteria for aquaculture settling basins can be found in Wheaton (1977), Stechey and Trudell (1990b), IDEQ (1997), and Summerfelt (1999).

Other technologies tested for solids capture from raceway effluents include microscreens, plate separators, swirl separators or hydrocyclones, media filters, air flotation and foam fractionation, and chemical flocculation (Summerfelt 1999; Cripps and Bergheim 2000). Of these, the technology with potential for practical use with the high flows and low concentrations from raceways is the use of microscreens. Advantages to microscreens include large capacities in a relatively small space and automatic operation. Disadvantages include relatively high costs, production of relatively dilute wastes when used individually, and only moderate efficiencies when treating dilute effluents. In an evaluation of microscreens on a commercial Atlantic salmon smolt and rainbow trout farm, the issues of dilute outflows from the initial separatory screens were addressed by adding an additional drum screen as a de-watering device, then using a sedimentation tank as a thickening tank for the sludge (Bergheim et al. 1998). The system removed 70–75% of the total suspended solids, 45–60% of the total phosphorus, and about 5% of the total nitrogen from the effluent water. The primary drawback to the system as far as application to raceway effluents is its cost. The estimated cost of effluent treatment, not including removal from the site, was $0.056

per salmon smolt or $0.79 per kg of fish. This amount is nearly half the total production cost of rainbow trout in some areas of the U.S. and would clearly not be economical.

Conclusions

In spite of the massive quantity of research describing various components of effluent management in raceways, the choice of system and approach is still best made on a site-specific basis. Many of the descriptive studies of effluents are diminishing in practical relevance due to rapid advances in fish nutrition and systems management. Of all the potential negative impacts of effluents from raceways, the most common and most visible still result from failure to control suspended and settleable solids from the facilities. Overall, of the nearly 700 raceway production systems in the U.S., very few have been identified as the cause of severe stream impairment, yet most contribute to some level of nutrient-related changes in stream habitat below their discharges. The degree of impact can be reduced significantly through 1) enhancements in feed quality and feeding efficiency, and 2) effective solids capture and handling.

Literature Cited

Alabaster, J. S. 1982. A survey of fish-farm effluents in some EIFAC Countries. Pages 5–20 in J. S. Alabaster, editor. Report of the EIFAC workshop on fish-farm effluents, Silkeborg, Denmark, 26–28 May 1981. EIFAC (European Island Fisheries Advisory Commission) Technical Paper No. 41. FAO, Rome, Italy.

Alanärä, A. 1994. The effect of temperature, dietary energy content and reward level on the demand feeding activity of rainbow trout *(Oncorhynchus mykiss)*. Aquaculture 126:349–359.

APHA (American Public Health Association). 1985. Standard methods for the examination of water and wastewater, 16th edition. Washington, D.C., USA.

Azevedo, P. A., C. Y. Cho, S. Leeson, and D. P. Bureau. 1998. Effects of feeding level and water temperature on growth, nutrient and energy utilization, and waste outputs of rainbow trout *(Oncorhynchus mykiss)*. Aquatic Living Resources 11:227–238.

Bailey, J. and A. Alanärä. 2001. A test of a feed budget model for rainbow trout, *Oncorhynchus mykiss* (Walbaum). Aquaculture Research 32:465–469.

Barinanga, M. 1990. Fish, money, and science in Puget Sound. Science 247:631.

Barrows, F. T. and R. W. Hardy. 2000. Feed manufacturing technology. Pages 354–359 in R. R. Stickney, editor. Encyclopedia of aquaculture. John Wiley & Sons, Inc., New York, New York, USA.

Bergheim, A. and A. R. Selmer-Olsen. 1978. River pollution from a large trout farm in Norway. Aquaculture 14:267–270.

Bergheim, A. and A. Sivertsen. 1981. Oxygen consuming properties of effluents from fish farms. Aquaculture 22:185–187.

Bergheim, A., S. J. Cripps, and H. Liltved. 1998. A system for the treatment of sludge from land based fish farms. Aquatic Living Resources 11:279–287.

Bergheim, A., H. Hustveit, A. Kittelsen, and A. R. Selmer-Olsen. 1984. Estimated pollution loadings from Norwegian fish farms. II. Investigations 1980–1981. Aquaculture 36:157–168.

Beveridge, M. C. M., M. J. Phillips, and R. M. Clarke. 1991. Quantitative and qualitative assessment of wastes from aquatic animal production. Pages 506–533 in D. E. Brune and J. R. Tomasso, editors. Aquaculture and water quality. World Aquaculture Society, Baton Rouge, Louisiana, USA.

Boaventura, R., A. M. Pedro, J. Coimbra, and E. Lencastre. 1997. Trout farm effluents: Characterization and impact on the receiving streams. Environmental Pollution 95:379–387.

Bodien, D. G. 1970. An evaluation of salmonid hatchery wastes. U.S. Department of the Interior, Federal Water Quality Administration, Northwest Region, Portland, Oregon, USA.

Bohl, M. 1988. National report for the Federal Republic of Germany on suspended solids from land-based fish farms. Pages 29–36 *in* M. Pursiainen, editor. National contributions on suspended solids from land-based fish farms. Monistettuja Julkaisuja, Helsinki, Finland.

Brett, J. R. and T. D. D. Groves. 1979. Physiological energetics. Pages 279–352 *in* W. S. Hoar, D. J. Randall, and J. R. Brett. Fish physiology volume VIII. Academic Press, New York, New York, USA.

Brisbin, K. J. 1971. Pollutional aspects of trout hatcheries in British Columbia. Prepared by Underwood, McLellan and Associates for the B.C. Department of Recreation and Conservation, Fish and Wildlife Branch, Victoria, British Columbia, Canada.

Burroughs, R. and H. Chenoweth. 1955. Evaluation of three types of fish rearing ponds. Research Report 39, U.S. Department of the Interior, Fish and Wildlife Service, Washington, D.C., USA.

Burrows, R. E. 1968. Salmonid husbandry techniques. Shearwater Fish Farming, Shearwater, England.

Buterbaugh, G. L. and H. Willoughby. 1967. A feeding guide for brook, brown, and rainbow trout. Progressive Fish-Culturist 29:210–215.

Butz, I. and B. Vens-Cappell. 1982. Organic load from the metabolic products of rainbow trout fed with dry food. Pages 73–82 *in* J. S. Alabaster, editor. Report of the EIFAC workshop on fish-farm effluents, Silkeborg, Denmark, 26–28 May 1981. EIFAC (European Island Fisheries Advisory Commission) Technical Paper No. 41. FAO, Rome, Italy.

Caddy, D. E. 1979. Water quality is key to trout profits. Fish Farmer 2(4):12–14.

Castledine, A. J. 1986. Aquaculture in Ontario. Ontario Ministry of Natural Resources, Ontario Ministry of Agriculture and Food, and Ontario Ministry of Environment, Queen's Printer for Ontario, Canada.

Cho, C. Y. and D. P. Bureau. 1998. Development of bioenergetic models and the Fish-PrFEQ software to estimate production, feeding ration and waste output in aquaculture. Aquatic Living Resources 11:199–210.

Cho, C. Y., J. D. Hynes, K. R. Wood, and H. K. Yashida. 1991. Quantitation of fish culture wastes by biological (nutritional) and chemical (limnological) methods; the development of high nutrient dense (HND) diets. Pages 37–50 *in* C. B. Cowey and C. Y. Cho, editors. Nutritional strategies and aquaculture waste, proceedings of the first international symposium on nutritional strategies and management of aquaculture waste. University of Guelph, Ontario, Canada.

Cho, C. Y., J. D. Hynes, K. R. Wood, and H. K. Yashida. 1994. Development of high nutrient-dense, low pollution diets and prediction of aquaculture wastes using biological approaches. Aquaculture 124:293–305.

Clark, E. R., J. P. Harman, and J. R. M. Forster. 1985. Production of metabolic and waste products by intensively farmed rainbow trout *Salmo gairdneri*. Journal of Fish Biology 27:381–394.

Clarke, S. 1990. Extruded diets and pollution reduction. Fish Farmer, International File

4(1):95.

Colt, J., K. Orwicz, and G. Bouck. 1991. Water quality considerations and criteria for high-density fish culture with supplemental oxygen. Pages 372–385 *in* J. Colt and R. J. White, editors. Fisheries Bioengineering Symposium 10. American Fisheries Society, Bethesda, Maryland, USA.

Cripps, S. J. 1995. Serial particle size fractionation and characterisation of an aquacultural effluent. Aquaculture 133:323–339.

Cripps, S. J. and A. Bergheim. 2000. Solids management and removal for intensive land-based aquaculture production systems. Aquacultural Engineering 22:33–56.

Dosdat, A., R. Métailler, E. Desbreyères, and C. Huelevan. 1998. Comparison of brown trout *(Salmo trutta)* reared in freshwater and seawater to freshwater rainbow trout *(Oncorhynchus mykiss)* II: Phosphorus balance. Aquatic Living Resources 11:21–28.

Einen, O., I. Holmefjord, T. Asgard, and C. Talbot. 1995. Auditing nutrient discharges from fish farms: theoretical and practical considerations. Aquaculture Research 26:701–713.

Ellis, J. E., D. L. Tackett, and R. R. Carter. 1978. Discharge of solids from fish ponds. Progressive Fish-Culturist 40:165–166.

Elliot, J. M. 1976. The energetics of feeding, metabolism and growth of brown trout (Salmo trutta L.) in relation to body weight, water temperature and ration size. Journal of Animal Ecology 45:923–948.

EPA (Environmental Protection Agency). 1974. Development document for proposed effluent limitation, guidelines and new source performance standards for the fish hatcheries and farms. Washington, D.C., USA.

Eskelinen, P. 1986. The phosphorus balance of rainbow trout. Canadian Translations in Fisheries and Aquatic Sciences, Number 5274. Vesihallituksen monistesarja 241:33–42.

Fauvre, A. 1977. Mise au point sur la pollution engendree les piscicultures. Piscicultures Francais 50:33–35.

Fivelstad, S., J. M. Thomassen, M. J. Smith, H. Kjartansson, and A. B. Sando. 1990. Metabolite production rates from Atlantic salmon *(Salmo salar* L.) and arctic char *(Salvelinus alpinus* L.) reared in single pass land-based brackish water and sea-water systems. Aquatech Systems A/S, Langoyneset, Norwegian Aquacultural Engineering 9:1–21.

Forster, R. P. and L. Goldstein. 1969. Formation of excretory products. Pages 313–350 *in* W. S. Hoar and J. C. Randall, editors. Fish physiology, volume 1. Academic Press, London, England.

Foster, M., R. Petrell, M. R. Ito, and R. Ward. 1995. Detection and counting of uneaten food pellets in a sea cage using image analysis. Aquacultural Engineering 14:251–269.

Foy, R. H. and R. Rosell. 1991a. Fractionation of phosphorus and nitrogen loadings from a Northern Ireland fish farm. Aquaculture 96:31–42.

Foy, R. H. and R. Rosell. 1991b. Loadings of nitrogen and phosphorus from a Northern Ireland fish farm. Aquaculture 96:17–30.

Frier, J-O., J. From, T. Larsen, and G. Rasmussen. 1995. Modeling waste output from trout farms. Nutritional strategies and management of aquaculture waste. Water Science and Technology 31:103–121.

Fromm, P. O. and J. R. Gillette. 1968. Effect of ambient ammonia on blood ammonia and nitrogen excretion rates in rainbow trout Salmo gairdneri. Comparative Biochemistry and Physiology 26:887–898.

Gagen, C.J. 1990. Fish, money, and science in Puget Sound. Science 248:290.

Garcia-Ruiz, R. and G. H. Hall. 1996. Phosphorus fractionation and mobility in the food and faeces of hatchery reared rainbow trout (Onchorhynchus mykiss). Aquaculture 145:183–193.

Gowen, R. J., D. P. Weston, and A. Ervik. 1991. Aquaculture and the benthic environment. Pages 187–205 in C. B. Cowey and C. Y. Cho, editors. Nutritional strategies and aquaculture wastes. Proceedings of the first international symposium on nutritional strategies and management of aquaculture waste. University of Guelph, Ontario, Canada.

Hardy, R. W. 1996. Alternate protein sources for salmon and trout diets. Animal Feed Science Technology 59:71–80.

Harris, J. 1981. Federal regulation of fish hatchery effluent quality. Pages 157–161 in L. J. Allen and E. C. Kinney, editors. Proceedings of the bio-engineering symposium for fish culture. FCS Publication 1. Fish Culture Section of the American Fisheries Society, Bethesda, Maryland, USA.

Haskell, D. C. 1955. Weight of fish per cubic foot of water in hatchery troughs and ponds. Progressive Fish-Culturist 17:117–118.

Heinsbroek, L. T. N. 1988. Waste production and discharge of intensive culture of African catfish, *Clarias gariepinus*, European eel, *Anguilla anguilla*, and rainbow trout, *Salmo gairdneri*, in recirculation and flow-through units in the Netherlands. Pages 55–61 in M. Pursiainen, editor. National contributions on suspended solids from land-based fish farms. Montistettuja Julkaisuja, Helsinki, Finland.

Henderson, J. P. and N. R. Bromage. 1988. Optimising the removal of suspended solids from aquacultural effluents in settlement lakes. Aquacultural Engineering 7:167–181.

Henderson, P. and N. Bromage. 1987. Diets to curb pollution. Fish Farmer Jul/Aug:39.

Hinshaw, R. N. 1973. Pollution as a result of fish culture activities. EPA report #PA-R3-73-009. U.S. Government Printing Office, Washington, D.C.

IDEQ (Idaho Division of Environmental Quality). 1997. Idaho waste management guidelines for aquaculture operations. Boise, Idaho, USA.

Iwama, G. K. 1991. Interactions between aquaculture and the environment. Critical Reviews in Environmental Control 21:177–216.

Jensen, R. 1972. Taking care of wastes from the trout farm. American Fish and US Trout News 16(5):4–6,21.

Juell, J-E., D. M. Furevik, and Å. Bjordal. 1993. Demand feeding in salmon farming by hydroacoustic food detection. Aquacultural Engineering 12:155–167.

Kazamzakeh, M. N. 1990. Fish feed extrusion technology; the case for the twin screw. Feed International, Feb. 22:4.

Kendra, W. 1991. Quality of salmonid hatchery effluents during a summer low-flow season. Transactions of the American Fisheries Society 120:43–51.

Ketola, H. G. 1982. Effects of phosphorus in trout diets on water pollution. Salmonid 6:12.

Ketola, H. G. 1985. Mineral nutrition: effects of phosphorus in trout and salmon feeds on water pollution. Pages 465–473 in C. B. Cowey, A. M. Mackie and J. G. Bell, editors. Nutrition and feeding in fish. Academic Press, London, England.

Ketola, H. G. and B. F. Harland. 1993. Influence of phosphorus in rainbow trout diets on phosphorus discharges in effluent water. Transactions of the American Fisheries Society 122:1120–1126.

Ketola, G. and M. E. Richmond. 1994. Requirement of rainbow trout for dietary phosphorus and its relationship to the amount discharged. Transactions of the American Fisheries Society 123:587–594.

Ketola, H. G., H. Westers, W. Houghton, and C. Pecor. 1991. Effect of diet on growth and survival of coho salmon and on phosphorus discharges in effluent water from a fish hatchery. Pages 402–409 *in* J. Colt and R. J. White, editors. Fisheries Bioengineering Symposium 10. American Fisheries Society, Bethesda, Maryland, USA.

Klapsis, A. and R. Burley. 1984. Flow distribution studies in fish rearing tanks. Part 1—design constraints. Aquacultural Engineering 3:103–118.

Klontz, G. W., I. R. Brock, and J. A. McNair. 1979. Aquaculture techniques: water use and discharge quality. Idaho Water Resource Research Institute, Moscow, Idaho, USA.

Knosche, R. 1971. Der Einfluss intensiver Fischproducktion auf das Wasser und Moglichkeiten zur Wasserreinigung Z. Binnenfisch D.D.R. 18:372–379.

Korzeniewski, K., A. Banat, and A. Moczulska. 1982. Changes in water of the Uniesc and Skotawa rivers, caused by intensive trout culture. Polish Archives of Hydrobiology 29:683–691.

Krieger, R. I., D. Marcy, J. H. Smith, and K. Tomson. 1987. Levels of nine potentially toxic elements in Idaho fish manure. Bulletin of Environmental Contamination and Toxicology 38:63–66.

Liao, P. B. 1970a. Pollutional potential of salmonid fish hatcheries. Water and Sewage Works 117:291–297.

Liao, P. B. 1970b. Salmonid hatchery waste water treatment. Water and Sewage Works 117:439–443.

Liao, P. B. and R. D. Mayo. 1974. Intensified fish culture combining water reconditioning with water pollution abatement. Aquaculture 3:61–85.

Liao, P. B., R. D. Mayo, and W. Williams. 1972. A study for development of hatchery water treatment systems. Report for the U.S. Army Corps of Engineers, Walla Walla, Washington District by Kramer, Chin & Mayo, Consulting Engineers, Seattle, Washington, USA.

Loch, D. D., J. L. West, and D. G. Perlmutter. 1996. The effect of trout farm effluent on the taxa richness of benthic macro-invertebrates. Aquaculture 147:37–55.

Lovell, T. 1989. Nutrition and feeding of fish. Van Nostrand Reinhold, New York, New York, USA.

Makinen, T. 1988. Report on suspended solids from fish farms in Finland. Pages 17–28 *in* M. Pursiainen, editor. National contributions on suspended solids from land-based fish farms. Monistettuja Julkaisuja, Helsinki, Finland.

Markmann, P. N. 1981. Biological effects of effluents from Danish fish farms. Pages 99–102 *in* J. S. Alabaster, editor. Report of the EIFAC workshop on fish-farm effluents, Silkeborg, Denmark, 26-28 May 1981. EIFAC (European Island Fisheries Advisory Commission) Technical Paper No. 41. FAO, Rome, Italy.

Merican, Z. O. and M. J. Phillips. 1985. Solid waste production from rainbow trout, *Salmo gairdneri* Richardson, cage culture. Aquaculture and Fisheries Management

1:55–69.

Mudrak, V.A. 1981. Guidelines for economical commercial fish hatchery waste water treatment systems. Pages 174–182 *in* L. J. Allen and E. C. Kinney, editors. Proceedings of the bio-engineering symposium for fish culture. FCS Publication 1. Fish Culture Section of the American Fisheries Society, Bethesda, Maryland, USA.

Mudrak, V. A. and K. R. Stark. 1981. Guidelines for economical commercial hatchery wastewater treatment systems. National Marine Fisheries Service Project Number Commercial 3–242–R. Pennsylvania Fish Commission, Philadelphia, Pennsylvania, USA.

Niemi, M. and I. Taipalinen. 1980. Hygienic indicator bacteria in fish farms. Research report. Helsinki National Board of Waters, Helsinki, Finland.

Odum, W. E. 1974. Potential effects of aquaculture on inshore coastal waters. Environmental Conservation 1:225–230.

Ogino, C. and H. Takeda. 1978. Requirements of rainbow trout for dietary calcium and phosphorus. Bulletin of the Japanese Society of Scientific Fisheries 44:1019–1022.

Olson, G. L. 1992. The use of trout manure as a fertilizer for Idaho crops. Pages 198–205 *in* J. Blake, J. Donald and W. Magette, editors. National livestock, poultry and aquaculture waste management. Publication 03-92. American Society of Agricultural Engineers, St. Joseph, Michigan, USA.

Persson, G. 1990. Eutrophication resulting from salmonid fish culture in fresh and salt waters: Scandinavian experiences. Pages 163–185 *in* C. B. Cowey and C. Y. Cho, editors. Proceedings of the first international symposium on feeding fish

in our waters: nutritional strategies in management of aquaculture waste, 5-8 June 1990. Guelph, Ontario, Canada.

Piper R. G., I. B. McElwain, L .E. Orme, J. P. McCraren, L. G. Fowler, and J. R. Leonard. 1982. Fish hatchery management. U.S. Dept. of Interior, Fish and Wildlife Service, Washington, D.C., USA.

Pitts, J.L. 1990. Fish, money, and science in Puget Sound. Science 248:290–291.

Purdom, C. E. 1982. Analyses of influent and effluent water in UK trout farms - an interim report. Pages 83–86 *in* J. S. Alabaster, editor. Report of the EIFAC workshop on fish-farm effluents, Silkeborg, Denmark, 26-28 May 1981. EIFAC (European Island Fisheries Advisory Commission) Technical Paper No. 41. FAO, Rome, Italy.

Querellou, J., A. Fauré, and C. Fauré. 1982. Pollution loads from rainbow trout farms in Brittany, France. Pages 87–97 *in* J. S. Alabaster, editor. Report of the EIFAC workshop on fish-farm effluents, Silkeborg, Denmark, 26-28 May 1981. EIFAC (European Island Fisheries Advisory Commission) Technical Paper No. 41. FAO, Rome, Italy.

Rodehutscord, M. 1996. Response of rainbow trout (*Oncorhynchus mykiss*) growing from 50 to 200 g to supplements of dibasic sodium phosphate in a semi-purified diet. Journal of Nutrition 126:324–331.

Rychly, J., and B. A. Marina. 1977. The ammonia excretion of trout during a 24-hour period. Aquaculture 11:173–178.

Selong, J. H. and L. A. Helfrich. 1998. Impacts of trout culture effluent on water quality and biotic communities in Virginia headwater streams. Progressive Fish-Culturist 60:247–262.

Seymour, E. A. and A. Bergheim. 1991. Towards a reduction of pollution from intensive aquaculture with reference to the farming of salmonids in Norway. Aquacultural Engineering 10:73–88.

Shanks, W. 1971. Hatchery water quality monitoring. Presented at the 22nd Northwest Fish Cultural Conference, 2-3 Dec. 1970. Portland, Oregon, USA.

Sloan, D. 2001. Verification of the state listings for the 1998 EPA 303d & 305b: Impaired waterbodies due to aquaculture. Report to the Joint Subcommittee on Aquaculture, aquaculture effluents task force. National Association of State Aquaculture Coordinators (NASAC), Raleigh, North Carolina, USA.

Smart, G. R. 1978. Investigations of the toxic mechanisms of ammonia to fish - gas exchange in rainbow trout (*Salmo gairdneri*) exposed to acutely lethal concentrations. Journal of Fish Biology 12:93–104.

Smith, C. E. and R. G. Piper. 1975. Lesions associated with chronic exposure to ammonia. Pages 497–514 *in* W. E. Ribelin and G. Migaki, editors. The pathology of fishes. University of Wisconsin Press, Madison, Wisconsin, USA.

Solbé, J. F. de L.G. 1982. Fish-farm effluents; a United Kingdom survey. Pages 29–56 *in* J. S. Alabaster, editor. Report of the EIFAC workshop on fish-farm effluents, Silkeborg, Denmark, 26–28 May 1981. EIFAC (European Island Fisheries Advisory Commission) Technical Paper No. 41. FAO, Rome, Italy.

Sparrow, R. A. H. 1981. Hatchery effluent water quality in British Columbia. Pages 162–166 *in* L. J. Allen and E. C. Kinney, editors. Proceedings of the bio-engineering symposium for fish culture. FCS Publication 1. Fish Culture Section of the American Fisheries Society, Bethesda, Maryland, USA.

Speece, R. T. 1973. Trout metabolism characteristics and the rational design of nitrification facilities for water re-use in hatcheries. Transactions of the American Fisheries Society 102:323–332.

Stechey, D. and Y. Trudell. 1990a. Aquaculture wastewater treatment: Wastewater characterization and development of appropriate treatment technologies for the New Brunswick smolt production industry: final report. Canadian Aquaculture Systems Bioengineering Technologies & Business Management Services, New Brunswick Fisheries and Aquaculture, Fredericton, New Brunswick, Canada.

Stechey, D. and Y. Trudell. 1990b. Aquaculture wastewater treatment: Wastewater characterization and development of appropriate treatment technologies for the Ontario trout production industry: Final report. Canadian Aquaculture Systems Bioengineering Technologies & Business Management Services, Queen's Printer for Ontario, Canada.

Storebakken, T. and E. Austreng. 1987. Ration level for salmonids. 2. Growth, feed intake, protein digestibility, body composition, and feed conversion in rainbow trout weighing 0.5–1.0 kg. Aquaculture 60:207–221.

Sugiura, S. H. and R. W. Hardy. 2000. Environmentally friendly feeds. Pages 299–310 *in* R. R. Stickney, editor. Encyclopedia of aquaculture. John Wiley and Sons, Inc., New York, New York, USA.

Sugiura, S. H., F. M. Dong, and R. W. Hardy. 2000. Primary responses of rainbow trout to dietary phosphorus concentrations. Aquaculture Nutrition 6:235–245.

Sugiura, S. H., F. M. Dong, C. K. Rathbone, and R. W. Hardy. 1998. Apparent protein digestibility and mineral availabilities in various feed ingredients for salmonid feeds. Aquaculture 159:177–202.

Sumari, O. 1982. A report on fish-farm effluents in Finland. Pages 21–28 *in* J. S. Alabaster, editor. Report of the EIFAC workshop on fish-farm effluents, Silkeborg, Denmark, 26–28 May 1981. EIFAC (European Island Fisheries Advisory Commission) Technical Paper No. 41. FAO, Rome, Italy.

Summerfelt, S. T. 1999. Waste handling systems. Pages 309–350 *in* F. W. Wheaton, volume editor. CIGR (International Commission of Agricultural Engineering) handbook of agricultural engineering, volume II, part II. Aquaculture engineering. American Society of Agricultural Engineers, St. Joseph, Michigan, USA.

Summerfelt, S. T., K. H. Holland, J. A. Hankins, and M. D. Durant. 1995. A hydroacoustic waste-feed controller for tank systems. Water Science and Technology 31:123–129.

Taylor, J. A. and C. J. Perrin. 1989. A database of literature describing associations between limnological variables and freshwater aquaculture activities. Report to British Columbia MAFF. Vancouver, British Columbia, Canada.

Timmons, M. B. and W. D. Youngs. 1991. Considerations on the design of raceways. *In* P. Giovannini, editor. Aquaculture systems engineering. ASAE Publication 02-91. American Society of Agricultural Engineers, St. Joseph, Michigan, USA.

Train, R. E., A. W. Briedenback, E. C. Beck, R. B. Schaffer, and D. F. Anderson. 1977. Development document for recommended effluent limitations guidelines and standards of performance for the fish hatcheries and farms point source category. Effluent Guidelines Division, Office of Water and Hazardous Materials, U.S. Environmental Protection Agency, Washington, D.C., USA.

UMA Engineering Ltd. 1979. Salmon rearing facilities waste water study—waste characterization. Prepared for the Department of Fisheries and Oceans, Vancouver, British Columbia, Canada.

Walden, C. C. and A. E. Birkbeck. 1974. Feasibility and design studies for water treatment at a trout hatchery. Prepared for British Columbia Department of Public Works by B.C. Research, Vancouver, British Columbia, Canada.

Warren, C. E., and G. E. Davis. 1967. Laboratory studies on the feeding, bioenergetics, and growth of fish. Pages 175–214 *in* S. D. Gerking, editor. The biological basis of freshwater fish production. John Wiley & Sons, New York, New York, USA.

Warrer-Hansen, I. 1978. Fish farms are having to watch their wastes. Fish Farming International 5(3):19.

Warrer-Hansen, I. 1982a. Methods of treatment of waste water from trout farming. Pages 113–121 *in* J. S. Alabaster, editor. Report of the EIFAC workshop on fish-farm effluents, Silkeborg, Denmark, 26-28 May 1981. EIFAC (European Island Fisheries Advisory Commission) Technical Paper No. 41. FAO, Rome, Italy.

Warrer-Hansen, I. 1982b. Evaluation of matter discharged from trout farming in Denmark. Pages 57–63 *in* J. S. Alabaster, editor. Report of the EIFAC workshop on fish-farm effluents, Silkeborg, Denmark, 26-28 May 1981. EIFAC (European Island Fisheries Advisory Commission) Technical

Paper No. 41. FAO, Rome, Italy.

Westerman, P. W., J. M. Hinshaw, and J. C. Barker. 1993. Trout manure characterization and nitrogen mineralization rate. Pages 35–43 *in* J.-K. Wang, editor. Techniques for modern aquaculture. Proceedings of an aquacultural engineering conference, 21-23 June 1993, Spokane, Washington. American Society of Agricultural Engineers, St. Joseph, Michigan, USA.

Westers, H. 1990. Fish hatchery effluents: problems and solutions. Presented at the 120th Annual Meeting of the American Fisheries Society, Pittsburgh, PA (USA), 26-30 August 1990. American Fisheries Society, Bethesda, Maryland, USA.

Westers, H. and K. M. Pratt. 1977. Rational design of hatcheries for intensive salmonid culture, based on metabolic characteristics. Progressive Fish-Culturist 39:157–165.

Wheaton, F. W. 1977. Aquacultural engineering. John Wiley & Sons, New York, New York, USA.

Weismann, D., H. Scheid and E. Pfeffer. 1988. Water pollution with phosphorus of dietary origin by intensively fed rainbow trout *Salmo gairdneri* Rich. Aquaculture 69:263–270.

Willoughby, H. 1968. A method for calculating carrying capacities of hatchery troughs and ponds. Progressive Fish-Culturist 30:173–174.

Willoughby, H., H. N. Larsen, and J. T. Bowen. 1972. The pollutional effects of fish hatcheries. American Fish and U.S. Trout News 17(3):6–20.

Windell, J. T., J. W. Foltz, and J. A. Sarokon. 1978. Effect of fish size, temperature and amount fed on nutrient digestibility of a pelleted diet by rainbow trout, *Salmo gairdneri*. Transactions of the American Fisheries Society 107:613–616.

Wong, K. B. and R. H. Piedrahita. 2000. Settling velocity characterisation of aquacultural solids. Aquacultural Engineering 21:233–246.

Yamamoto, T., T. Shima, H. Furuita, N. Suzuki, and M. Shiraishi. 2001. Nutrient digestibility values of a test diet determined by manual feeding and self-feeding in rainbow trout and common carp. Fisheries Science 67:355–357.

Youngs, W. D. and M. B. Timmons. 1991. A historical perspective of raceway design. Pages 160–169 *in* Engineering aspects of intensive aquaculture. Northeast regional agricultural engineering service publication NRAES-49. Cooperative Extension, Ithaca, New York, USA.

Zeigler, T. 1988. Effluent management of trout farms. Presented at the 1988 Annual Convention of the Ontario Trout Farmers Association, Toronto, Ontario, Canada.

Impacts of Cage and Net-pen Culture on Water Quality and Benthic Communities

Robert R. Stickney

Texas Sea Grant College Program, 2700 Earl Rudder Fwy South, Suite 1800,

College Station, Texas 77845 USA

ABSTRACT

Metabolic waste products from organisms cultured in cages and net-pens can affect water quality through the release of nutrients and can also reduce dissolved oxygen concentrations. Feces and unconsumed feed that fall to the sediments under aquaculture cages and net-pens can be present in sufficient quantity to create anoxic areas that extend beyond the shadow of the structures that are the source of those contaminants. A significant body of literature on both types of impact has appeared in recent years from studies conducted in both fresh and marine waters.

Eutrophication of coastal areas as a result of nutrient addition and the impacts of anoxic zones under cages and net-pens have become major issues in many places around the world. In some instances, the situation has become so severe that fish production has been impacted through what has been called self-pollution.

In freshwater ponds and lakes, cages may be used when the water body is difficult to harvest or where the intent is to separate the culture species from other aquatic organisms. Instances where impacts on water quality and the benthic environment were as significant as what has been seen at some marine cage and net-pen sites have been reported. Another concern in freshwater cage culture has been infiltration by dissolved nutrients into ground water.

Aquaculturists have become intensely aware of these problems in recent years and have actively been involved in the development of methods to deal with them. Most solutions involve proper siting coupled with responsible culture practices. By siting facilities in areas that provide proper water exchange rates, nutrients are quickly and widely dispersed and waste products do not accumulate on the bottom under the cages or net-pens. By employing the proper food formulations and feeding practices, fecal production can be reduced and feed wastage can be largely avoided. Stocking densities can also be adjusted to avoid overloading the system. By taking the proper precautions and employing best management practices, cage and net-pen aquaculture can co-exist with the natural aquatic ecosystem.

For all practical purposes, extraction of food from the sea through commercial fishing has peaked (Chapter 1), while demand continues to increase both because the human population continues to grow, and because the buying power of people in developing countries is increasing. The increasing demand for seafood can only be met through increases in aquaculture production. While there are those who argue that at least part of the demand can be met by finding ways to adapt krill and fish that currently are used in the production of fish

meal into products that are acceptable human foods, most attempts to do that have been failures. Also, even if a significant portion of those species could be used as human food, demand would ultimately exceed supply and the aquaculture option would once again be the only viable one.

For many years, in fact for centuries in some parts of the world, aquaculture was practiced virtually in the absence of criticism. During those centuries, the majority of aquaculture production was in freshwater environments and was limited to small farming operations. A notable exception was oyster culture, which has historically been a highly extensive form of aquaculture with limited, if any negative perceived environmental impact.

With the expansion of aquaculture into public waters during the latter part of the 20th century, criticisms began to appear. Largely ignored by the aquaculture community at first, the objections of the critics became a major issue when lawsuits were filed and attempts were made to impose increasingly restrictive regulations on the aquaculture industry. At the same time, the list of objections increased. Included on the list with respect to salmon culture, for example, is everything from visual pollution to excessive noise and odors (Stickney 1990). Major objections have been associated with degraded environmental quality, disease transmission, use of antibiotics, and interaction of escapees with wild populations. A symposium was held at Aquaculture '01 in Orlando, Florida during January 2001 which addressed each issue in detail. A book based on symposium presentations is currently under development.

Most of the objections surround the rearing of aquatic animals, primarily finfish, in cages or net-pens. Cages have been, in general, relatively small structures with rigid frames covered with wire mesh or netting. They have been used, until recently, most often in freshwater environments. Cages provide an opportunity for culturists to conduct their activities in water bodies that are difficult to seine. Examples are deep ponds or lakes, and reservoirs that have standing trees in them. In lakes where a variety of species are present, the use of cages will allow the culturist to isolate the target species from others. This may occur in public or private waters. In most parts of the world there are regulations involving the use of cages in public freshwater bodies. Where permitted at all, the deployment of cages in public waters often requires permits, may

involve payment of fees, and is controlled with respect to numbers and sizes of cages per unit of area.

Net-pens are primarily found in coastal waters. Most consist of net bags suspended from frames that float at the surface. They lack rigid structure below the surface. Such net-pens may be 10 to 20 m on a side at the surface and 10 or more meters deep. As marine aquaculture has moved from protected coastal waters to exposed and even open ocean locations, the need for more rigid structures has been recognized. Today, marine structures that are actually cages of hundreds to thousands of cubic meter volumes are being utilized in some situations.

Cages have been employed for the culture of various species in freshwater, including channel catfish, tilapia, rainbow trout, and various species of carp. Cages of one type or another (including small cages, called happas, used for holding fry) can be found in many countries throughout the world, though cage culture does not tend to dominate in any nation where freshwater species are being reared. In the marine environment, however, cage and net-pen culture can be the dominant types of production units. Atlantic salmon are currently being cultured in cages and net-pens in Canada, Chile, Ireland, Japan, Norway, Scotland, and the United States. Sea bass and sea bream are cultured in cages or net-pens in countries bordering on the Mediterranean Sea, while red sea bream and yellowtail are cultured in Japan. Culture of species such as tuna and cobia is being practiced on a small scale in some countries, and there are other species that could be included in a more exhaustive list.

Serious impacts on water quality and the sediments occurred in Japan late in the last century. As a result, Japan has altered its cage culture practices to limit impacts, though it has been suggested that new technology will be required that will promote the decomposition of organic matter that accumulates below the net-pens (Tsutsumi 1995). As Taiwan turns to cage aquaculture, lessons learned from the collapse of its shrimp culture industry in the early 1990s are guiding the process wherein the goal is to make marine aquaculture sustainable (Anonymous 2000).

The focus of this paper is on two related topics associated with the environmental impacts of cage and net-pen culture: water quality and the benthic community. Scientific investigations that have been conducted to evaluate those impacts are presented, after which policies and practices for dealing with the problems are described.

Water Quality

Changes in water quality can result from the release of ammonia directly into the water and from the dissolution of various nutrients from waste feed and feces. Water quality impacts may or may not occur in conjunction with impacts on the benthic environment (Tlusty et al. 1999), or the impacts on water quality may be limited compared with those on the benthos (Doughty and McPhail 1994). Marine fish culture may range from little measurable impact on water quality (Arulampalam et al. 1998) to significant negative impact (Tsutsumi 1995). Freshwater lakes have been successfully employed for fish culture without significant impacts on water quality (Anonymous 1997), though there are also instances where clear degradation of water quality can be documented. For example, effects of salmonid cage culture operations in freshwater lochs in Scotland on water quality, particularly with respect to increased phosphorus level, have been observed (Kelly 1995).

Water quality became a major issue with regard to development of a cage culture of salmonids in iron mine pit lakes in Minnesota in the early 1990s. The typical mine pit lake is smaller than 100 ha and over 64 m deep. They fill through groundwater inflow and have no surface outflows (Axler et al. 1992, 1994). Fish culture was initiated in a few of the more than 200 such lakes. As a result of cage culture activities, dramatic increases in dissolved phosphorus and nitrogen levels occurred. Algal blooms occurred in the previously highly oligotrophic waters, followed by oxygen depletions (Axler et al. 1994; Yokom 1994). Aeration was required to maintain acceptable dissolved oxygen levels (Axler et al 1992; Yokom 1994). Within 18 mo after culture in one pit mine was terminated under orders from the Minnesota Pollution Control Agency in 1993, water quality had almost completely recovered. Management during the recovery period consisted only of aeration during one summer (Axler et al. 1998).

Examples of studies on water quality in conjunction with net-pen or cage culture include research conducted in Australia (McGhie et al. 2000), Greece (Klaoudatos et al. 1996), Japan (Tsutsumi 1995), Korea (Kim et al. 1998), Malaysia (Choo 1994), Norway (Johannessen et al. 1994), South Africa (Anonymous 1997), Sweden (Troell 1996), the United Kingdom (O'Connor et al. 1993; Doughty and McPhail 1994; Black et al. 1996a, 1996b), the United States (Findlay et al. 1995), Zimbabwe (Troell and Berg 1997), and the western Mediterranean (Mazzola et al. 1999).

Impacts on the Sediments and Benthic Community

Fewer studies on the impacts of cage and net-pen culture on water quality were uncovered in the scientific literature than those associated with impacts on the sediments. The practice of farming aquatic animals in cages and net-pens results in some loss of feed along with feces from the facilities and, in many cases, the accumulation of both on the bottom directly below or near the cages or net-pens. Local current dynamics greatly influence accumulation rates.

Sedimentation rates as high as 52.1 kg/m^2

per yr were reported under salmon net-pens in one Puget Sound, Washington location (Weston and Gowen 1988), while the accumulation rate under salmon pens in Scotland was a more modest 10 kg/m^2 per yr (Gowen and Bradbury 1987). Phillips et al. (1985) estimated that for every metric ton of rainbow trout produced in cages, 150–300 kg of waste feed and 250–300 kg of dry weight feces are discharged.

With the rapid development of cage and net-pen culture in the 1980s, studies on the impacts of that type of aquaculture on the sediments and benthic community began to appear in significant numbers. Mahnken (1993) cited a number of such studies that focused on the impacts on sediment redox potential, nutrient dynamics and deposition, and benthic community structure. The conclusion that was drawn from those studies was that the impact varied greatly, ranging from no discernible effect to the creation of anaerobic conditions and an absence of benthic macrofauna. Various other studies have been conducted in several countries throughout much of the world since Mahnken (1993) produced his review. Mahnken (1993) concentrated on information associated with salmon culture, though various studies conducted on other species prior to 1993, including some on salmonids, were not included in his review. Examples of work not included in the previous review are studies conducted in Australia (Woodward et al. 1992), Canada (Chang and Thonney 1992; Tlusty et al. 1999), Croatia (Katavic and Antolic 1999), Denmark (Holmer and Kristensen 1992), Indonesia (Costa-Pierce and Roem 1990), Norway (Frogh et al. 1991; Raa and Liltved 1991), the Philippines (Delos Reyes 1993), and the United Kingdom (Maloney et al. 1991).

Various models to predict the impacts from cage and net-pen culture on the environment have been developed. Examples are Beveridge (1984), Panchang and Richardson (1992), Silvert (1992), Axler et al. (1993), Hargrove (1994), Kelly (1995), McDonald et al. (1996), Huang and Wen (1998) and Lupatsch and Kissil (1998). While a number of models exist, according to Stewart (1997) none of them provides estimates of the limits on the assimilative capacity of different locations or the consequences if limits are exceeded.

Gowen et al. (1990) indicated that there are four environmental impacts associated with intensive aquaculture in natural environments. Those are hypernutrification (or eutrophication), benthic enrichment, increased biochemical oxygen demand, and changes in the bacterial flora. The work of Silvert (1992), which attempted to characterize the effects on environment listed by Gowen et al. (1990), provides an example of the types of models that have been developed. Silvert (1992), examined local impacts (benthic deposition), regional impacts (eutrophication), and internal impacts (dissolved oxygen depletion). For example, regional impacts were described in the following formula:

$$DN = R_N T/AZ,$$

where DN is the nitrogen level in an inlet housing the aquaculture site, R_N is the amount of soluble nitrogen released into the water by the fish being produced, T is the flushing time, and AZ is the area of the inlet times its mean depth. Other mathematical expressions were developed for the other impacts.

One of the goals for the models developed by Silvert (1992) was to provide tools that could be used by managers. For example, to minimize regional impacts from an aquaculture operation, agencies could employ the following expression:

$$Y/A = ZC_{crit}/gT,$$

where Y/A is fish production per unit area, Z is mean depth, T is flushing time, C_{crit} and g are dependent on the established maximum acceptable loading for organic matter and the rate at which fish release fine particulates and soluble compounds.

While most biological studies deal with impacts on benthic fauna, effects on sea grasses have also been reported (Delgado et al. 1999). Effects on sea grasses after cessation of cage fish farming were apparent even after water quality conditions had returned to normal.

The use of disease treatment chemicals in cage and net-pen culture has been objected to because of the possibility of developing resistant bacteria or imparting direct toxicity to benthic organisms. Aquaculturists are sensitive about the use of antibiotics and other chemicals in conjunction with their operations. Not only do the fish farmers wish to avoid developing antibiotic resistant bacteria that would make treatment ineffective, the cost of treating fish is not trivial. Culturists typically treat only when they are confident that a disease is present, and only treat for the prescribed period. While even the best management practices cannot ensure that no chemicals reach the sediments, the studies conducted to date seem to indicate that development of antibiotic resistance in sediment microbes is not as significant a problem as some have feared.

Following therapy with oxytetracycline, Kerry et al. (1996a) examined the sediments under and adjacent to marine salmon cages in Ireland for the presence of the antibiotic and for the frequency of oxytetracycline-resistant microorganisms. Concentrations of the antibiotic were higher immediately under the cages than 10 m down current. The chemical was not detected in other areas outside of the cage footprint. Frequencies of oxytetracycline-resistant microorganisms in sediments containing the antibiotic ranged from less than 1% to 8.5%. There was no correlation between chemical concentration and frequency of disease resistance. In another study, Kerry et al. (1994) found that 33 d after drug therapy had been terminated

the frequency of resistance in microorganisms under the cages was not significantly higher than in samples from sediments free of the antibiotic. In the United States, Capone et al. (1996) also looked at oxytetracycline levels in sediments under and adjacent to salmon cages. They found the antibiotic in the sediments under and as far as 30 m away from the cages.

Traces of oxytetracycline were found in oysters and dungeness crabs *Cancer magister* collected at the salmon farm studied by Capone et al. (1996). About half of the red rock crabs *Cancer productus* sampled contained levels in excess of those approved by the U.S. Food and Drug Administration in seafoods that are commercially sold.

Microcosm studies have been conducted on oxytetracycline residence in sediments (Capone et al. 1996b) and on the development of oxytetracycline-resistant bacteria (Kerry et al. 1996b). One-fourth to one-half of the chemical remained in the microcosm after 60 d. It was determined that the extent of bacterial cell division was an important factor in the results.

Romet-30· (sulfadimethoxine, an antibacterial compound) was also evaluated in the Capone et al. (1996) study. It appeared to have a very short life in marine sediments (just detectable after 2 d as compared with a 36-d half-life for oxytetracycline). In at least one study of the effects of the antibiotic Ivermectin on polychaetes under an Atlantic salmon farm in Ireland, there was no evidence of adverse impact though the farm was orally administering the drug to control sea lice (Costelloe et al. 1998).

With time, the accumulation of waste feed and feces under cages and net-pens can lead to the development of anoxic sediments. Mahnken (1993) cited studies in which accumulation rates have ranged from 0 cm to as much as 40 cm deep under salmon farms. When anoxia occurs, the microbiological community changes from aerobic to anaerobic species, heterotrophic organisms are eliminated, and the sediments may begin to emit methane, carbon dioxide, and hydrogen sulfide (Samuelson et al. 1988).

Various measures of benthic impacts have been employed. Diver observations, collection of grab samples for benthos sampling, and chemical analysis of sediments have commonly been used. Acoustic sounding has also been employed (Dougall and Black 1999). Mazzola et al. (1999) reported that sediment protein and lipid composition are good descriptors of impacts. Lipid analysis was used by Henderson et al. (1997) to assess sediment characteristics in Scotland.

Pearson and Rosenberg (1978) proposed a model of organic enrichment (Fig. 1) that predicts development of an anoxic zone in the area of highest impact (such as immediately under cages or net-pens). This zone, which has also been termed the azoic zone, has been reported to exist in a number of studies beginning in the early 1980s (Stewart 1984; Brown et al. 1987; Lumb 1989). Outside of the azoic zone is a region that is still heavily impacted, though not anoxic. It is colonized by pollution-tolerant opportunistic species. That community may be low in diversity, but may have high numbers and biomass. The next region is a transition zone in which a more complex community becomes established. There may actually be higher productivity in that region than exists under background conditions. That zone finally grades into conditions reflective of reference areas. Impacts on the sediments and benthic community are usually restricted to an area no more than 30 m from the edge of the cages or net-pens (Frid and Mercer 1989; McGhie et al. 2000).

While most studies have looked at impacts on the environment outside of the cages or net-pens, Black et al. (1996a) pointed out that enrichment of the sediments can have a negative impact on the animals being reared, in that case, salmon. Fish health on several salmon farms in Scotland could be correlated

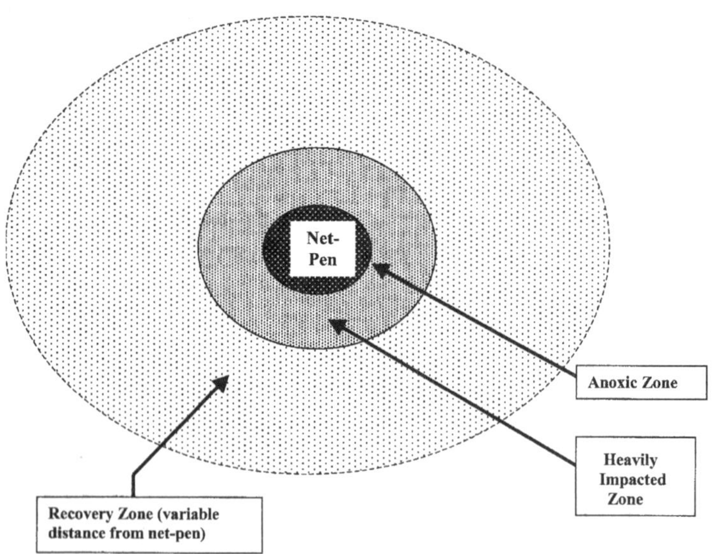

FIGURE 1. *Impacts of marine cage or net-pen culture on benthic habitat (based on Pearson and Rosenberg 1978).*

with the incidence of low current velocities and the log mean hydrogen sulfide concentration.

Seasonal variations can occur. Accumulation of waste products was found to increase during winter, for example, under Newfoundland salmon net-pens (Tlusty et al. 1999). That was associated with reduced digestion efficiency by the fish. In a study of nutrient accumulation in the sediments under tilapia cages in a lake, the general pattern was for concentrations to be highest prior to winter turnover (Troell and Berg 1997). The authors concluded that tropical lakes may be better able to process nutrients deposited in the sediments as organic wastes than temperate lakes.

Return of the sediments and benthic community to reference conditions following removal of cages or net-pens requires the passage of time. Even when oxygen levels in the sediments return to normal, carryover effects of organic matter deposition can be shown through examination of the redox potential (McGhie et al. 2000). At one Puget Sound, Washington site that was monitored after production of salmonids in cages ceased, rapid increases in abundance and diversity of the benthic community occurred within 3 mo (Mahnken 1993). Further recovery approaching an equilibrium stage occurred over the next 3 to 25 mo, with rare species recovery being slow.

Methods of Reducing Impacts

Modification of feeding methods and feed formulations have been recommended as means of reducing the amount of waste that is produced in cage and net-pen culture (Hennessy et al. 1996; New 1996). Such modifications might include changes in feeding practices to reduce the amount of feed that falls through the cage or net-pen and modifying feed formulations to make them more fully digestible and reduce the level of dietary phosphorus, thereby reducing solid waste production (Gavine et al. 1995). New (1996)

indicated that research is needed to develop optimum feeding rates, frequency and timing as well as to develop methods for accurate biomass determination (as fish are typically fed on the basis of estimated biomass).

The use of low-fiber, high-protein feeds, while possibly leading to lower fecal production, often results in increased ammonia production. Phosphorus release to the environment can be reduced by feeding the mineral in forms that are highly digestible and not providing that nutrient at levels above the nutritional requirement of the target species. Significant changes in the phosphorus content of trout diets have demonstrated that it is possible to reduce phosphorus waste associated with cage culture operations (Gavine et al. 1995). Wastes from facilities using dry feeds are significantly less than from facilities that feed wet feeds (New 1996). In Japan, Thailand (Immink 1996), and other nations, ground trash fish are used as feed, and feed loss from cages is high.

O'Connor et al. (1993) evaluated harrowing of sediments under a salmon farm after a period wherein the farm was out of production. A survey conducted 3 mo after harrowing showed significantly improved conditions compared with pre-harrowing evaluations. It does not seem as though much additional research on this particular approach has been conducted.

The suggestion has been made, and legislation considered or implemented in some nations, to eliminate negative environmental impacts from waste feed and feces. One proposed approach involves virtually diapering the cages or net-pens to collect those wastes. While that approach is fraught with technical problems and thought to be economically prohibitive, particularly when large structures are involved, a cage system in which wastes falling toward the bottom are collected appears to have been successfully employed in Switzerland (Anonymous 1999).

The most common approach to deal with the accumulation of wastes is to properly site the facility so currents will carry away and disperse the wastes as was suggested for Puget Sound, Washington (Parametrix 1990) and British Columbia, Canada (Levings 1994) net-pen facilities. Apparently, relatively strong currents are not always required to achieve the desired result. At one site in Greece, impact on the benthos was low though current speeds were very low. That result was associated with the presence of large numbers of wild fish feeding around the cages (Dougall and Black 1999). Each prospective culture site should be evaluated independently since tide range, depth, proximity to shore and shoreline configuration, and perhaps other physical factors, in addition to current speed and direction can influence dispersal of particulate matter from cages and net-pens.

Klaoudatos et al. (1996) examined the water quality and rate of sedimentation associated with a 200-ton sea bass cage culture operation in Greece and concluded that the environmental impact was very low. A similar result was found at a coastal cage culture facility in Malaysia (Choo 1994). Certainly, siting considerations should take carrying capacity into consideration as has been recommended by Immink (1996) for Thailand cage culture operations and Fresi et al. (1993) for Mediterranean mariculture operations.

Polyculture is another approach that has been used to reduce waste impacts. Troell (1996) and Troell et al. (1997) reported results associated with growing the alga *Gracilaria chilensis* on ropes in conjunction with a salmon cage operation. Algae located 10 m from fish cages grew more rapidly than at distances of 150 m and 1 km. Dissolved ammonia removal was 50% in winter and 90–95% in spring. Polyculture with mussels did not result in efficient waste reduction. The authors estimated that 1 ha of algae cultivated near fish cages could remove at least 5% of the dissolved inorganic nitrogen released from the fish and

---

TABLE 1. *Approaches to reducing environmental impacts from cage and net-pen culture activities.*

- Determine permitting requirements and begin the permit application process early so permits will be in place when the facility is completed and ready for stocking.

- Evaluate prospective sites to determine flushing rates, direction in which waste products will be carried as a function of tidal flow and wind generated currents, and identify any physical conditions such as water depth, that might affect dispersal of nutrients and solids.

- When negative benthic impacts occur, remove fish from the location and take remedial steps such as harrowing the bottom and allowing sufficient time for recovery prior to restocking; then stock at densities that do not recreate the problem.

- Consider establishing operations in suitable offshore environments to reduce environmental impacts.

- Use feeds designed for high digestibility and low rates of nitrogen excretion.

- Stock fish at densities that are conducive to stress reduction as a means of avoiding disease problems.

- Keep handling of fish to a minimum to avoid handling stress that could lead to disease epizootics.

- Employ treatment chemicals only when a disease has been detected and only provide chemicals in the minimum amounts and for the minimum duration that have been shown effective.

- Consider polyculture as a means to reduce the accumulation of nutrients and solids in the immediate vicinity of the cages or net-pens.

27% of the dissolved phosphorus. An added economic benefit would also be obtained from the practice when the edible algae are sold.

One way in which marine aquaculturists have addressed the problems associated with rearing fish in cages and net-pens is to move their facilities into more open water; in some cases, out of sight of land. Not only does open ocean culture defuse the visual pollution issue (which is usually raised by adjacent land owners who object to the presence of cages and net-pens on the basis that they interfere with the aesthetics of their view), but it also usually means increased flushing rates. Open ocean areas are less likely to experience water quality and benthic impacts than protected inshore areas due to an increased rate of water exchange and rapid dispersion of solids. The presence of currents implies that dispersal would increase with increasing water depth. Open ocean facilities also serve as fish attracting devices, thus waste feed will be consumed. Issues that are not relieved when facilities are sited in open water include interference with navigation and encroachment on commercial fishing grounds.

The major impediments to open ocean aquaculture, however, are related to survival of facilities during storms. While new cage types have been developed, including submerged and submersible cages, that can withstand most storms, a considerable amount of work needs to be done in the area of cage engineering. Currently, the costs of cages that can withstand storms is so much higher than their inshore counterparts that profits are difficult to achieve.

Obtaining permits from regulators has been a fact of life for aquaculturists in much of the developed world for several years. In many developing nations, however, the realization that aquaculture cannot be allowed to expand in the absence of regulations is a relatively recent occurrence. Once governments recognized that uncontrolled expansion of aquaculture cannot only create

environmental problems but also eventually becomes self-limiting, the race was on to develop appropriate guidelines and restrictions. As inland and coastal waters become fully developed in terms of the number and type of aquaculture facilities that can be supported without undue environmental impacts, the pressure to move to open ocean aquaculture will only increase.

In conclusion, there is no question that aquaculture in cages and net-pens can have severe environmental impacts. At the same time, the literature on the subject demonstrates that environmental impacts do not always occur. Proper siting, taking the carrying capacity of the environment into consideration, applying best management practices with respect to use of antibiotics and other chemicals, developing and employing feeds and feeding techniques that will reduce feed loss and fecal production, and using feeds that keep nutrient losses to the environment at a minimum are all factors that will keep impacts to a minimum.

Environmental impacts cannot be completely avoided. It will ultimately come down to whether some level of impact is acceptable in order for aquaculture to help fill the gap between seafood supply and demand. Neither extreme—eliminating aquaculture if there is any environmental impact or allowing aquaculture to expand in an uncontrolled manner—is acceptable to society. Responsible cage and net-pen culture will require scientists, engineers, regulators, and producers to work in concert. There has been at least one encouraging development in the state of Washington, USA where a state court has ruled that for at least one salmon net-pen farming operation the activity in Puget Sound did not pose a significant risk to water or sediment quality and did not violate existing water pollution laws (Elliot and Tremaine 2000). Table 1 presents a summary of some of the practices that can be employed by cage and net-pen culturists to reduce environmental impacts.

Literature Cited

Anonymous. 1997. Trout culture impact on a South African river. Fish Farming International 24(3):35.

Anonymous. 1999. Swiss lake farm avoids pollution. Fish Farming International 26(7):26.

Anonymous. 2000. Taiwan's offshore potential: First symposium outlines country's cage culture. Fish Farming International 27(1):26.

Arulampalam, P., F. M. Yusoff, M. Shariff, A. T. Law, and P. S. S. Rao. 1998. Water quality and bacterial populations in a tropical marine cage culture farm. Aquaculture Research 29:617–624.

Axler, R., C. Larsen, C. Tikkanen, and M. McDonald. 1992. Water quality changes in mine pit lakes used for salmonid culture. Pages 32–33 *in* Aquaculture '92: Growing toward the 21st Century. World Aquaculture Society, Baton Rouge, Louisiana, USA.

Axler, R., C. Owen, J. Ameel, E. Ruzycki, and H. Henneck. 1994. Water quality issues associated with aquaculture: A case study in Minnesota mine pit lakes. Lake and Reservoir Management 9:53.

Axler, R., C. Tikkanen, M. McDonald, C. Larsen, and G. Host. 1993. Fish bioenergetics modeling to estimate waste loads from a net-pen aquaculture operation. Pages 596–604 in J-K. Want, editor. Techniques for modern aquaculture. American Society of Agricultural Engineers, St. Joseph, Michigan, USA.

Axler, R., S. Yokom, C. Tikkanen, M. McDonald, H. Runke, D. Wilcox, and B. Cady. 1998. Restoration of a mine pit lake from aquacultural nutrient enrichment. Restoration Ecology 6:1–19.

Beveridge, M. C. M. 1984. Cage and pen fish farming. Carrying capacity models and environmental impact. FAO Fisheries technical paper no. 255. Food and Agriculture Organization of the United Nations, Rome, Italy.

Black, K. D., M. C. B. Kiemer, and I. A. Ezzi. 1996a. Benthic impact, hydrogen sulphide and fish health: Field and laboratory studies. Pages 16–26 in K. D. Black, editor. Aquaculture and sea lochs. Scottish Association for Marine Science, Oban, UK.

Black, K. D., M. C. B. Kiemer, and I. A. Ezzi. 1996b. The relationships between hydrodynamics, the concentration of hydrogen sulphide produced by polluted sediments and fish health at several marine cage farms in Scotland and Ireland. Journal of Applied Ichthyology 12:15–20.

Brown, J. R., J. Gowen, and D. S. McLusky. 1987. The effect of salmon farming on the benthos of a Scottish sealoch. Journal of Experimental Marine Biology and Ecology. 109:39–51.

Capone, D. G., D. P. Weston, V. Miller, and C. Shoemaker. 1996. Antibacterial residues in marine sediments and invertebrates following chemotherapy in aquaculture. Aquaculture 145:55–75.

Chang, B. D. and J. P. Thonney. 1992. Overview and environmental status of the New Brunswick salmon culture industry. Bulletin of the Aquaculture Association of Canada 92-3:61–63.

Choo, P. S. 1994. A study of the water quality at a coastal cage-culture site in Penang, Malaysia. Fishery Bulletin Department of

Fisheries (Malaysia), 86:1–9.

Costa-Pierce, B. A. and C. M. Roem. 1990. Waste production and efficiency of feed use in floating net cages in a eutrophic tropical reservoir. Pages 257–271 in B. A. Costa-Pierce and O. Soemarwoto, editors. Reservoir fisheries and aquaculture development for resettlement in Indonesia. ICLARM technical report. no. 23. International Center for Living Aquatic Resource Management, Manila, Philippines.

Costelloe, M., J. Costelloe, B. O'Connor, and P. Smith. 1998. Densities of polychaetes in sediments under a salmon farm using Ivermectin. Bulletin of the European Association of Fish Pathology 18:22–25.

Delgado, O., J. Ruiz, M. Perez, J. Romero, and E. Ballesteros. 1999. Effects of fish farming on seagrass (Posidonia oceanica) in a Mediterranean bay: Seagrass decline after organic loading cessation. Oceanology Acta 22:109–117.

Delos Reyes, M. R. 1993. Fishpen culture and its impact on the ecosystem of Laguna de Bay, Philippines. ICLARM Conference Proceedings 26:74–84.

Dougall, N. M. and K. D. Black. 1999. Determining sediment properties around a marine cage farm using acoustic ground discrimination: RoxAnn,. Aquaculture Research 30:451–458.

Doughty, C. R. and C. D. McPhail. 1994. Monitoring the environmental impacts and consent compliance of freshwater fish farms. Aquaculture Research 26:557–565.

Elliot, R. and D. W. Tremaine. 2000. Court rules in favor of salmon net-pen farms. Waterlines 10(2):5.

Findlay, R. H., L. Watling, and L. M. Mayer. 1995. Environmental impact of salmon net-pen culture on marine benthic communities in Maine: A case study.

Estuaries 18:145–179.

Fresi, E., S. Cataudella, A. Lazzari, M. DiBitetto, and C. Lorenzi. 1993. Environmental aspects of land-based and mariculture systems. Page 303 in M. Carrillo, L. Dahle, J. Morales, P. Sorgeloos, N. Svennebvig, and J. Wyban, editors. From discovery to commercialization. Special publication no. 19 of the European Aquaculture Society. European Aquaculture Society, Oostende, Belgium.

Frid, C. L. J. and T. S. Mercer. 1989. Environmental monitoring of caged fish farming in macrotidal environments. Marine Pollution Bulletin 20:379–383.

Frogh, M., M. Schaanning, N. DePauw, and J. Joyce. 1991. Benthic degradation in bottom sediments from salmon cage farming at 66°N in Norway. Page 111 in N. Depauw and J. Joyce, editors. Aquaculture and the environment. Special publication no. 14 of the European Aquaculture Society. European Aquaculture Society, Oostende, Belgium.

Gavine, F. M., M. J. Phillips, and A. Murray. 1995. Influence of improved feed quality and food conversion ratios on phosphorus loadings from cage culture of rainbow trout, Oncorhynchus mykiss (Walbaum), in freshwater lakes. Aquaculture Research 26:483–495.

Gowen, R. J. and N. B. Bradbury. 1987. The ecological impact of salmonid farming in coastal waters: A review. Oceanography and Marine Biology Annual Review 25:563–575.

Gowen, R. J., H. Rosenthal, T. Mäkinen, and I. Ezzi. 1990. Environmental impact of aquaculture activities. Pages 257–283 in N. dePauw and R. Billard, editors. Special publication no. 12 of the European Aquaculture Society. European Aquaculture Society, Bredene, Belgium.

Hargrove, B. T., editor. 1994. Modelling benthic impacts of organic enrichment from marine aquaculture. Canadian Technical Report Fisheries and Aquatic Science 1949:1–136.

Henderson, R. J., D. A. M. Forrest, K. D. Black, and M. T. Park. 1997. The lipid composition of sealoch sediments underlying salmon cages. Aquaculture 158:69–83.

Hennessy, M. M., L. Wilson, W. Struthers, and L. A. Kelly. 1996. Waste loadings from two freshwater Atlantic salmon juvenile farms in Scotland. Water, Air, and Soil Pollution 86:235–249.

Holmer, M. and E. Kristensen. 1992. Impact of marine fish cage farming on metabolism and sulfate reduction of underlying sediments. Marine Ecolology Progress Series 80:191–201.

Huang, X. and W. Wen. 1998. A study on restriction of marine environment on capacity of cage-culture in Gongwan Bay. Tropical Oceanology 17:57–64.

Immink, A. J. 1996. An assessment of marine fish cage culture in the Ranong mangrove ecosystem, Thailand. Institute of Aquaculture, Stirling University, Scotland.

Johannessen, P. J., H. B. Botnen, and O. F. Tvedten. 1994. Macrobenthos: Before, during and after a fish farm. Aquaculture and Fishery Management 25:55–66.

Katavic, I. and B. Antolic. 1999. On the impact of a sea bass (Dicentrarchus labrax L.) cage farm on water quality and macrobenthic communities. Acta Adriatic 40(2):19–32.

Kelly, L. A. 1995. Predicting the effect of cages on nutrient status of Scottish freshwater lochs using mass-balance models. Aquaculture Research 26:469–477.

Kerry, J., R. Coyne, D. Gilroy, M. Hiney, and P. Smith. 1996a. Spatial distribution of oxytetracycline and elevated frequencies of oxytetracycline resistance in the sediments beneath a marine salmon farm following tetracycline therapy. Aquaculture 145:31–39.

Kerry, J., M. Hiney, R. Coyne, D. Cazabon, S. NicGabhainn, and P. Smith. 1994. Frequency and distribution of resistance to oxytetracycline in micro-organisms isolated from marine fish farm sediments following therapeutic use of oxy-tetracycline. Aquaculture 123:43–54.

Kerry, J., M. Slattery, S. Vaughan, and P. Smith. 1996b. The importance of bacterial multiplication in the selection, by oxytetracycline-HCl, of oxytetracycline-resistant bacteria in marine sediment microcosms. Aquaculture 144:103–119.

Kim, D. S., J-W. Choi, and J-G. Je. 1998. Community structure of meiobenthos for pollution monitoring in mariculture farms in Tongyong coastal area, southern Korea. Journal of the Korean Fisheries Society 31:217–225.

Klaoudatos, S. D., A. J. Conides, and M. V. Chatziefstathiou. 1996. Environmental impact assessment (EIA) studies in floating cage culture systems in Greece. Pages 525–532 in J. Taussik and J. Mitchell, editors. Partnership in coastal zone management. Samara Publishing Ltd., Cardigan, UK.

Levings, C. D. 1994. Some ecological concerns for net-pen culture of salmon on the coasts of the Northeast Pacific and Atlantic oceans, with special reference to British Columbia. Journal of Applied Aquaculture 4:65–141.

Lumb, C. M. 1989. Self pollution by Scottish salmon farms? Marine Pollution Bulletin 20:375–379.

Lupatsch, I. and G. W. Kissil. 1998. Predicting aquaculture waste from gilthead seabream (Sparus aurata) culture using a nutritional approach. Aquatic Living Resources 11:265–268.

Mahnken, C. V. W. 1993. Benthic faunal recovery and succession after removal of a marine fish farm. Doctoral dissertation. University of Washington, Seattle, Washington, USA.

Maloney, D., M. Gillooly, M. Costello, J. Wilson, N. DePauw, and J. Joyce. 1991. Water quality and sediment nutrients around salmon cages in a freshwater lake. Pages 202–203 in N. Depauw and J. Joyce, editors. Aquaculture and the environment. Special publication no. 14 of the European Aquaculture Society. European Aquaculture Society, Oostende, Belgium.

Mazzola, A., S. Mirto, and R. Danovaro. 1999. Initial fish-farm impact on meiofaunal assemblages in coastal sediments of the western Mediterranean. Marine Pollution Bulletin 38:1126–1133.

McDonald, M. E., C. A. Tikkanen, R. P. Axler, C. P. Larsen, and G. Host. 1996. Fish simulation culture model (FIS-C): A bioenergetics based model for aquacultural wasteload application. Aquaculture Engineering 15:243–259.

McGhie, T. K., C. M. Crawford, I. M. Mitchell, and D. O'Brien. 2000. The degradation of fish-cage waste in sediments during fallowing. Aquaculture 187:351–366.

New, M. B. 1996. Responsible use of aquaculture feeds. Aquaculture Asia 1(1):3–4ff.

O'Connor, B., J. Costelloe, P. Dinneen, and J. Faull. 1993. The effect of harrowing and fallowing on sediment quality under a salmon farm on the west coast of Ireland. ICES Council meeting papers. International Council for the Exploration of the Sea, Copenhagen, Denmark.

Panchang, V. and J. Richardson. 1992. A review of mathematical models used in assessing environmental impacts of salmonid net-pen culture. Journal of Shellfisheries Research 11:204–205.

Parametrix. 1990. Fish culture in floating net-pens. Washington Department of Fisheries, Olympia, Washington, USA.

Pearson, T. H., and R. Rosenberg. 1978. Macrobenthic succession in relation to organic enrichment and pollution of the marine environment. Oceanography and Marine Biology (New York) Annual Review 16:229–311.

Phillips, M. J., M. C. M. Beveridge, and J. F. Muir. 1985. Waste output and environmental effects of rainbow trout culture. ICES C.M. 1985/F:21. International Council for the Exploration of the Sea, Copenhagen, Denmark.

Raa, J. and H. Liltved. 1991. An assessment of the compatibility between fish farming and the Norwegian coastal environment. Pages 51–59 N. Depauw and J. Joyce, editors. Aquaculture and the environment. Special publication no. 14 of the European Aquaculture Society. European Aquaculture Society, Oostende, Belgium.

Samuelson, O. B., A. S. Ervik, and E. W. Solheim. 1988. A qualitative and quantitative analysis of sediment gas and diethylether extract of the sediment from salmon farms. Aquaculture 74:277–285.

Silvert, W. 1992. Assessing environmental impacts of finfish aquaculture in marine waters. Aquaculture 107:67–79.

Stewart, J. E. 1997. Environmental impacts of aquaculture. World Aquaculture 28:47–52.

Stewart, K. I. 1984. A study of the environmental impact of fish cage culture on an enclosed sea loch. Master's thesis. University of Stirling, Scotland.

Stickney, R. R. 1990. Controversies in salmon aquaculture and projections for the future of the aquaculture industry. Pages 455–461 in Proceedings of the Fourth Pacific Congress on Marine Science and Technology, Tokyo, Japan.

Tlusty, M. F., M. R. Anderson, and V. A. Pepper. 1999. Assuring sustainable salmonid aquaculture in Bay d'Espoir, Newfoundland. Bulletin of the Aquaculture Association of Canada 98-2:35–37.

Troell, M. 1996. Intensive fish cage farming—impacts, resource demands and increased sustainability through integration. Stockholm University, Stockholm, Sweden.

Troell, M. and H. Berg. 1997. Cage fish farming in the tropical Lake Kariba, Zimbabwe: Impact and biogeochemical changes in sediment. Aquaculture Research 28:527–544.

Troell, M., C. Halling, A. Nilsson, A. H. Buschmann, N. Kautsky, and L. Kautsky. 1997. Integrated marine cultivation of Gracilaria chilensis (Gracilariales, Bangiophyceae) and salmon cages for reduced environmental impact and increased economic output. Aquaculture 156:45–61.

Tsutsumi, H. 1995. Impact of fish net-pen culture on the benthic environment of a cove in South Japan. Estuaries 18:108–115.

Weston, D. P. and R. J. Gowen. 1988. Assessment and prediction of the affects of salmon net-pen culture on the benthic environment. Technical report 414, Technical appendix A. Washington Department of Fisheries, Olympia, Washington, USA.

Woodward, I. O., J. B. Gallagher, M. J. Rushton, P. J. Machin, and S. Mihalenko. 1992. Salmon farming and the environment of the Huon estuary, Tasmania. Technical Report Sea Fisheries Research Laboratory, Division of Sea Fisheries (Tasmania) 45:1–58.

Yokom, S. 1994. Recovery of a mine pit lake following removal of aquacultural loading: Observed changes and model predictions. Lake and Reservoir Management 9:125–126.

Recirculating Systems, Effluents, and Treatments

SHULIN CHEN

Department of Biological Systems Engineering,
Washington State University, Pullman, Washington 99194 USA

STEVE SUMMERFELT

Freshwater Institute, P.O. Box 1889, Shepherdstown, West Virginia 25443 USA

Tom Losordo

Department of Zoology, North Carolina State University, Raleigh, North Carolina 27695 USA

RON MALONE

Department of Civil and Environmental Engineering, Louisiana State University,
Baton Rouge, Louisiana 70803 USA

ABSTRACT

Recirculating aquaculture systems have internal water treatment components that continuously process fish excretory products so that an adequate water quality can be maintained as the culture water is reused. As a result, the recirculating aquaculture technology offers many advantages over conventional aquaculture systems in terms of site selection, environmental control and water conservation. Disadvantages, on the other hand, include being energy intensive and the requirement for a higher level of management skills. From an environmental perspective, recirculating aquaculture systems typically produce less effluent volume than flow through systems. This chapter summarizes the characteristics of recirculating aquaculture systems, fish excretion pertinent to system design and environmental impact, major wastewater treatment components in a recirculating aquaculture system, waste generation from a recirculating system, and processes for processing and disposal of the waste.

Recirculating Systems as a New Aquaculture Technology

As seafood demand increases worldwide and ocean catch declines, aquaculture has become the fastest growing sector of agriculture in many parts of the world. Major limitations to the development of aquaculture are availability of water resources and adverse environmental impacts. As human population increases and industries grow, water becomes a scarce resource that has many conflicting uses. The fact that traditional aquaculture production systems usually require relatively large amounts of water limits production of fish in many regions. Environmentally, effluents from traditional aquaculture systems, which are characterized by high volume and relatively low pollutant concentrations (Phillips et al. 1991), are difficult to treat economically with conventional wastewater treatment technologies. Consequently, it seems to be inevitable that aquaculture will have to face more stringent restrictions over development in many areas of the world, as concern over environmental impact and competition for resources grows (Phillips et al. 1991). In fact, concern over the impact of aquaculture wastes and effluents has already brought aquaculture under the scrutiny of many regulatory agencies in many developed countries (Losordo and Timmons 1994). These limitations are a formidable challenge to the development of aquaculture.

The concerns for water conservation and the environmental impacts of aquaculture activities have promoted the development of technologies for water reuse (or recirculation). With its internal water conditioning units, a recirculating aquaculture system (RAS) continuously processes the culture water used in the system, removing fish excretion products, and maintaining a healthy environment for fish. In a RAS system, water treatment and water quality control processes are designed into the systems to allow high degree of control of the environment.

Most efficient RASs can recycle 95% of the water, producing a very small amount of effluent. Additionally, wastes from RAS can generally be more easily collected and concentrated. Because of the internal water treatment capability, a RAS typically carries a high fish biomass density.

Besides water conservation and environmental benefits, a RAS provides flexibility in site selection, can be isolated from contaminations in source water, and can be constructed and operated for a specific market. The RAS technology virtually eliminates the significant water resource and appropriate climatic conditions required by traditional fish culture systems. A properly designed RAS can be placed almost anywhere for producing a quality-controlled product continuously throughout the year (Losordo 1998). This feature allows for the culture of warm water species in cold climatic conditions, or marine species grown inland. Thus RASs allow fresh product to be available at locations and times when the product would not ordinarily be available. Recirculating aquaculture systems can also be integrated with vegetable production so that nutrients from fish culture

(ammonia, nitrate, phosphorus, etc.) can be effectively utilized by plants (Rakocy et al. 1992; Rakocy 1997).

RAS systems have disadvantages as well. Besides higher energy consumption, a RAS usually requires a higher initial investment than that of traditional aquaculture production systems. Additionally, design and application of RASs is typically more complicated than other systems. In the last two decades, however, significant efforts have been devoted towards RAS research and development. As a result, considerable progresses have been made in recirculating technologies. In general, RASs have attracted worldwide interests, and have been used in a variety of applications.

This chapter summarizes the characteristics of recirculating aquaculture systems, fish excretion pertinent to system design and environmental impact, major wastewater treatment components in a recirculating aquaculture system, waste generation from a recirculating system, and methods for processing and disposal of the waste.

Major Pollution Parameters and their Environmental Impacts

Major environmentally significant substances contained in effluents from RASs include phosphorus (P), nitrogen (N), and organic matter. Phosphorous is an essential element for aquatic life and is often regarded as a limiting production factor in freshwater aquatic environments. Biologically available P in aquaculture discharge can stimulate the primary productivity of receiving waters and accelerate the eutrophication process. Phosphorus in RAS discharge water is present as either soluble P (e.g., orthophosphate) and or as insoluble P. Orthophosphate originates from P excreted in the fish urine and from the conversion of insoluble P contained in feces and uneaten feed to soluble P by microbial actions.

Nitrogen is another important element

supporting aquatic life. Like P, excess concentrations of N also increase aquatic productivity and possible eutrophication. The nitrogen compounds of concern to fish health include ammonia and nitrite, whereas ammonia and nitrate are a concern to receiving waters. Ammonia is a direct product of fish excretion and has high solubility in water. Ammonia exists in two forms in water: the un-ionized fraction (NH_3) and the ionized fraction (NH_4^+). The ratio of the two fractions is controlled by pH and temperature. The higher the pH and temperature, the higher the un-ionized fraction. Ammonia can be biologically oxidized by nitrifying bacteria to nitrite then to nitrate under aerobic conditions. The un-ionized fraction of ammonia and nitrite is highly toxic to aquatic life. For example, when the un-ionized ammonia concentration exceeds 0.0125 to 0.025 mg/L, growth rates of rainbow trout are reduced and damage to gill, kidney, and liver tissue may occur (Idaho DEQ 1998). Both ammonia and nitrate can be used directly by algae and other plants as nutrients. Leaching of nitrate to ground water can cause groundwater contamination.

Organic substances are also important in evaluating the environmental impacts of aquaculture effluent. Biological decay of organics consumes oxygen and may reduce the dissolved oxygen concentration in receiving waters. The impact of organic decay on oxygen concentration can be more directly represented by biochemical oxygen demand (BOD)—the amount of oxygen necessary to biologically oxidize the readily decomposable organic matter. In the measurement of standard BOD_5 of a sample, the sample is sealed in a filled bottle, the bottle is incubated in dark at 20 C for 5 d, and the measured dissolved oxygen reduction in the bottle can be converted to BOD contained in the water sample in mg (of BOD)/L (of water).

Another parameter that is often used to evaluate effluent quality is total suspended solids (TSS), which is defined according to the standard analysis method (APHA 1995) as the concentration of particles greater than 2 microns in size. The impacts of TSS on the receiving water include contributing to BOD, increasing the turbidity, and altering the physical habitat of the receiving water.

The primary source of P, N, and organic matter in aquaculture effluent is fish feed. Feed ingredients of animal origin (fish meal, meat, and bone meal) are rich sources of P. Phosphorous is also available from cereal grains and other plant protein sources. However, most of the P is present as phytate that is not well utilized by fish. The concentration and availability of P in the feed have the greatest influence on P retention in the fish and conversely on the amount excreted. Nitrogenous compounds, primarily ammonia, are the result of protein catabolism. Quality and quantity of dietary protein and the protein to energy ratio of the diet influence nitrogen retention and excretion.

Fish Waste Excretion and Basic Components of RAS

Fish Excretion

The majority of the waste materials from a RAS is related to fish excretion. The amount of waste excretion is related to feeding rate since virtually all the wastes generated from an intensive aquacultural system originate from the feed. Assuming a typical feed conversion ratio of 1 to 2 and neglecting the impact of uneaten feed, 80% of feed (on a dry mass basis) input to an aquacultural system will eventually be wasted as fish excretion products (Hopkins and Manci 1989), including CO_2, ammonia, and feces. Relating excretion rate to feeding rate is also convenient since feeding rate is a key operational parameter in an aquacultural operation.

The amount of feces excreted by fish is typically reported as TSS. The direct TSS excretion rate varies from report to report, ranging from 0.30 to 0.52 kg per kg-feed for trout and 0.18 to 0.69 kg per kg-feed for catfish (Chen et al. 1997). Clearly, TSS excretion rate will vary with species, temperature, types of feed and feeding rates. However, values in the range of 0.2 to 0.4 kg-TSS per kg-feed appear to be typical. In addition to the environmental impact externally to a RAS, particulate matters can cause fish health problems within the RAS and can increase BOD while decomposing.

Another parameter often used in measuring fish excretion is BOD_5. BOD_5 excretion rate is generally related to the feeding rate. The excretion rate of BOD_5 varies more widely and is typically lower than that of TSS. Of the BOD_5 excreted, over half is in particulate form and the rest is dissolved in water.

Fish excretion of nitrogen is also determined by feeding. For instance, in typical salmonid diets, approximately 7.2–7.7% of the feed is nitrogen. Of the N in feed, 67–75% will be lost to the environment, either as excretion products or as uneaten feed (Chen et al. 1997). Nitrogen excretion occurs in two forms, the dissolved form such as ammonia and the particulate form such as organic N in feces. Ammonia comprises between 70 to 90% of the nitrogenous catabolites with a typical excretion rate of 3% of the feeding rate. Besides its potential environmental impacts, ammonia is toxic to fish, thus has to be removed from a RAS. In a study of nutrient budgets in tanks with different stocking densities of hybrid tilapia, Siddiqui and Al-Harbi (1999) found that the largest amount of total nitrogen (59–72%) contained in feed was lost through aquaculture effluent.

Phosphorus excretion is also significant. Ketola and Harland (1993) reported that the retention of P in several salmonid diets ranged from 14–22%, meaning that approximately 80% of dietary P was discharged into the water. Phosphorous is excreted in soluble and particulate forms: the form of P consumed by the fish will affect the excretion amount of each form. Another report by Siddiqui and Al-Harbi (1999) indicated that phosphorous in the effluent could represent up to 60–62% of the feed input. The N and P excreted in particulate form can be converted to dissolved form through biological decomposition processes, during which organic nitrogen will be transformed to ammonia and organic P to orthophosphate during decay of TSS.

One of the major factors that contributes to the differences in reported excretion values is the amount of uneaten feed that ends up as feces. Under laboratory conditions, the amount of uneaten feed is usually minimized, whereas, under commercial operating conditions the amount of unconsumed feed is less controllable. Consequently, the waste excretion rates obtained under laboratory conditions are typically lower than those obtained under large-scale culture conditions.

Basic components of a RAS

The most important function of a RAS is to provide adequate life support for fish production. The main indicators of the proper rearing conditions include: 1) clean water with proper chemistry; 2) sufficient oxygen, 3) appropriate temperature range, and 4) the absence of pathogens. Accumulation of fish excretory products within a RAS creates unhealthy conditions that may result in reduced growth rates, lower feed conversion efficiency, poor fish health, and mortality. Due to the high biomass density within the system, several unit processes are required to meet the biological needs of the fish and to process the wastes. The major components of a RAS include a culture unit for fish, an aeration unit for oxygenation, a solids and liquid separation unit for the removal of suspended particles, a biofiltration unit to convert ammonia to nitrate, and a disinfection unit for control of pathogens. There are many variations in terms of design configuration and operation of these engineering components. The following section provides a summary of the major types of systems and associated components.

Types of Recirculating Systems

In RASs, water exiting the culture unit is passed through complementary water treatment processes before being reused in the same or different culture units. The water treatment processes each act to change the character of the water being recirculated. The rate that the water is exchanged through the culture tank is often set to transport dissolved oxygen into the fish culture environment, as well as to export waste metabolites produced by the fish. Careful selection of this water exchange rate through the culture tank and proper design of the complementary water treatment processes are required to maintain a healthy fish culture environment in a RAS. When designed and managed properly, commercial recirculating technology can now provide a controlled environment with uniformly healthy water quality.

Water is most often recirculated within systems that use either circular or rectangular tanks. However, an improved understanding of the dynamics of algal production and waste removal within pond systems has led to the innovative application of recirculating pond-based systems, known as partitioned aquaculture systems (PAS). This section will summarize the types and organization of treatment processes for both tank-based and pond-based RASs.

Tank-based RASs

There are at least three common classifications of tank-based water reuse and RASs–serial-reuse, partial-reuse, and fully-recirculating.

Serial-reuse systems are well suited for raceway culture situations, where water discharged from one or more tanks flows into one or more tanks located downstream.

However, a large water supply at the correct culture temperature and with sufficient water head must be available for application of serial-reuse technology. Water flowing through the serial-reuse raceways transports dissolved oxygen to the fish and flushes the waste metabolites. Serial-reuse systems often incorporate quiescent zones and hydraulic drops of 30–120 cm at the end of each raceway to capture some settleable solids, replenish dissolved oxygen, and strip dissolved carbon dioxide before the water enters the next raceway (Idaho DEQ 1998). Ammonia is controlled by flushing action. However, acid-base equilibrium and thus the water's pH controls the fraction of the total ammonia-nitrogen that is present in the toxic un-ionized form (i.e., NH_3). In regions where the water contains relatively low levels of alkalinity and pH values less than 7.0, the fraction of toxic un-ionized ammonia-nitrogen is lower and the total ammonia-nitrogen concentration can be allowed to increase to levels well above 1 mg/L while water is reused through a series of 10 or more raceways. As a general guide, serial-reuse systems can produce approximately 6 kg of fish annually for every 1 L/min of water flow (or 50 lb of fish annually for every 1 gpm of water flow) (MacMillan 1992). Serial-reuse systems in regions with slightly acidic waters could potentially produce more fish per unit water flow.

Partial-reuse systems use the same basic water treatment processes and have the same pumping requirement and general layout as fully-RASs, except that partial-reuse systems do not make use of biofilters and therefore have to control accumulation of total ammonia-nitrogen by flushing the system with make-up water at levels of 10–50% of the total recirculating flow per day. Compared to serial-reuse raceway systems, partial-reuse systems use much smaller flows to achieve production goals (e.g., 40–50 kg of fish annually for every 1 L/min of make-up water flow) and can significantly increase waste capture—about 80% overall capture of particulate wastes—by concentrating wastes into much smaller discharges (Summerfelt et al. 2000). Partial-reuse systems require much larger make-up water flows than fully-RASs. The high volume of make-up water flow maintains the partial-reuse system water temperature at the same temperature as the make-up water, unless extreme differences in ambient air and water temperatures are encountered. Without biofiltration and in an open system, water treatment usually consists only of processes for solids removal, aeration, and/or oxygenation. However, as the fraction of un-ionized ammonia declines with declining pH levels, the amount of carbon dioxide stripped from the system with aeration can be controlled to keep the pH down to a predetermined level (within reason) and, thus, maintain safe levels of un-ionized ammonia in the partial-reuse system (Summerfelt et al. 2000).

Fully-RASs are generally operated to reuse from 95% of the recirculating water flow (as in open-reuse systems designed for coldwater fish production) to more than 99.9% of the flow (as in RASs designed for warmwater fish production, such as tilapia, African catfish, and eel). At a minimum, traditional fully RASs will require clarifiers or filters to reduce particulate solids, biological filters to abate dissolved

wastes (primarily ammonia), and aerators to add dissolved oxygen and decrease dissolved carbon dioxide (Fig. 1). Less traditional fully-RASs such as aquaponic systems (which treat the water using plants that are also marketed as produce) and systems using organic detrital algae soup (ODAS) or activated sludge-type treatments are also used for commercial finfish culture. These three types of tank-based fully-RASs have their own distinct features, advantages and disadvantages, and some of these are discussed below.

Traditional Fully-RASs (i.e., Traditional Treatment Processes and Fixed-Film Biofilters)

The water exchange rate through each culture tank and the entire RAS, as well as the types of water treatment processes selected for use in a RAS, will depend largely on the water quality required by the fish being cultured and on the economics and scale of the water flow to be treated. However, tank-based RASs will generally require some mix of the following water treatment processes (Summerfelt et al, in press):

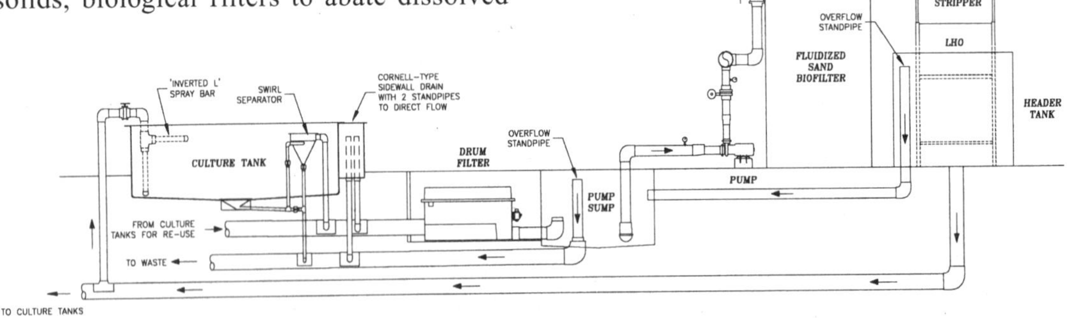

FIGURE 1. *This profile view (courtesy of PRAqua Technologies, Ltd., Nanaimo, British Columbia) shows the path water takes through the major components within one type of fully-RAS (from Summerfelt et al., in press). After pumping the water through the biofilter, the water flows by gravity through all of the treatment components and fish culture tanks until the water returns to the pump sump, where it is pumped again for further reuse. Make-up water is added in the pump sump through a float valve (not shown) and excess water overflows the system from the bottom discharge of each swirl separator and/or from the pump sump.*

- Sedimentation units, granular filters, or mechanical filters to reduce particulate solids (Table 1);

- Biological filters to abate dissolved wastes (Table 2);

- Aerators or strippers to add DO and decrease dissolved carbon dioxide or nitrogen gas to levels closer to atmospheric saturation;

- Oxygenation units to increase DO concentrations above atmospheric saturation levels;

- Advanced oxidation units (i.e., ultraviolet (UV) filters or units to add ozone) to disinfect, oxidize organic wastes and nitrite, or supplement the effectiveness of other water treatment units;

- pH controllers to add alkaline chemicals for maintaining acceptable pHs;

- Heaters or chillers to maintain the water temperature at a desired level.

Water treatment processes are sometimes multifunctional in that certain processes can change more than one characteristic of the

TABLE 1. *Principal Recirculating Clarification Technologies*

Technology	Removal Mechanism	Effective Removal (Microns)	Issues	Comments
Settling Basins	Gravity	>100	Low cost, significant labor associated with draining, poor fine solids removal, requires a lot of space	Most appropriate for low water reuse systems and/or light loads where fine solids can be flushed
Lamella Separator	Gravity	>50	Reduced space requirements when compared to settling basins, subject to biofouling problems	Most appropriate for low HRT and/or moderate loads where fine solids can be flushed
Hydrocyclones	Gravity	>1,000	Compact, low head, moderate flow rates, limited to coarse solids removal	Primarily used as a concentration device with double drains
Double Drains	Gravity	>1,000	Subject to clogging, requires round tank	Accelerates removal of heavy solids from water column in round tanks
Microscreens	Straining	>40	Low head loss, high flow capacities, poor fine solids capture, capital expense, mechanical upkeep required failure, moderate backwash water loss, highly resistant to biofouling	Popular both in the United States and Europe, can require complementary fine solids removal device for high HRT and/or display applications
Upflow Sand Filters	Gravity, Straining, Interception, Adsorption	>20	Compact, low flows, moderate head loss, good fine solidss capture, high backwash water loss	Popular technology for small baitfish systems and crab shedding systems, backwashing water losses limit finfish applications
Floating Bead Filters	Gravity, Straining, Interception, Adsorption	>30	Capital expense, moderate head loss, low backwash water loss, moderate hydraulic capacity, good fine solids control	Emerging technology, popular in the United States for moderately sized systems
Pressurized Sand Filters	Gravity, Straining, Interception, Adsorption	>1	High head loss, high backwash water loss, superb fine solids capture, susceptible to biofouling	Used in combination with ozone in display or brood stock applications where organic loading is very light, biofouling problems prohibit use in most growout applications
Cartridge Filters	Straining	>.1	Rapid clogging, superb fine solids capture, high pressure loss, require frequent manual cleaning, susceptibility to biofouling	Practical application limited to low volume, pristine applications usually experimental scale or display
Bag Filters	Straining	>1	Rapid clogging, superb fine solids capture, require frequent manual cleaning	Most frequently observed in smaller experimental systems

TABLE 2. *Overview of common biofiltration format used in recirculating sysems*

Process	Specific Surface Area (m^2/m^3)	Optimum Substrate (mg-TAN/l	Biofilm Control Mechanisms	Comments
Rotating Biological Filters	200-500	>3	Gentle hydraulic washing, sloughing	Inherently provides for secondary gas exchange; low headloss characteristics; many designs subject to mechanical failure; poor specific area characteristics, large foot print
Trickling Filter	200-500	>3	Hydraulic washing, sloughing	Good gas exchange characteristics; high headloss; poor specific area characteristics
Submerged Bed	200-500	<1	Endogenous respiration or intermittent draining	Low headloss; subject to biofouling; large foot print
Coarse Sand Fluidized Bed	2,500	1-3	Continuous mechanical abrasion between sand particles	High flow requirements; moderate headloss; good specific surface area; high biofouling resistance; substrate sensitive performance
Fine Sand Fluidized Bed	10,000+	<0.3	Endogenous respiration	Superb specific area characteristics; excellent low substrate performance,; moderate headloss; subject to sand loss under moderate loading; oxygen supplementation may be required
Upflow Sand Filters	2,500	<0.5	Mechanical abrasion between sand particles induced by intermittent hydraulic backwashing	Bioclarifier; good specific surface area; small foot print; high backwash water loss; low areal hydraulic flux-rates; moderate headloss
Mechanical Bead Filters	1,100	0.3-2	Aggressive mechanical abrasion and hydraulic shear induced by propeller or paddle	Bioclarifier; small foot print; high biofouling resistance; moderate headloss, backwashing management required
Bubble-washed bead Filters	1,100	0.3-1	Gentle hydraulic shear and abrasion between particles due to air injection	Bioclarifier; good specific surface area; good biofouling resistance; low to moderate backwashing water loss, backwashing management required
Hydraulically-washed Bead Filters	1,100	0.3-0.5	Gentle hydraulic shear and abrasion between particles induced by hydraulic expansion	Bioclarifier; good specific surface area; provides excellent solids capture; subject to biofouling at high substrate levels; moderate to high backwashing water loss, backwashing management required
Micro-bead Filters	4,000	0.3-3	Continuous hydraulic shear induced by gentle hydraulic expansion	Excellent specific surface area, low headloss, high flow demands, hydraulics limit bed depth and diameters
Moving Bed Filters	500	0.5-3	Continuous hydraulic shear induced by gentle hydraulic expansion	Low surface area, good high substrate performance, low head operation

water. Two examples of multifunctional processes used in RASs are trickling biofilters and dual-drain circular culture tanks. Air ventilated through trickling biofilters can add dissolved oxygen and strip dissolved carbon dioxide, while simultaneously nitrifying bacteria growing on the trickling filter media convert ammonia to nitrate (Nijhof 1995). Circular tanks installed with a bottom center drain and an elevated drain are used for fish culture at the same time that the tank's bottom drain is concentrating settleable solids in a relatively small discharge (Timmons et al. 1998).

Systems Using ODAS Or 'Activated Sludge-Type' Treatments

Some commercial RASs for tilapia and

shrimp utilize a completely non-conventional wastewater treatment approach compared to the approach that was just described. These non-conventional systems are not designed to rapidly remove solids, but are rather designed to convert dissolved and particulate wastes into microbial protein using a combination of photosynthesis and heavy aeration within the culture tanks and within suspended growth-type biofilters. The bacteria, micro-algae, fungi, and protozoa that are produced are in turn consumed by filter-feeding tilapia and supplement the tilapia's standard feed (Serfling 2000). This type of water treatment has been nicknamed 'organic detrital algae soup' treatment (ODAS).

According to Serfling (2000), optimal waste conversion and tilapia production requires total suspended solids concentrations of ODAS of 70–130 mg/L. By removing some solids in settling ponds (or tanks), commercial RASs that incorporate ODAS treatment primarily do not require many of the more expensive water treatment processes previously described. The ODAS systems have reportedly been used successfully for commercial culture of tilapia and for marine shrimp and *Macrobrachium* prawns (Serfling 2000). Several variations of this ODAS system have been used in greenhouses throughout North America. However, RASs that incorporate ODAS treatment would clearly not be capable of producing fish species that do not tolerate high levels of suspended solids.

Aquaponic Systems

RASs can incorporate vegetable hydroponics as a water treatment component where the plants are used to recover valuable nutrients that would otherwise accumulate or be discharged into the environment (Rakocy and Hargreaves 1993). Sale of the vegetable can generate additional revenues and enhance total system profit potential (Rakocy and Hargreaves 1993; Jenkins and Wade 1997; Adler et al. 2000).

Sand and gravel beds, nutrient film technique (NFT), and floating rafts have all been used for hydroponic production of plants (Rakocy and Hargreaves 1993). Achieving effective plant production in aquaponic systems requires effective solids removal before the hydroponic unit, maintenance of aerobic conditions in the plant's root zone, and sufficient macronutrients (N, P, K, Ca, Mg, and S) and micronutrients (Cl, Fe, Mn, Zn, Cu, Mo, and B) available for plant growth. Unfortunately, it can be difficult to meet the production requirements of both fish and vegetables in integrated aquaponic systems (Rakocy and Hargreaves 1993): water temperatures ideal for fish can be too warm for certain vegetables; nutrient levels typical of recirculating aquaculture system contain insufficient quantities of certain macronutrients and micronutrients for plant growth; chemotherapeutics used to control fish pathogens can kill plants; and, pesticides used to control insects can kill fish. Also, the value of the vegetable crop produced within integrated aquaponic systems can be much greater than the value of the fish crop produced, which can place most of the focus on vegetable production to maintain profitability (Rakocy and Hargreaves 1993; Jenkins and Wade 1997). Most commercial aquaponic systems have been relatively small operations that have serviced niche markets.

Pond-based RASs

In traditional pond culture, the pond is both the culture environment and the waste treatment process—which is largely via algal photosynthesis—and fish production levels of about 5,000 kg/ha are commercially achievable (Brune and Wang 1998). The partitioned aquaculture system (PAS) recirculates water in outdoor pond-based systems and has increased catfish production in ponds to approximately 17,000 kg/ ha in an experimental system by

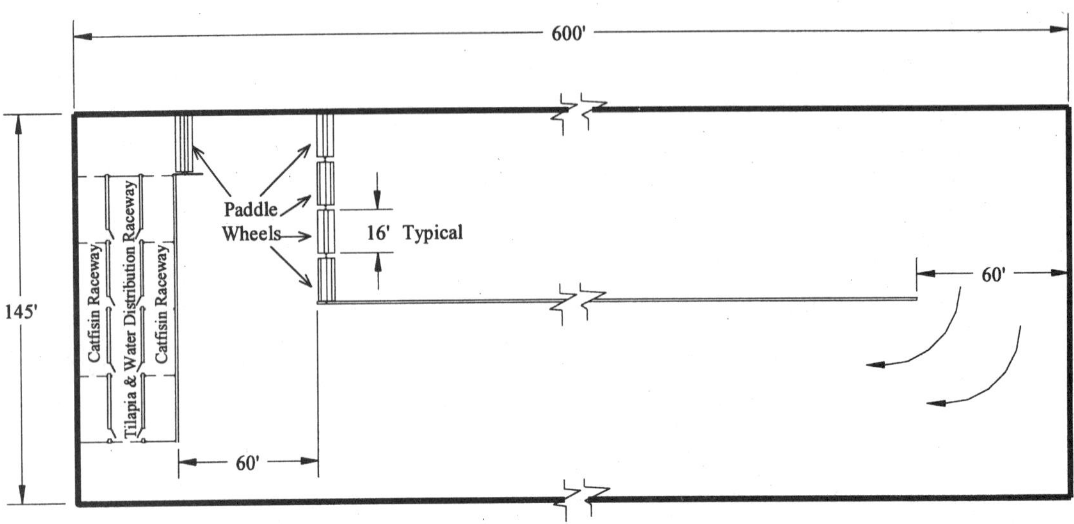

FIGURE 2. *The partitioned aquaculture system separates the fish culture into high density raceways and the waste treatment into a pond configuration that produces long narrow channels, and uses paddle wheel circulators to maintain uniform and controlled water velocities through both the raceways and the pond channels (courtesy of Dave Brune, Clemson University, Clemson, South Carolina, USA*

increasing the ammonia removal rate and net oxygen production through management of algal productivity and biomass concentration (Drapcho and Brune 2000). To achieve this, the PAS separates the fish culture into high density raceways and the waste treatment into a pond configuration that produces long narrow channels, allowing for maintenance of uniform and controlled water velocities through both the raceways and the pond channels (Fig. 2). The water flow through the raceway and through the pond channels is maintained by separate paddle wheel circulation devices that are capable of moving large water volumes against very low head. A quiescent zone is located immediately after the raceway to capture settleable wastes, which may be vacuumed from the PAS. Confining the fish to raceways simplifies fish grading and harvest operations, eliminates predation by birds (when the raceway is covered with netting), and allows for improved feeding and more efficient allocation of energy for oxygen control (Brune and Wang 1998). The

PAS design promotes efficient algae growth and gas exchange in the pond channels by producing uniform mixing within the pond channels. The water mixing combined with filter-feeder co-culture (tilapia or shellfish) allows for improved algal growth and control of algal density, which helps to maintain proper dissolved oxygen concentrations, reduce the accumulation of total ammonia-nitrogen, as well as decrease the probability of an algal die-off and off flavor produced in the fish flesh by unwanted algal populations in the pond (Brune and Wang 1998; Drapcho and Brune 2000). Because waste solids are captured and removed from the PAS for subsequent land application, the PAS produces little discharge to surface waters and only requires enough makeup water to meet evaporation and seepage losses. By recirculating water in ponds, the PAS technology builds upon the largest and least expensive production technology in the U.S. by maximizing the best features of ponds while minimizing some of the problems encountered with intensive pond production (Brune and Wang 1998).

RAS Water Use Compared to Other Aquaculture Systems

In the past, intensive fish farming practices used water resources for two reasons: 1) to carry oxygen to the fish, and 2) to receive the waste produced in the system (metabolic by-products and other materials) and disperse or carry these wastes away so that they did not accumulate in or around the fish farm to harmful and undesirable levels (Phillips et al. 1991). Intensive fish farming practices were often developed without much consideration for either the real or perceived effect of their waste on the farm's surrounding environment. The large effluent volume discharged from flowing water fish farms will contain dilute concentrations of wastes relative to many effluent standards. However, even though the waste concentration in most aquaculture effluents is relatively low, the cumulative waste load discharged to receiving watersheds can be significant due to the large flowrates involved (Summerfelt 1998). Therefore, the recent increased emphasis to reduce, manage and control effluents, as well as the growing competition for, and conservation of, water resources has created a more difficult regulatory, economic, and social environment under which existing aquaculture facilities must operate (MacMillan 1992; Ewart et al. 1995). Also, it is clearly much more difficult to locate and permit new aquaculture facilities with today's scarce untapped water resources and more stringent regulations. For these reasons, when compared to other types of land-based aquaculture production systems, application of recirculating aquaculture technologies greatly reduces the amount of water required to achieve a given level of fish production (Table 3). The reduced discharge volumes make it possible to use different practices for sludge processing and disposal,

TABLE 3. *Water consumption requirements reported for intensive fish culture systems.*

	Annual production (kg/yr) Makeup water required (L/min)	Recirculation rate on a flow basis (%)	Reference
Coldwater systems			
Single-pass	0.8–1.4 kg/yr per 1 L/min	--	Pillay (1982)
	9.7 kg/yr per 1 L/min	--	Alabaster (1982)
Serial-reuse raceways	6 kg/yr per 1 L/min	--	MacMillan (1992)
	--	--	Mudrak (1981)
Partial-reuse	40–50 kg/yr per 1 L/min	80-85%	Summerfelt et al. (2000)
Fully-recirculating	160 kg/yr per 1 L/min	6%	Heinen et al. (1996)
Warmwater systems			
Serial-reuse raceways	16 kg/yr per 1 L/min	--	Ray (1981)
Ponds	294 kg/yr per 1 L/min	--	Losordo (1991)
Recirc through wetlands	~145 kg/yr per 1 L/min	78%	Massingill et al. (1998)
Fully-recirculating	3,900 kg/yr per 1 L/min	>> 99%	Losordo (1991)
	5,000 kg/yr per 1 L/min		Losordo et al. (1994)

such as storage tanks, lagoons and land application. Retaining 100% of their discharge is not an option available for raceways and net pens.

Waste Production from RASs

Because of the internal wastewater treatment units employed in a RAS, the discharge from a RAS is typically associated with solids removal. In other words, the amount of effluent produced is often determined by the amount of liquid that is necessary to facilitate the removal of the solids from the RAS. The effluent is typically intermittent in nature and usually has relatively high solids concentration and is called sludge. This sludge stream can be further processed to produce two streams, the first one is more concentrated sludge and the second one is less concentrated effluent.

Waste Production

Amount of solids and associated sludge volume are two major factors in waste management for RASs. The solids mass production rate is determined by the waste excretion rate of the fish within the system, as well as the system's internal waste treatment capability. The volume of sludge generated, on the other hand, is controlled not only by the amount of solids produced, but also the degree to which the TSS is concentrated in the sludge stream. When solids are removed from a system either through backwashing a filter or cleaning a sedimentation tank, the resultant sludge is usually dilute, and the solids content is likely below 1%.

Because of the high-energy utilization efficiency of aquatic animals, fish culture produces the least amount of waste compared with large animals for a unit weight gain. However, considering daily waste generation on a live weight base, fish produce a comparable amount of wastes as do other animals, except fish culture produces a much higher sludge volume. A comparison is given in Table 4 in which a feeding rate of 2% of body weight per day was assumed for a catfish system (Chen et al. 1997).

The amount of effluent generated by an aquaculture operation is largely determined by the type of culture system. Typical RASs operate with a water exchange rate less than 10% of the system volume per day. As discussed earlier, the discharge is usually related to solids removal. Typically, the effluent is further treated before final disposal.

Characterization of Aquacultural Wastes

Information on sludge properties is also critical for waste treatment and disposal. Some preliminary data are currently available that provides characterization of waste stream from RAS.

Sludge composition. RAS sludge can be characterized both by the concentration of

TABLE 4. *Waste generation (/day/1000kg-Live Weight) comparison between recirculating catfish systems and other commercial animal operations (BOD: Biochemical oxygen demand, TKN: total Kjeldahl nitrogen).*

Animals	BOD$_5$ (kg)	Solids (kg)	TKN (kg)	Sludge Volume (L)
Fish	0.8–1.3	3.9–6.3	0.2–0.32	65–630
Beef Cattle	1.6	9.5	0.32	30
Dairy Cows	1.4	7.9	0.51	51
Poultry	3.4	14	0.74	37
Swine	3.1	8.9	0.51	76

TABLE 5. *The chemical compositions of aquacultural sludge*

Parameter	Aquacultural Sludge [a]			Municipal Sludge [b]	
	Range	Mean	Standard deviation	Range	Typical
TS (%)	1.4–2.6	1.8	0.35	2.0–8.0	5.0
BOD_5 (mg/L)	1,588–3,867	2,756	212	2,000–30,000	6,000
TKN (N, % of TS)	3.7–4.7	4.0	0.5	1.5–4	2.5
TP (P, % of TS)	0.6–2.6	1.3	0.7	0.4–1.2	0.7
pH	6.0–7.2	6.7	0.4	5.0–8.0	6.0

[a] Ning 1996.

[b] Data from Metcalf and Eddy, Inc. 1991.

waste constituents or by the ratios of a given constituent to total solids (TS) in the sludge. The TS concentration in RASs is mainly contributed by TSS. The BOD_5/TS ratio can be used as a measure of the degree of sludge stabilization. A high BOD_5/TS ratio implies a sludge that will rapidly decay and potentially cause oxygen depletion and odor problems if not properly managed. Similarly, the nutrient content of the sludge can be described by total Kjeldahl nitrogen (TKN) and total phosphorus (TP) to TS ratio. Higher ratios represent better fertilizer values or strong pollution potential. The characteristics of the aquacultural sludge from a RAS using a plastic beads filter is illustrated in Table 5 with reference to municipal sludge.

Compared with typical municipal sludge, aquacultural sludge has lower solid and BOD_5 concentrations. Of the solids contained in the aquacultural sludge, more than 80% is volatile, 20% higher than that of municipal sludge. Aquacultural sludge has higher nitrogen and phosphorus contents. As shown in Table 5, the average TKN content of the aquacultural sludge was 4.0% of the dry solid mass, whereas, the typical value for municipal sludge is only 2.6%. The average value of TP was 1.3% of the dry solid mass, while the typical municipal sludge contains only 0.7%. It has been found (Olson 1991) that fish wastes contain a higher percentage of nitrogen than cattle, pig, and sheep wastes. The high TP and TKN contents in the fish sludge originate from the feed, and most fish feeds contain 7.2% to 7.7% of nitrogen by weight. Of the nitrogen in food, 67% to 75% will be lost to the aquatic environment. The phosphorus content of the commercial fish diet ranges from 1.2% to 2.5% (Iwama 1991).

Physical characteristics. Major physical sludge characteristics that affect the design of a treatment system include particle settling velocity, density, and size distribution. It was observed (Ning 1996) that the solids particles in aquacultural sludge from a bead filter settled out fairly quickly with an average zone settling velocity of 1.37×10^{-3} m/s (8 cm/min). The wet density of the sluge was measured as 1.004 g/mL, which is close to the typical value of a municipal sludge. The particle size distribution of the aquacultural sludge showed that particles less than 60 microns and greater than 1,000 microns accounted for 15% and 17 % of the total dry weight, while particles ranging from 60–105, 105–500, and 500–1,000 microns represented 8%, 29%, and 31%, respectively.

Biochemical characteristics. A major biochemical characteristic of sludge is the decay rate constant that represents how fast the waste material decays under given conditions. A laboratory study (Ning 1996) found that oxygen availability and temperature had significant impacts upon decay rate constants.

From 10 C to 30 C, the anaerobic decay rate constants of BOD_5 varied from 0.004/d to 0.037/d. For the same temperature range, the aerobic decay rate constants varied from 0.188/d to 0.329/d. The impact of temperature on digestion rates was more significant for anaerobic digestions than for aerobic digestions. The results also show that aerobic digestions were much more efficient than anaerobic digestions; the respective reaction rates for aerobic digestions were, on average, approximately 10 times higher than anaerobic rates.

It was also found that aquacultural sludge had digestion rate constants comparable to domestic sludge (Ning 1996). For example, the maximum aerobic digestion rate constant of BOD_5 for aquacultural sludge at 20 C was 0.14 to 0.32/d, whereas, the reported value for municipal sludge was 0.05 to 0.3/d (Metcalf and Eddy, Inc. 1991).

Effluent Treatment Processes

Sludge management is typically the focus of effluent treatment of a RAS. A proper waste management strategy is now considered critical for maintaining the legality, profitability, and sustainability of an aquaculture facility. Treatment and ultimate disposal are the two major steps. For small-scale RASs, discharge to publicly-owned treatment works is the most popular option. For large-scale RASs, an on-farm management system is necessary. The following discussions describe typical processes.

Gravity Thickening

The purpose of thickening is to increase solids content and reduce the volume of water that has to be disposed. A thickening process is typically employed between the point of effluent discharge from the RAS and the sludge storage and stabilization units. In virtually all applications, it is always desirable to increase solids concentrations to a higher level to improve the economics of treatment and disposal. Clarification, often in settling tanks or ponds, is a common thickening process during which particles settle to the bottom by gravity. When the sediments are removed from the settling unit, solids concentrations are usually at 2 to 5%.

Sludge Stabilization and Storage

A stabilization process is necessary in environmentally sensitive areas where offensive odors need to be minimized. There are typically two major benefits that are related to sludge stabilization: decay of organics and volume reduction. Stabilization processes can provide for complete oxidation of readily degraded organics, resulting in a sludge that is unoffensive in nature. Stabilized sludge poses few problems when disposed through land application or land filling. In addition, stabilization can also reduce sludge volumes by 50–75% (Reynolds 1982).

Sludge storage is required in cold climate. During wintertime, frozen ground, along with wet winter weather, decreases waste utilization on land and increases surface runoff potential. Therefore, land application of waste in these regions is limited only to certain times of the year when the vegetation and crops on land are actively growing. As a result, waste has to be stored during the winter months.

Lagoons are the most feasible technology for stabilizing aquacultural sludge. Anaerobic lagoons have been used to treat waste discharges from all phases of the vast agricultural industry and have also been considered suitable treatment processes for aquacultural wastes (Ning 1996). In an anaerobic lagoon, the organic loading is so high that no appreciable oxygen concentration exists. Sludge introduced into the lagoon ranges from that containing relatively light solids concentrations (approximately 0.1% solids) to slurries containing just enough water to transport the solids into the lagoon. Anaerobic lagoons function successfully over a wide solids-loading range with little maintenance. The major parameters used for anaerobic lagoon design are volatile suspended solids (VSS). Whenever possible, a two-stage lagoon system should be used for sludge stabilization. A two-stage lagoon system is typically designed in such a way that the first stage is anaerobic, and the second stage is facultative. The main objective of an anaerobic lagoon is BOD reduction though organic decay, while the main objective of a facultative lagoon is nitrogen reduction, as well as additional BOD removal. Typical effluent quality from a facultative lagoon is not adequate for direct discharge, but is more appropriate for irrigation.

Clearly, a lagoon system can also be used for sludge storage in winter. However, the design for the two functions is usually different. Storage capacity is the major design criteria for storage lagoons (or ponds) where sludge and precipitation volume is more important. Organic (VSS) loading rate, on the other hand, is the major design criteria for anaerobic lagoons where the amount of organics and the decomposition rate of the organics are more important.

Sludge Disposal

Land Application. Currently, the most often used aquacultural sludge disposal process is direct land application. Application methods include using sprinklers and tank trucks. Because high-rate land application of animal manure as a waste has been proven to cause adverse environmental impacts (Overcash et al. 1983), a better approach for aquacultural sludge management is to use the waste only according to its fertilizer value for crops. The high nitrogen content (4–6%) of aquacultural waste makes it valuable to crops as a fertilizer, but limitations for such application have also been identified (Olson 1991). The first is odor, which prohibits this option in populated areas. The second is the propensity for the applied sludge to form a crust. If the sludge is not thoroughly plowed into the soil, some plant seedlings may be unable to push through the crust. The third limitation is the expense of hauling and spreading. The fourth is the slow nitrogen release rate. Since about 90% of the total nitrogen is in organic form, only one-third of the nutrients can be used in the first year (Olson 1991). This makes application in high rainfall areas questionable, since runoff of the unused nitrogen may cause water quality problems in local surface waters.

The guidelines for application rates of aquacultural sludge on cropland have not yet been established. However, it would be reasonable to manage aquacultural waste following guidelines similar to those for managing other types of waste. Most animal waste management plans are based on nitrogen. The fundamental premise is that the rate of animal waste applied on land should not provide more plant available nitrogen than crops need, in order to avoid contaminating groundwater with nitrate. Studies on animal manure indicate that crops typically remove between 100 to 200 kg of nitrogen/ha (Overcash et al. 1983). Therefore, a similar rate for application may eliminate nitrogen accumulation in the soil and avoid adverse impacts on the environment. The high nitrogen content that makes aquacultural waste valuable to crops as a fertilizer could, of course, also make over-application of nitrogen more likely. Olson (1991) tested three application rates of trout manure in a greenhouse (111, 222, and 336 kg N/ha) for growing spring wheat. Satisfactory results were obtained from application rates of both 222 and 336 kg N/ha. Subject to further experimental verifications, Olson (1991) recommends 222 kg N/ha as a design criterion for aquacultural sludge application on land. Such information would be useful for estimating the amount of land needed for sludge land application for a given operation. The characteristics of aquaculture sludge, however, may vary substantially depending upon waste management systems and many other factors (Westerman et al. 1993; Twarowska et al. 1997). Thus, the nutrient values and corresponding rate for land application may be different for aquaculture sludge from different operations and for different crops.

Composting. Another process that is often used for agricultural waste management is composting, an aerobic treatment occurring in the thermophilic temperature range. During the composting process, the carbonaceous constituents are readily removed through microbial conversion to carbon dioxide and water. These carbon losses are substantial resulting in waste volume and weight reduction. The aerobic environment leads to odorless operation due to the nature of end products of the predominant microorganisms. Nitrogen will also be reduced through principally ammonia volatilization. However, phosphorus, potassium, and other minerals will not be removed during composting. A challenge in composting aquacultural waste is the dilute nature of the sludge; bulking agents are typically needed to bring the solids content up to 40%.

Application Examples

This section will briefly describe the waste treatment system incorporated into a large-scale demonstration of recirculating aquaculture technology located on the campus of North Carolina State University. The production facility, currently producing mostly tilapia, consists of a 7.5-m^3 quarantine tank, a 15-m^3 nursery system, and four 60-m^3 growout tanks with associated recirculating treatment systems components. Each system has similar solid waste capture components consisting of an in-tank particle trap, an external sludge collector, and a drum screen filter. The facility is described in detail by Hobbs et al. (1997), while the operational characteristics of the main production tank system is described in

Losordo et al. (2000).

Waste Collection and Storage Components

The locations of the major fish tanks and waste handling and treatment components of the system are shown in plan view in Fig. 3. The waste handling and treatment components include particle traps and sludge collectors, drum filters, settling basin, and waste storage pond.

Particle traps and sludge collectors. Solid

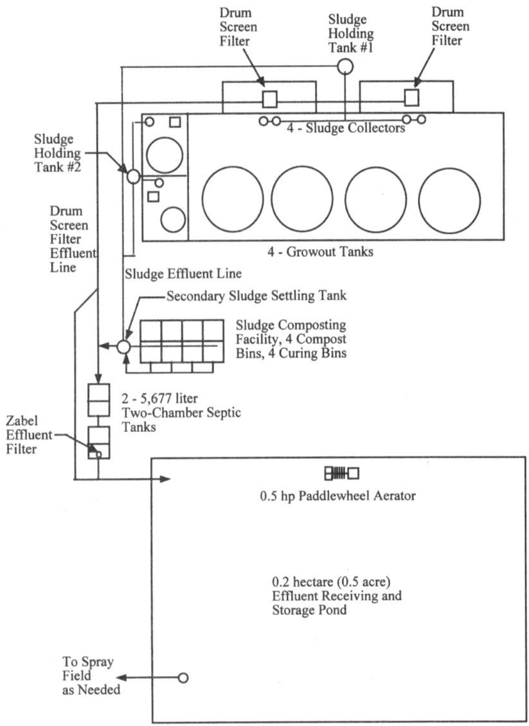

FIGURE 3. *North Carolina State University Fish Barn recirculating aquaculture waste treatment schematic.*

waste collection in the NC State Fish Barn system begins within the culture tank each of which is fitted with a particle trap at (center) bottom. Approximately 20% of the TS that are put into the system as feed is removed by the particle trap and associated sludge collector in a small (30 L/min) flow-stream (Twarowska et al. 1997). It was also determined (Twarowska et al. 1997) that 80% of waste solids flowing from the particle trap were settled in the small "sludge collector" mounted external to the fish culture tank. These conical bottom sludge collectors are 38 cm in diameter and 100 cm high with a working volume of 90 L. Clarified effluent leaves the top of the sludge collectors and is processed further in the recycle flow-stream by a drum screen filter. The settled sludge is approximately 7% solids by dry weight (70.7 g TS/L; Twarowska et al. 1997) and is discharged from the sludge collectors every 2 hr through an automated ball valve. The sludge discharge interval is dictated by the biological activity of the sludge. At water temperatures above 24 C, gas production within the sludge will float the sludge to the top of the sludge collector if not dumped at intervals of 2 hr or less. A volume of approximately 60 L of sludge and water are wasted with each sludge dump. The sludge from the four growout tank sludge collectors is stored in a small 1,325-L plastic storage tank located below ground and outside of the facility. Similarly, the sludge collectors from the quarantine and nursery tank systems discharge into an identical but separate storage tank. While the sludge is retained within these storage tanks, the supernatant is allowed to flow by gravity from the top of the storage container directly to the secondary sludge settling tank located adjacent to the composting facility (Fig. 3).

Drum screen filter. All of the water leaving the fish culture tanks is processed through a drum screen filter fitted with a 40μm screen. Suspended solids are retained on the inside of the screen and automatically cleaned from the screen with a high pressure backwash. Approximately 31% of the TSS that are input to the system as feed are removed by the drum screen filter. The solid waste stream from the drum screen filter backwash is comparatively dilute. The TSS concentration of the drum screen filter effluent measured by Twarowska et al. (1997) averaged 1.41 g/L (0.14% dry weight solids). The total flow from four drum screen filters within the facility averages 25,000 L/d and flows by gravity directly to the septic tank settling components.

Waste Treatment Processes

Composting The composting area is located approximately 36 m from the fish production facility. The composting facility is covered, contains eight bays, and has overall outside dimensions of 7.54 m long and 3.84 m wide. Each bay has dimensions of 1.8 m x 1.8 m x 1.1 m working depth. Each bay has a removable gate on the outside wall so that composted material can be removed with a small "front-end loader." Four of the composting bays are designated as primary composting areas and have a combination aeration and drainage chamber built into the base of each bay. The floor of each of the composting bays is covered with 2.54-cm diameter gravel. Low-pressure air is provided in an upflow direction through the composting material with a 1.3-hp regenerative blower. The blower is operated for 30 min and is off for 30 min. Any liquid underflow is collected and routed back to a secondary sludge settling tank. Additionally, at the top of each composting bay, there is a hose outlet where sludge from the secondary sludge settling tank can be pump sprayed on composting material in any of the four composting bays.

On a weekly or biweekly basis (as needed), the sludge in the secondary sludge settling tank adjacent to the compost facility (Fig. 3) is pumped onto a fresh layer (15–20 cm thick) of composting material (usually waste wood chips from a local tree trimming service). As mentioned previously, any underflow liquid flows back to the secondary settling basin and eventually overflows into the septic tank settling basins. After pumping down the level

TABLE 6. *Water quality characteristics of effluent at various points in the waste treatment system[a] of recirculating aquaculture systems at the North Carolina State University Fish Barn.*

Parameter	TKN[b] (mg / L)	NH$_3$-N (mg / L)	NO$_2$N (mg / L)	NO$_3$N (mg / L)	TP (mg / L)	PO$_4$-P (mg / L)	COD (mg / L)	TS (%)	TSS (mg / L)
Septic Tank 1 Inflow	50.3	2.96	5.35	109.0	28.6	5.98	1043	0.22	752
Septic Tank 2 Inflow	47.5	2.42	31.17	78.5	22.7	11.50	690	0.18	364
Septic Tank 2 Outflow	37.7	3.42	44.00	36.4	17.6	12.20	409	0.16	205
Receiving Pond	8.94	0.12	1.93	8.2	4.95	3.68	153	0.11	44

[a] Results presented above were sampled 4 wk after startup of the waste handling system. Flow from the system into the receiving pond for the sample period was 15.5 m^3 per d.

[b] TKN = Total Kjeldahl Nitrogen, TP = Total Phosphorus, PO$_4$-P = Orthophosphate, COD = Chemical Oxygen Demand, TS = Total Solids, TSS = Total Suspended Solids.

of the secondary settling basin, solids in the primary settling basins are then pumped to refill the secondary settling basin. After a particular bay is filled with composted material and sludge, the material is allowed to compost for up to 4 wk. Four of the composting facility bays are used for compost material "curing." At the end of the active composting period, determined by the internal temperature of the compost, the material is turned over while moving it to a curing bay for another period of up to 4 wk. Any dead fish from the production facility are also added to the active composting bay.

Septic tank settling basins. Effluent flows directly from the drum screen filters to a series of two "septic tank" settling basins. These tanks are designed for use in domestic wastewater treatment when combined with a leaching field. Both septic tanks have overall outside dimensions of 301 cm long x 152 cm wide x 168 cm high. Each septic tank has two chambers. The first chamber has inside dimensions of 193 cm long x 137 cm wide x 160 cm deep. The second chamber has inside dimensions of 91 cm long x 137 cm wide x 160 cm deep. The operating water depth in each septic tank is approximately 135 cm.

There are four circular openings that allow flow from the first chamber to the second chamber. These circular openings are 10 cm in diameter and are located approximately 30 cm from the bottom of the settling basin. Partially clarified effluent flows from the first septic tank to the second tank. Prior to being discharged (by gravity) to the pond, the effluent from the second chamber of the second septic tank flows through an effluent filter (Zabel Model A300, Zabel Environmental Technology, Jefferson-town, Kentucky, USA). This filter retains particles larger than 0.8 mm. Important characteristics of the water quality at the influent and effluent locations of each septic tank and in the pond can be found in Table 6.

Effluent Receiving Pond. Effluent from the Zabel filter flows directly to the 0.2-ha (0.5-acre) storage pond. The pond is a highly compacted clay-lined "levee" style pond commonly used in aquaculture production systems. The pond has a 0.5-hp paddle wheel type aerator that is operated manually to ensure the pond remains aerobic. While plans are in place to stock the pond with a filter-feeding species, the pond has not yet been stocked with fish. A tractor driven portable pump is available to pump water from the pond for irrigation of the adjacent fields. No water is discharged from this aquaculture site.

The example described above should not be used to estimate the waste volume or characteristics from a recirculating production system. These characteristics will be defined by the system in use. What should be noted from this example and particularly from the data in Table 6 is that RASs by themselves cannot eliminate environmental impacts. Keep in mind that approximately 40% of the solid waste produced in this fish production system is not included in the waste stream measured. That portion of the waste solids is collected in the sludge collector and treated in the composting facility.

Waste from this and other RASs can be broadly categorized as low volume and high strength. However, the data presented in this example shows that a significant portion of the waste can be captured for further treatment with simple settling tank technology. The inflow versus outflow concentrations of septic tank 1 and septic tank 2, respectively, indicate that 28% of the total nitrogen, 38% of the TP, 60% of the COD, and 73% of the TSS were retained within the septic tank system. While we expect the effluent pond concentrations to increase with time, in this case, the water within the pond can be easily disposed of on a cover crop. In the southern United States, fescue or bermudagrass over-seeded with ryegrass would be acceptable. For this example system in North Carolina, with an average daily flow of 22.7 m^3 (6,000 gpd), an irrigated crop area of 1.2 to 2.0 ha (3–5 acres) would be sufficient. In more urban settings, a more intensive waste treatment system would be needed.

Literature Cited

Adler, P. R., J. K. Harper, E. W. Wade, F. Takeda, and S. T. Summerfelt. 2000. Economic analysis of an aquaponic system for the integrated production of rainbow trout and plants. International Journal of Recirculating Aquaculture 1:15–34.

Alabaster, J. S. 1982. Report of the European Inland Fisheries Advisory Commission's (EIFAC) workshop on fish-farm effluents. Silkeborg, Denmark, 26-28 May, 1981. EIFAC Technical Paper 41. Food and Agricultural Organization of the United Nations, Rome, Italy.

APHA (American Public Health Association). 1995. Standard methods for the examination of water and wastewater, 19th edition American Public Health Association, Washington, D.C., USA.

Brune, D. E. and J.-K. Wang. 1998. Recirculation in photosynthetic aquaculture systems. Aquaculture Magazine 24(3):63–71.

Chen, S., D. E. Coffin, and R. F. Malone. 1997. Sludge production and management for recirculating aquacultural systems. Journal of World Aquaculture Society 28:303–315.

Drapcho, C. M. and D. E. Brune. 2000. The partitioned aquaculture system: impact of design and environmental parameters on algal productivity and photosynthetic oxygen production. Aquacultural Engineering 21:151–168.

Ewart, J., J. A. Hankins, and D. Bullock. 1995. State policies for aquaculture effluents and solid wastes in the Northeast region. NRAC Bulletin No. 300. Northeast Regional Aquaculture Center, University of Massachusetts, Dartmouth, Massachusetts, USA.

Heinen, J. M., J. A. Hankins, and P. R. Adler. 1996. Water quality and waste production in a recirculating trout culture system with feeding of a higher-energy or a lower-energy diet. Aquaculture Research 27:699-

710.

Hobbs, A. O., T. Losordo, D. DeLong, J. Regan, S. Bennet, R. Gron, and B. Foster. 1997. A commercial, public demonstration of recirculating aquaculture technology: The CP&L / EPRI Fish Barn at North Carolina State University. Pages 151–158 in M.B. Timmons and T. Losordo, editors. Advances in aquacultural engineering. Northeast Regional Agricultural Engineering Service, NREAS-105, Ithaca, New York, USA.

Hopkins, T.A., and W.E. Manci. 1989. Feed conversion, waste and sustainable aquaculture, the fate of the feed. Aquaculture Magazine 15(2):30.

Idaho DEQ (Department of Environmental Quality). 1998. Idaho waste management guidelines for aquaculture operations. Idaho Department of Health and Welfare, Division of Environmental Quality, Twin Falls, Idaho, USA.

Iwama, G. K. 1991. Interactions between aquaculture and the environment. Critical Reviews in Environmental Control 21:177–216.

Jenkins, M. R. and E. M. Wade. 1997. A preliminary economic analysis of a low-cost, small-scale, natural gas powered, integrated recycle aquaculture-hydroponic system for tilapia. Pages 163–173 in M. B. Timmons and T. Losordo, editors. Advances in aquacultural engineering. Northeast Regional Agricultural Engineering Service, Ithaca, New York, USA.

Ketola, H. G. and B. F. Harland. 1993. Influence of phosphorous in rainbow trout diets on phosphorous discharges in effluent water. Transactions of American Fisheries Society 122:1120–1126.

Losordo, T. M. 1991. Engineering considerations in closed recirculating systems. Pages 58–69 in P. Giovannini, session chairman. Aquaculture systems engineering. Proceedings of the World Aquaculture Society and the American Society of Agricultural Engineers at the World Aquaculture Society's 22nd Annual Meeting, 16-20 June, 1991, San Juan, Puerto Rico. American Society of Agricultural Engineers, Saint Joseph, Michigan, USA.

Losordo, T. M. 1998. Recirculating aquaculture production systems: the status and future, part I. Aquaculture Magazine 24(1):38–45.

Losordo, T. M., and M. B. Timmons. 1994. An introduction to water reuse systems. Pages 1-7 in M. B. Timmons and T. M. Losordo, editors. Aquaculture water reuse systems: Engineering design and management. Elsevier, Amsterdam.

Losordo, T. M., A. O. Hobbs, and D. P. DeLong. 2000. The design and operational characteristics of the CP&L / EPRI fish barn: a demonstration of recirculating aquaculture technology. Aquacultural Engineering 22:3–16.

Losordo, T. M., P. W. Westerman, and S. K. Liehr. 1994. Water treatment and waste-water generation in intensive recirculating fish production systems. Bulletin of National Research Institute of Aquaculture, Suppl. 1:27–36.

MacMillan, R., 1992. Economic implications of water quality management for a commercial trout farm. Pages 185–190 in J. Blake, J. Donald, and W. Magette, editors. National livestock, poultry, and aquaculture waste management. American Society of Agricultural Engineers, St. Joseph, Michigan, USA.

Massingill, M. J., E. M. Kasckow, R. J. Chamberlain, J. M. Carlberg, and J. C. Van Olst. 1998. Pg. 47. Pages 72-79 in G.

S. Libey and M. B. Timmons, editors. The Second International Conference on Recirculating Aquaculture. Virginia Polytechnic Institute and State University, Roanoke, Virginia, USA.

Metcalf and Eddy, Inc., 1991. Wastewater engineering, treatment/disposal/reuse, third edition. McGraw Hill, New York, New York, USA.

Mudrak, V. A. 1981. Guidelines for economical commercial fish hatchery wastewater treatment systems. Pages 174–182 *in* L. J. Allen and E. C. Kinney, editors. Proceedings of the Bio-engineering Symposium for Fish Culture. American Fisheries Society, Fish Culture Section, Bethesda, Maryland, USA.

Nijhof, M. 1995. Bacterial stratification and hydraulic loading effects in a plug-flow model for nitrifying trickling filters applied in recirculating fish culture systems. Aquaculture 134:49–64.

Ning, Z. 1996. Characteristics and digestibility of aquacultural sludge. Master's thesis. Louisiana State University, Baton Rouge, Louisiana, USA.

Olson, G. L. 1991. The use of trout manure as a fertilizer for idaho crops. Pages 198–205 *in* J. Blake, J. Donald, and W. Magette, editors. National livestock, poultry, and aquaculture waste management. American Society of Agricultural Engineers, St. Joseph, Michigan, USA.

Overcash, M. R. F. J. Humenik and J. R. Miner, 1983. Livestock waste management, volume II, CRC Press, Inc., Boca Raton, Florida, USA.

Pillay, T. V. R. 1982. Aquaculture and the environment. Fishing News Books, Oxford, England.

Phillips, M. J., M. C. M. Beveridge, and R. M. Clarke. 1991. Impact of aquaculture on water resources. Pages 568–591 *in* D.

E. Brune and J. R. Tomasso, editors. Aquaculture and water quality. Advances in world aquaculture. volume 3. The World Aquaculture Society, Baton Rouge, Louisiana, USA.

Rakocy, J. E. 1997. Integrating tilapia culture with vegetable hydroponics in recirculating systems. Pages 163–184 *in* B. A. Costa-Pierce and J. E. Rakocy, editors. Tilapia aquaculture in the Americas, volume one. The World Aquaculture Society, Baton Rouge, Louisiana, USA.

Rakocy, J. E. and J. A. Hargreaves. 1993. Integration of vegetable hydroponics with fish culture: A review. Pages 112–136 *in* J.-K. Wang, editor. Techniques for modern aquaculture. American Society of Agricultural Engineers, Saint Joseph, Michigan, USA.

Rakocy, J. E., T. M. Losordo, and M. P. Masser. 1992. Recirculating aquaculture tank production systems, integrating fish and plant culture. Southern Regional Aquaculture Center Publication No. 454. Southern Regional Aquaculture Center, Stoneville, Mississippi, USA.

Ray, L. 1981. Channel catfish production in geothermal water. Pages 192–195 *in* L. J. Allen and E. C. Kinney, editors. Proceedings of the Bio-engineering Symposium for Fish Culture. American Fisheries Society, Fish Culture Section, Bethesda, Maryland, USA.

Reynolds, T. D. 1982. Unit operations and processes in environmental engineering. Brooks/Cole Engineering Division, Monterey, California, USA.

Serfling, S. A. 2000. Closed-cycle, controlled environment systems: The Solar Aquafarms story. Global Aquaculture Advocate 3(3):48–51.

Siddiqui, A. Q. and A. H. Al-Harbi. 1999. Nutrient budgets in tanks with different stocking densities of hybrid tilapia. Aquaculture 170:245–252.

Summerfelt, S. T. 1998. An integrated approach to aquaculture waste management in flowing water tank culture systems. Pages 87–97 *in* G. S. Libey and M. B. Timmons, editors. The Second International Conference on Recirculating Aquaculture. Virginia Polytechnic Institute and State University, Roanoke, Virginia, USA.

Summerfelt, S. T., J. Bebak-Williams, and S. Tsukuda. In press. Controlled systems: water reuse and recirculation. *in:* G. Wedemeyer, editor. Fish hatchery management, second edition. American Fisheries Society, Bethesda, Maryland, USA.

Summerfelt, S. T., J. Davidson, T. Waldrop, and S. Tsukuda. 2000. A partial-reuse system for coldwater aquaculture. Pages 167–175 *in*: The Third International Conference on Recirculating Aquaculture. Virginia Polytechnic Institute and State University, Roanoke, Virginia, USA.

Timmons, M. B., S. T. Summerfelt, and B. J. Vinci. 1998. Review of circular tank technology and management. Aquacultural Engineering 18:51–69.

Twarowska, J. G., P. W. Westerman, and T. M. Losordo. 1997. Water treatment and waste characterization evaluation of an intensive recirculating fish production system. Aquacultural Engineering 16:133–147.

Westerman, P. W., J. M. Hinshaw, and J. C. Barker. 1993. Trout manure characterization and nitrogen mineraliation rate. Pages 3543 *in* J. K. Wang, editor. Techniques for modern aquaculture. American Society of Agricultural Engineers, Saint Joseph, Michigan, USA.

Environmental Interactions of Bivalve Shellfish Aquaculture

Paul G. Olin

University of California Cooperative Extension, Sea Grant Extension Program, 2604 Ventura Avenue,

Santa Rosa, California 95403 USA

ABSTRACT

The most widely cultured filter feeding bivalve mollusks in the United States are oysters, clams, and mussels. A variety of culture systems are used including bottom culture, floating bags, rack and bag systems, long lines, and trays. The environmental interactions vary significantly between these systems and to a lesser extent between the cultured species. Potential impacts include changes in water quality parameters such as dissolved oxygen, nutrient levels, turbidity, suspended sediments, organic particulates, and transmittance. In most cases, these changes result in improved water quality, and shellfish growers are acutely aware of the importance good water quality plays in their ability to produce a safe and healthy product.

At the community ecology level, shellfish gear and culture practices can influence the diversity and abundance of fish and invertebrates in surrounding waters and benthic substrates. The complex three-dimensional habitat created by cultured shellfish is colonized by a wide diversity of vertebrate and invertebrate fauna that interact with organisms in adjacent waters. Historically, shellfish aquaculture has resulted in the intentional and unintentional introduction of nonindigenous species, but today the use of hatcheries and strict regulatory programs protect against most unintentional introductions.

At a societal level, shellfish culture has been criticized for altering aesthetic views and impeding navigation. At the same time, societal benefits accrue from employment in largely rural areas and the rigorous attention to water quality that has highlighted and corrected many problems that would otherwise have gone undetected.

Bivalve aquaculture differs significantly from the culture of most finfish and crustaceans in that cultured shellfish exploit naturally occurring phytoplankton at the base of the estuarine food chain; thus obviating the need for external feed inputs. For this reason, shellfish aquaculture does not result in additional nutrient loading, but rather, a transfer of nutrients from the water column to benthic sediments through deposition of feces and pseudofeces, and a net removal of nutrients when shellfish are harvested. Nutrient deposition can alter sediment chemistry and benthic communities, and in severe cases result in localized hypoxia. Harvesting shellfish with dredges can cause reversible changes in sediment structure and disrupt benthic communities. Proper siting and management of shellfish farms can minimize environmental impacts associated with concentrated nutrient deposition and harvesting, provide economic opportunities, and result in a net benefit to water quality and the surrounding ecosystem.

For thousands of years shellfish harvested from North American waters provided an important source of nutrition and a popular food for native Americans and the European immigrants that followed. In the last few decades, wild shellfish harvest has declined dramatically due primarily to diseases, overharvesting, and harvest closures resulting from pollution. This is true in North America and throughout the world; thus, the popularity of shellfish and the resulting high demand has led to their being farmed in coastal waters adjacent to nearly every continent. The primary bivalves cultured in North America include oysters, clams, and mussels. The primary farmed species in the U.S. and the regions

TABLE 1. *Commonly cultured bivalves in the United States*

Common Name	Species	Culture Locations
Eastern oyster	*Crassostrea virginica*	Atlantic, Pacific, Gulf
Pacific oyster	*Crassostrea gigas*	Pacific
Quahog or hard clam	*Mercenaria mercenaria*	Atlantic, Gulf
Manila Clam	*Venerupis japonica*	Pacific
Atlantic mussel	*Mytilus edulis*	Atlantic, Gulf
Pacific mussel	*Mytilus galloprovincialis*	Pacific

where they are grown can be found in Table 1.

As the human population increases and agricultural production expands to meet demand, there are increasing concerns about how agriculture, including aquaculture, is impacting the environment and other natural resources. Modern agricultural production relies extensively on advanced technology including formulated feeds, fertilizers, and pesticides to produce a safe food supply for an ever-increasing population. This is true in terrestrial crop production as well as animal agriculture, and the runoff of chemicals and nutrients contributing to degradation and eutrophication of coastal waters is of paramount concern. One notable exception to this scenario is the production of filter feeding bivalve shellfish, which require few external inputs when they are farmed in coastal embayments and estuaries. Their unique filter feeding capacity allows shellfish to filter naturally produced phytoplankton and organic particulates from the water column. Bivalves are capable of selectively filtering phytoplankton and organic detrital particles and prior to consumption, they expel unwanted particles as mucous-coated pellets referred to as pseudofeces. Other particles are consumed and digested, and then either assimilated or excreted as fecal pellets. The end result is a transfer of organic matter from the water column to sediments beneath cultured bivalves, and a net removal of assimilated nutrients from coastal waters when the shellfish are harvested (Folke and Kautsky 1989; Shpigel et al. 1993; Dame 1996).

A variety of techniques are used to culture bivalve shellfish including bottom culture, floating bags, lantern nets, rack and bag systems, long lines, and trays (Figs. 1-5). The degree to which shellfish production impacts surrounding water quality, benthic substrate, and other organisms is largely determined by the water depth, tidal exchange, and shellfish density. Potential impacts include changes in water quality parameters such as dissolved oxygen, nutrient levels, turbidity, suspended sediments, organic particulates, and light transmittance. In most cases these changes result in improved water quality, and shellfish growers are acutely aware of the important role that good water quality plays in their ability to

FIGURE 1. *Intertidal oyster beds, Totten Inlet, Washington, USA.*

produce a safe and healthy product.

Shellfish aquaculture can influence the diversity and abundance of fish and invertebrates in surrounding waters and benthic substrates. The complex three-dimensional habitat created by cultured shellfish is colonized by a wide diversity of vertebrate and invertebrate fauna. These organisms include both predators and prey, and many remain in the estuarine system when the shellfish are harvested. Changes in sediment composition, and the diversity and abundance of benthic invertebrates can occur when organic levels rise beneath culture gear as a result of deposition of feces and pseudofeces. Sediment structure

FIGURE 2. *Longline oyster culture gear, Willappa Bay, Washington, USA.*

and benthic communities can also be disrupted through harvesting practices such as tonging, dredging, and raking. These alterations have been shown to be temporary and benthic communities revert to pre-culture conditions in a relatively short time frame when shellfish grounds are allowed to lie fallow. Historically, shellfish aquaculture has resulted in the intentional and unintentional introductions of nonindigenous species, but today the use of hatcheries and strict regulatory programs largely protect against unintentional introductions.

In the Unites States 110 million Americans or about half the U.S. population now live near coastal and estuarine shores. Coastal counties

FIGURE 3. *Oyster dredge lifting harvested oysters on barge.*

are growing three times faster than counties elsewhere in the nation. With this increasing concentration of people and the changing economies that result, it is not surprising that conflicting uses often arise in nearshore coastal waters. Shellfish gear has been criticized for altering scenic vistas, and recreational users often frown upon commercial activity in coastal waters. Perceptions often vary as a result of cultural and regional differences. Floating long lines and aquaculture structures are generally viewed with satisfaction in Japan where they are valued as a source of highly prized fresh seafood. In a similar fashion, Americans value the pastoral vistas and rolling green hillsides that support dairy cows and cattle that provide

FIGURE 4. *Rack and Bag oyster culture, Tomales Bay, California, USA.*

meat, milk and cheese. It is important to recognize different uses and values in order to minimize conflicts over the utilization of coastal resources. For example, site selection is critical to ensure that floating aquaculture structures do not impede navigation. While aquaculture does compete with other coastal uses, significant benefits accrue to society through sustainable production of fresh shellfish, and opportunities for economic development and rural employment.

The Interstate Shellfish Sanitation Commission regulates shellfish growers under the National Shellfish Sanitation Program. Part of the regulatory program covers certification of shellfish growing waters, and a growing area can be classified as approved, conditionally approved, restricted, conditionally restricted, prohibited, or closed. Growing areas are routinely monitored for coliform bacteria as an indicator of sewage contamination, and growing water classification can be downgraded if unacceptably high levels of bacteria are present. Shellfish growers are acutely aware of the danger this represents in a business where the greatest capital is clean water, and there are many instances where their rigorous attention to water quality has highlighted and corrected problems that would otherwise have gone undetected.

Aquaculture in coastal waters is usually conducted on public property through shellfish leases managed by state natural resource agencies. The terms and conditions of these leases vary from state to state, but generally require growers to obtain regulatory permits from state and federal environmental protection agencies, the Army Corp of Engineers, and public health agencies. The situation in Washington State differs in that growers often hold private title to shellfish grounds in the intertidal zone. However, they operate in a similar regulatory framework.

Shellfish aquaculture has the potential to impact the environment and the economy in common with almost any endeavor whether it is a cornfield, a pasture, or a small municipality. It is important to recognize this potential and address issues of concern to ensure shellfish aquaculture provides a sustainable source of food, and that growers are good stewards of natural resources. Proper siting and management of shellfish farms can minimize adverse environmental impacts associated with concentrated nutrient deposition and harvesting, and have a net benefit on water quality and the surrounding natural resources.

Shellfish Farming Techniques

Oyster aquaculture, particularly along the West Coast of the United States has become a completely integrated farming operation (Conte et al. 1996). Cultured oysters are produced in hatcheries where broodstock are maintained and induced to spawn. Larval oysters are fed cultured phytoplankton until they are competent to settle and attach to a substrate. Some oysters are settled on pieces of shell or cultch. This cultch with attached oysters is then placed on intertidal and subtidal bottom where it is left until the oysters reach market size, usually in one to three years depending on location and temperature (Fig. 1). Some seeded cultch is strung on lines that are suspended from stakes or rails (Fig. 2).

FIGURE 5. *Floating oyster culture bags, Tomales bay, California, USA.*

After reaching market size, oysters are harvested by hand, using tongs, or with mechanical or suction dredges (Fig. 3). The harvested oyster clusters are separated into singles, or processed at a shucking plant where the meats are packaged in containers for sale. Oysters destined for the half shell market are often settled as larvae on sand grains or very small pieces of shell. Referred to as cultchless oysters, these individual oysters are often grown in high-density polyethylene (HDPE) mesh bags placed on the bottom, suspended off the bottom on racks, or placed in floating bags attached to long lines (Figs. 4, 5 respectively).

Farmed clams are also produced in hatcheries where they are held in trays or upwellers during a nursery period before being planted in a growout area or placed in HDPE bags for growout. Grounds where clams are planted are frequently covered with small plastic mesh screening as predator protection. Manila clam grounds often have the substrate modified through the addition of gravel and oyster shell that is usually spread from a barge (Toba et al.1992).

Mussel seed is produced in hatcheries and also collected from the wild when spat collector lines are strung in areas known to have good larval densities. These mussels are grown in a nursery phase, and juveniles are then placed in mussel socking or on lines for growout. Mussel growout lines and socks are often suspended between buoys or attached as vertical drop lines suspended from a surface or submerged line between buoys.

These general culture techniques illustrate the areas where shellfish culture can influence environmental variables and organisms in surrounding waters and adjacent shorelines. Water quality can be altered by oysters acting as biofilters, while activities associated with seeding, maintaining, and harvesting shellfish beds can temporarily resuspend sediments and alter benthic communities. Sediment chemistry can be influenced by deposition of organic waste, and increased sedimentation can result from altered current velocities through shellfish culture gear. Cultured shellfish also create habitat that is utilized by other organisms.

Water Quality

Water quality is the most critical element in siting a shellfish farm, and growers must ensure that waters are free of contamination from sewage, industrial chemicals, and non-point source pollution from rural, urban, and residential runoff. Through their filter feeding abilities bivalve shellfish act as biofilters removing phytoplankton, suspended sediments, and organic particulates from the water column. Filtering capacity depends on the species, temperature, and food availability (Haven and Morales-Alamo 1970; Mohlenberg and Riisgard 1979; Doering and Oviatt 1986). Filtration rates are summarized in Table 2.

The essential ecological function of oysters in the Chesapeake Bay is convincingly made by Mann (2000), wherein he describes the decline of oyster filtration as being responsible for increased turbidity that reduced light penetration and resulted in declines of submerged aquatic vegetation (SAV). The loss of SAV that stabilized bay sediments deteriorated water quality to the detriment of oysters and thus, two important habitats and the ecological functions they supported in the bay were substantially reduced.

The importance of this biofiltration to a functioning estuarine ecosystem is further illustrated by Newell (1988) who calculated that in 1870, prior to extensive harvesting and disease, oysters in the Chesapeake Bay filtered the entire bay volume in 3.3 da. The remnant oyster populations in the Chesapeake Bay in 1988 required 325 da to filter a comparable volume. In the Providence River section of Narragansett Bay, Rice (2001) found that the population of northern quahogs was able to effectively filter 21.3% of the tidal prism in a

TABLE 2. *Bivalve filtration rates.*

Species	Filtration Rate L/h	Size[a]	Temp. (C)	Reference
Crassostrea gigas	3.8	5.0 g	20	Kobayashi et al. 1997
	2.4	5.0 g	20	Powell et al. 1992
	2.4	1.0 g	13–17	Ropert and Goulletquer 2000
Crassostrea virginica	9.9–24.3	12 cm	N/A	Galtsoff 1964
	26	8–10 cm	25	Nelson 1938
	2.7–3.6	1.5 g	20	Shumway and Cucci 1996
Mercenaria mercenaria	8.2	4.8 g	18–20	Coughlan and Ansell 1964
Mya arenaria	2.8	1.0 g	13	Mohlenberg and Riisgard 1979
Mytilus edulis	9.1	1.4 g	13	Mohlenberg and Riisgard 1979
	3.9	1.0 g	10	Vahl 1973
	1.9	1.0 g	15	Thompson and Bayne 1974
Ostrea edulis	0.5–0.9	0.2 g	12	Shumway and Cucci 1986

[a]Denotes dry weight (g) or shell length (cm) of animal.

single tide cycle. As mentioned earlier, this filtration of algae and detrital particles improves light penetration and conditions for the growth of SAV such as eelgrass that provides important habitat for many fish and invertebrates (Cohen et al. 1984; Peterson and Black 1991; Gottleib and Schweighofer 1996). In many estuaries where wild shellfish populations have been depleted, cultured shellfish acting as biofilters may already be duplicating this important ecological function.

While positive associations between bivalve shellfish and SAV exist, another study in Oregon documented declines in eelgrass abundance associated with oyster culture (Everett et al. 1995). These researchers concluded that stake culture methods adversely impacted eelgrass through increased sedimentation and physical disturbance associated with planting and harvest, while eelgrass declines associated with rack culture were attributed to erosion and possibly shading. Eelgrass often thrives on oyster beds employing ground culture techniques (Fig. 6). These studies highlight the importance of site selection and site specific culture methods to ensure that shellfish aquaculture is compatible with protection of valuable natural resources such as eelgrass.

FIGURE 6. *Intertidal oyster beds and eelgrass, Samish Bay, Washington, USA.*

The complexity of the ecological interactions between shellfish, the habitat they create and SAV is nicely demonstrated in a series of experiments conducted by Peterson and Heck (1999, 2001a, 2001b) involving the mussel *Modiolus americanus* and the seagrass *Thalassia testudinum*. This work demonstrated that mussels were an effective benthopelagic coupler, increasing pore water ammonium and phosphorous concentrations which resulted in decreased C:N and C:P ratios in leaf tissue. Field experiments documented that the length and width of seagrass blades significantly increased in response to the presence of mussels and increased sediment nutrient levels. In addition, the presence of mussels was associated with decreased epiphytic fouling on seagrass blades with the potential to increase photosynthetic efficiency. Field experiments using mussel mimics and nutrient supplements documented the relative contributions of increased nutrients and reduced epiphytic loads to increased seagrass productivity, with the greatest positive response attributed to nutrient enrichment. Habitat structure provided by mussel mimics significantly increased densities of epiphytic grazers and reduced epiphytic fouling.

Many coastal waters receive excessive nutrient loading from municipal sewage discharges and are listed as impaired water bodies. Shellfish aquaculture has been proposed in some areas as having the potential to alleviate some of the eutrophication that can result from this nutrient overloading. In an interesting study, Rice (2001) calculated that every kilogram of shellfish meats harvested from growing waters results in the net removal of 16.8 g of nitrogen and that if an individual harvested and consumed 5,600 oysters annually, they would mitigate the impact of their own nitrogen excretion. Mussel farms have been proposed in Sweden as a means of reducing the negative effects of eutrophication and to increase extraction of nutrients from eutrophic Swedish coastal waters (Haamer et al. 1999). Natural oyster reefs in South Carolina were also shown to be processing nitrogen and phosphorous at high rates and appeared to play an important role in cycling nutrients from suspended organic particulates and plankton into oyster tissue and dissolved forms (Dame et al. 1989).

There is no net increase of nutrients in coastal waters as a result of shellfish culture given that all nutrition for growth and survival results from exploiting naturally occurring phytoplankton and suspended detrital particles. Nutrients consumed by oysters are assimilated as oyster tissue and shell, excreted as ammonia in the water, or transferred to the sediments via deposition of waste.

Sediment Conditions

It is recognized that bivalve mollusks increase the amount of organic material in underlying sediments through deposition of feces and pseudofeces. This constitutes a transfer of nutrients from the water column to underlying sediments, and in extreme cases, high-density oyster cultivation has increased organic levels in the substrate to the point that anoxic conditions occurred (Castel et al. 1989). In a mesocosm experiment, Doering and Oviatt (1986) observed a 58% increase in sedimentation in tanks containing quahogs. Studies on benthic substrate composition and chemistry beneath oyster culture gear on the southwest coast of France documented a 3- to 4-fold increase in meiofaunal abundance, accompanied by a 42% reduction in benthic macrofauna (Castel et al. 1989; Nugues et al. 1996). These changes in benthic species diversity and abundance are thought to result from increased predation on macrofauna, and their greater sensitivity to reduced oxygen levels often associated with localized organic enrichment.

In some Pacific Northwest estuaries, populations of the burrowing shrimps *Neotrypaea californiensis* and *Upogebia pugettensis* have increased to such densities that bottom sediments have been destabilized and oysters planted on the bottom sink into the soft substrate. These population increases are thought to have resulted from a decline in predator populations such as sturgeon, increased salinities resulting from fresh water diversions, or the effect of large scale climactic events on shrimp recruitment. As a result, many former oyster grounds are now less productive due to sediment destabilization. Under agreement with the Washington Department of Ecology, oyster growers in Washington State spray designated oyster beds in the intertidal zone with the insecticide carbaryl to control burrowing shrimp (Feldman et al. 2000). Growers and state agencies signed a memorandum of understanding to pursue integrated pest management and seek alternative control measures in 2001. By removing the shrimp from their beds and planting them with live oysters, a more stable habitat is created, and a different and more diverse benthic community is established (Dumbauld et al. 2001)

In the culture of Manila clams, growers apply gravel and shell to intertidal sediments, which is reported to reduce polychaetes in favor of bivalves and nemerteans (Simenstad and Fresh 1995). Spencer et al. (1996, 1997) documented increased sedimentation and organic content in sediments when predator exclusion netting was placed over Manila clam beds. This resulted in dramatic shifts in benthic species abundance, and the original community was substantially altered during the 2-yr growout period. The environment and species complex was comparable to control areas within 12 mo after harvest, demonstrating the resilience of these benthic communities to disturbance (Spencer et al. 1998). In another study, it required in excess of 18 mo for benthic impacts of an intensive mussel farm to disperse through a combination of microbial activity and wave action (Davies et al. 1980). In a study of hard clam culture in the Indian River Lagoon in Florida, little effect on benthic communities was observed adjacent to or beneath culture bags, which was attributed to the relatively low culture densities involved (Mojica and Nelson 1993).

It is evident that bivalve shellfish culture has the potential to alter sediment chemistry and composition, with resulting changes in the diversity and abundance of benthic invertebrates. It is also evident that these changes in benthic communities are reversible. It is the responsibility of shellfish growers and managing resource agencies to ensure that culture practices promote the long-term sustainability of both natural resources and shellfish production.

Species Interactions

Bivalve shellfish compete with other filter feeding organisms for the plankton and particulate detrital particles they consume. There is little research to document these competitive interactions, but given the increased productivity of most coastal waters receiving nutrients from point and non-point sources in the watershed, it is unlikely that natural productivity is a limiting factor in most estuarine systems. In many cases, cultured bivalves are simply replacing native populations whose numbers have declined from overharvest and disease. High stocking densities of bivalves that filter out larvae may influence other invertebrate populations if they are recruitment limited (Baldwin et al. 1995). On the other hand, some invertebrates may benefit from the settlement substrate provided by shellfish and the gear used to culture them.

Epibenthic habitat created by oyster and mussel shell is rapidly utilized by a number of vertebrate and invertebrate species (Doty et al. 1990; Armstrong et al. 1992; Suchanek 1992; Dumbauld et al. 1993, 2000; Feldman et al. 2000). In one study of mussel rafts in Spain, López-Jamar et al. (1984) found that a rich epifaunal community associated with the rafts constitutes the main food resource for three demersal fish species in the area. The habitat created by cultured shellfish is also used by fish and invertebrates that are important prey species for juvenile salmon, flatfish, and Dungeness crabs (Simenstad and Fresh 1995; Feldman et al. 2000).

Cultured mussels provide a complex habitat that is colonized by over 100 species (Tenore and Gonzalez 1976). While not approaching the more than 300 species found in natural beds of *Mytilus californiensis* along the California coast, like their wild counterparts these cultured mussels support a complex ecosystem providing habitat for many species including both predators and prey (Suchanek 1992).

In studies to evaluate the effect of trampling on rocky shore marine communities Brosnan and Crumrine (1994) found that foliose algae and barnacles were damaged but algal turf increased in relative abundance at trampled sites. These results may not be representative of impacts associated with shellfish culture as these researchers trampled their study sites 250 times/mo for 1 yr, a level of impact more closely associated with a popular beach or tidepool area than a harvested shellfish bed. Additional studies have also documented changes in invertebrate and plant communities on trampled intertidal mudflats and rocky shorelines, although once again at use levels far greater than those typically found on commercial shellfish beds as the site was visited by 2,000 religious pilgrims a month between May and September (Chandrasekara and Frid 1996; Fletcher and Frid 1996).

Considerable attention has been given to the effects mechanical shellfish harvesting techniques have on the bottom and benthic communities. Harvesting practices using towed or suction dredges to collect shellfish on the bottom significantly impact benthic communities (Trianni 1995; Jennings and Kaiser 1998; Kaiser et al. 1998). Suction dredging and mechanical drags or rakes remove and relocate sediments creating furrows and surface irregularities. These physical alterations are corrected over time by tidal, wind, wave, and current action, and disturbed areas recover so they are similar to adjacent grounds within a few days or months (Hall and Harding 1997). The time to recover is based on localized weather conditions, sediment composition, and harvest gear. In this same study, the benthic infaunal community recovered within 3 mo. One study by Kaiser et al. (1996) evaluating recovery times following clam harvest found that restoration of the benthic community following suction dredging took 7 mo. In another study, 9–12 mo were required for recovery of benthic infauna (Spencer et al. 1998). It is likely that the rate of recovery of benthic communities will depend on sediment structure and be highly seasonal in relation to reproductive cycles of the organisms present.

In North America, natural resource management policies protect SAV habitat like eelgrass beds as critical biologically diverse habitat. Shellfish culture employing racks and bags has been shown to reduce eelgrass community coverage in areas where this gear is deployed (Pregnall 1993; Everett et al. 1995). As a result, natural resource management agencies have adopted policies to restrict activities that would negatively impact SAV (Wyllie-Echeverria et al. 1994).

Methods of Reducing Impacts

Bivalve shellfish aquaculture can improve water quality and provide important habitat for fish and invertebrates. There can also be shifts in benthic communities associated with organic enrichment of sediments and harvesting practices. Sediment enrichment and changes in benthic infauna are largely reversible in relatively short time frames. Conflicting uses in the coastal zone dictate that shellfish growers must site and manage facilities in such a way that alterations to view planes and obstructions to navigation are minimized. Shellfish growers should select culture sites that have adequate water flow and adopt culture practices to ensure that unacceptable levels of organic materials do not accumulate beneath shellfish beds. Growers along the east and west coasts of the United States have recently begun a process to establish an environmental policy and environmental codes of practice to address these issues (PCSGA 2001).

In addition to providing employment and economic opportunities, perhaps one of the greatest benefits of bivalve aquaculture is the rigorous attention to water quality and monitoring that accompanies it. In essence, shellfish farmers are conducting privately funded water quality monitoring programs that provide an early warning in the event that water quality is compromised. In many cases this has resulted in improved management of coastal septic systems and municipal waste treatment and reduced pollution loads to coastal waters. This protection and improvement of coastal water quality benefits multiple natural resources and the people who use and value them.

Literature Cited

Armstrong, D. A., O. Iribarne, P. A. Dinnel, K. A. McGraw, J. A. Schaffer, R. Palacios, M. Fernandez, K. Feldman, and G. Willaims. 1992. Mitigation of Dungeness crab, *Cancer magister*, losses due to dredging in Grays Harbor by development of intertidal shell habitat: pilot studies during 1991. Publication #FRI-UW-9205. Fisheries Research Institute, University of Washington, Seattle, Washington, USA.

Baldwin, B., J. Borichewski, and R. A. Lutz. 1995. Predation on larvae: Filtration by adult bivalves directly and indirectly kills bivalve larvae. 23rd Benthic Ecology Meeting, Rutgers State University, New Brunswick, New Jersey, USA.

Brosnan, D. M. and L. L. Crumrine. 1994. Effects of human trampling on marine rocky shore communities. Journal of Experimental Marine Biology and Ecology 177:79–97.

Castel, J., P. J. Labourg, V. Escaravage, I. Auby, and M. E. Garcia. 1989. Influence of seagrass beds and oyster parks on the abundance and biomass patterns of meio- and macrobenthos in tidal flats. Estuarine, Coastal and Shelf Science 28:71–85.

Chandrasekara, W. U. and C. L. J. Frid. 1996. Effects of human trampling on tidalflat infauna. Aquatic Conservation: Marine and Freshwater Ecosystems 6:299–312.

Cohen, R. R. H., P. V. Dresler, E. J. P. Phillips, and R. L. Corey. 1984. The effect of the Asiatic clam, *Corbicula fluminea*, on phytoplankton of the Potomac River, Maryland. Limnology. Oceanography 29:170–180.

Conte, F. S., S. C. Harbell, and R. L. RaLonde. 1996. Oyster culture: Fundamentals and technology of the west coast industry. Western Regional Aquaculture Center Publication No. 94–101. University of Washington, Seattle, Washington, USA.

Coughlan, J. and D. A. Ansell. 1964. A direct method for determining the pumping rate of siphonate bivalves. Journal of Conservation and International Explorations Mer 29:205–213.

Dame, R. F. 1996. Ecology of marine bivalves: an ecosystem approach. CRC Press, Boca Raton, Florida, USA.

Dame, R. F., J. D. Spurrier and T. G. Wolaver. 1989. Carbon, nitrogen, and phosphorus processing by an oyster reef. Marine Ecology Progress Series 54:249–256.

Davies, G. P., J. Dare, and D. B. Edwards. 1980. Fenced enclosures for the protection of seed mussels (*Mytilus edulis*) from predation by shore crabs (*Carcinus Maenus*). Fisheries Research Technical Report. No. 56. Ministry of Agriculture,

Fisheries, and Food, London, England.

Doering, P. H. and C. A. Oviatt. 1986. Application of filtration rate models to field populations of bivalves: an assessment using experimental mesocosms. Marine Ecology-Progress Series 31:265–275.

Doty, D., D. Armstrong, and B. Dumbauld. 1990. Comparison of carbaryl pesticide impacts on Dungeness crab (*Cancer magister*) versus benefits of habitat derived from oyster culture in Willapa Bay, Washington. Publication #FRI-UW-9020. Fisheries Research Institute, University of Washington, Seattle, Washington, USA.

Dumbauld, B., D. Armstrong, and T. McDonald. 1993. Use of oyster shell to enhance intertidal habitat and mitigate loss of Dungeness crab (*Cancer magister*) caused by dredging. Canadian Journal Fisheries Aquatic Science 50:381–390.

Dumbauld, B. R., K. M. Brooks, and M. H. Posey. 2001. Response of an estuarine benthic community to application of the pesticide carbaryl and cultivation of Pacific oysters (*Crassostrea gigas*) in Willapa Bay, Washington. Marine Pollution Bulletin 42:826–844.

Dumbauld, B. R., E. P. Visser, D. A. Armstrong, L. Cole-Warner, K. L. Feldman, and B.E. Kauffman. 2000. Use of oyster shell to create habitat for juvenile Dungeness crab in Washington coastal estuaries: Status and prospects. Journal Shellfish Research 19:379–386.

Everett, R., G. Ruiz, and J. T. Carlton. 1995. Effect of oyster mariculture on submerged aquatic vegetation: an experimental test in a Pacific Northwest estuary. Marine Ecology Progress Series 125:205–217.

Feldman, K. L., D. A. Armstrong, B. R. Dumbauld, T. H. DeWitt, and D.C. Doty. 2000. Oysters, crabs and burrowing shrimp: Review of an environmental

conflict over aquatic resources and pesticide use in Washington State's (USA) coastal estuaries. Estuaries 23:141–176.

Fletcher, H. and C. L. J. Frid. 1996. Impact and management of visitor pressure on rocky intertidal algal communities. Aquatic Conservation: Marine and Freshwater Ecosystems 6:287–297.

Folke, C. and N. Kautsky. 1989. The role of ecosystems for a sustainable development of aquaculture. Ambio 1989 18:234–243.

Galtsoff, P. F. 1964. The American oyster: *Crassostrea virginica*. Fishery bulletin, volume 64. U.S. Department of the Interior, Washington, D.C., USA.

Gottleib, S. J. and M. E. Schweighofer. 1996. Oysters and the Chesapeake Bay ecosystem: A case for exotic species introduction to improve environmental quality? Estuaries 19:639–650.

Haamer, J., A. S. Holm, L. Edebo, O. Lindahl, F. Norén, and B. Hernroth. 1999. Farming mussels strategically to recycle nutrients and create balance in the ecosystem. A review of knowledge and suggestions for action. Fiskeriverket Rapport 6:5–29.

Hall, S. J. and M. J. C. Harding. 1997. Physical disturbance and marine benthic communities: The effects of mechanical harvesting of cockles on non-target benthic infauna. Journal of Applied Ecology 34:497–517.

Haven, D. and R. Morales-Alamo. 1970. Filtration of particles from suspension by the American oyster, *Crassostrea virginica*. Biological Bulletin 139:248–264.

Jennings, A. C. and M. J. Kaiser. 1998. The effects of fishing on marine ecosystems. Advances in Marine Biology 34:209–224.

Kaiser, M. J., D. B. Edwards, and B. E. Spencer. 1996. A study of the effects of commercial clam cultivation and harvesting on benthic infauna. Aquatic Living Resources 9:57–63.

Kaiser, M. J., I. Lang, S. D. Utting, and G. M. Burnell. 1998. Environmental impacts of bivalve mariculture. Journal of Shellfish Research 17:59–66.

Kobayashi, M., E. E. Hoffman, E. N. Powell, J. M. Klinch, and K. Kusaka. 1997. A population dynamics model for the Japanese oyster, *Crassostrea gigas.* Aquaculture 149:285–321.

López-Jamar, E., J. Inglesias, and J. J. Otero. 1984. Contribution of infauna and mussel-raft epifauna to demersal fish diets. Marine Ecology-Progress Series 15:13–18.

Mann, R. 2000. Restoring the oyster reef communities in the Chesapeake Bay: A commentary. Journal of Shellfish Research 19:335–339.

Mohlenberg, F. and H. U. Riisgard. 1979. Filtration rate, using a new indirect technique, in thirteen species of suspension feeding bivalves. Marine Biology 54:1 43–147.

Mojica, R. and W. Nelson. 1993. Environmental effects of hard clam (*Mercenaria mercenaria*) aquaculture in the Indian River Lagoon, Florida. Aquaculture 29:135–139.

Nelson, T. C. 1983. The feeding mechanism of the oyster. I. On the pallium and brachial chambers of *Ostrea virginica, O. edulis,* and *O. angulata.* Journal of Morphology 63:1–61.

Newell, R. I. E. 1988. Ecological changes in the Chesapeake Bay: Are they the result of overharvesting the American oyster, *Crassostrea virginica*? Pages 536–546 *in* Understanding the estuary: Advances in Chesapeake Bay research, conference proceedings. Chesapeake Research Consortium, Publication No. 29, Baltimore, Maryland, USA.

Nugues, M. M., M. J. Kaiser, B. E. Spencer, and D. B. Edwards. 1996. Benthic community changes associated with intertidal oyster cultivation. Aquaculture Research 27:913–924.

PCSGA (Pacific Coast Shellfish Growers Association). 2001. Environmental policy statement. PCSGA, Olympia, Washington, USA.

Peterson C. H. and R. Black. 1991. Preliminary evidence for progressive sestonic food depletion in an incoming tide over a broad tidal sand flat. Estuarine, Coastal, and Shelf Science 32:405–413.

Peterson, B. J. and K. L. Heck, Jr. 1999. The potential for suspension feeding bivalves to increase seagrass. Journal of Experimental Marine Biology and Ecology 240:37–52.

Peterson, B. J. and K. L. Heck, Jr. 2001a. Positive interactions between suspension-feeding bivalves and seagrass—a facultative mutualism. Marine Ecology Progress Series 213:143–155.

Peterson, B. J. and K. L. Heck, Jr. 2001b. An experimental test of the mechanism by which suspension feeding bivalves elevate seagrass productivity. Marine Ecology Progress Series 218:115–125.

Powell, E. N., E. E. Hoffman, and S. M. Rays. 1992 Modeling oyster populations. I. A commentary on filtration rate. Is faster always better? Journal of Shellfish Research 11:387–398.

Pregnall, M. 1993. Regrowth and recruitment of eelgrass (*Zostera marina*) and recovery of benthic community structure in areas disturbed by commercial oyster culture in the South Slough Estuarine Research Reserve, Oregon. Master's thesis. Bard

College, Annandale-on-Hudson, New York, USA.

Rice, M. A. 2001. Environmental impacts of shellfish aquaculture: Filter feeding to control eutrophication. Pages 1–12, *in* H. O. Halvorson, editor. Proceedings of the Workshop on Aquaculture and the Marine Environment. Northeast Rhode Island Agricultural Experimentation Station and Rhode Island Cooperative Extension Service. Publication No. 3861, Kingston, Rhode Island, USA.

Ropert, M. and P. Goulletquer. 2000. Comparative physiological energetics of two suspension feeders: Polychaete annelid *Lanice conchilega* (Pallus 1766) and Pacific cupped oyster *Crassostrea gigas* (Thunberg 1795). Aquaculture 181:171-189.

Shpigel, M., A. Neor, D. M. Popper, and H. Gordin. 1993. A proposed model for "environmentally clean" land-based culture of fish, bivalves and seaweeds. Aquaculture 117:115–128.

Shumway, S. E. and T. L. Cucci. 1986. The effects of the toxic dinoflagellate *Protogonyaulax tamarensis* on the feeding and behavior of bivalve mollusks. Aquatic toxicology 10:9–27.

Shumway, S. E. and T. L. Cucci. 1996 Natural environmental factors. Pages 467–513 *in* V. S. Kennedy, R. I. E. Newell, and A. F. Eble, editors. The Eastern oyster. *Crassostrea virginica*. Maryland Sea Grant College, College Park, Maryland, USA.

Simenstad, C. and K. Fresh. 1995. Influence of intertidal aquaculture on benthic communities in Pacific Northwest estuaries: Scales of disturbance. Estuaries 18:43–70.

Spencer, B. E., M. J . Kaiser, and D. B. Edwards. 1996. The effects of Manila clam cultivation on an intertidal benthic community: The early cultivation phase. Aquaculture Research 27:261–276.

Spencer, B. E., M. J. Kaiser and D. B. Edwards. 1997. Ecological effects of intertidal Manila clam cultivation: Observations at the end of the cultivation phase. Journal of Applied Ecology 34:444–452.

Spencer, B. E., M. J. Kaiser and D. B. Edwards. 1998. Intertidal clam harvesting: Benthic community change and recovery. Aquaculture ecological effects of intertidal Manila clam cultivation: Observations at the end of the cultivation phase. Aquaculture Research 29:429–437.

Suchanek, T. H. 1992. Extreme biodiversity in the marine environment : Mussel bed communities of *Mytilus californianus*. The Northwest Environmental Journal 8:150–152.

Tenore, K. and N. Gonzalez. 1976. Food chain patterns in the Ria de Arosa, Spain, an area of intense mussel aquaculture. Pages 601–609 *in* G. Persoone and E. Jaspers, editors. Population dynamics, volume 2. Universa Press, Ostend, Belgium.

Thompson, R. J. and B. L. Bayne. 1974. Some relationships between growth, metabolism, and food in the mussel, *Mytilus edulis*. Marine Biology 27:317–326.

Toba, D., D. S. Thompson, K. K. Chen, G. J. Anderson, and M. B. Miller. 1992. Guide to Manila clam culture in Washington. Washington Sea Grant Program, University of Washington, Seattle, Washington, USA.

Trianni, M. S. 1995. The influence of commercial aquaculture activities on the benthic infauna of Arcata Bay. Master's thesis. Humboldt State University, Arcata, California, USA.

Vahl, O. 1973. Pumping and oxygen consumption rate in *Mytilus edulis L.*, of different sizes. Ophelia 12:45–52.

Wyllie-Echeverria, S, A. M. Olsen, and M. J. Hershman, editors. 1994. Seagrass science and policy in the Pacific Northwest: Proceedings of a seminar series (SMA 94-1, EPA 910/R-94-004). U.S. Environmental Protection Agency, Cincinnati, Ohio, USA.

Manipulations of Diets and Feeding to Reduce Losses of Nutrients in Intensive Aquaculture

DELBERT M. GATLIN III

Department of Wildlife and Fisheries Sciences, Texas A&M University System, 2258TAMUS, College Station, Texas 77843-2258 USA

RONALD W. HARDY

Hagerman Fish Culture Experiment Station, University of Idaho, Hagerman, Idaho 83332 USA

ABSTRACT

Potential negative impacts of aquaculture effluents due to the contribution of nutrients such as phosphorus, nitrogen, and organic matter of dietary origin have received considerable attention in recent years. In an effort to reduce nutrient enrichment, a variety of nutritional manipulations have been developed to minimize waste production and maximize nutrient utilization and growth efficiency of fish. Many such manipulations have enhanced the economical competitiveness and environmental sustainability of commercial aquaculture. This paper reviews various nutritional strategies including advancements in diet formulation, ingredient processing, feed manufacturing, and feeding strategies that have contributed substantially to reducing the excretion of enriching nutrients and thus enhancing nutrient utilization and production efficiency in aquaculture.

Nutrient losses by fish in aquaculture systems have received considerable attention in recent years due to the negative effects these losses may have on organisms within the culture systems as well as on receiving waters. Interest in this area has heightened as various sectors of the aquaculture industry have expanded and intensified production (Goddard 1996). Aquacultural production in culture systems such as cages, net-pens, and raceways has received the most attention due to the ready exchange of culture water with the surrounding environment that can lead to eutrophication and sedimentation (Black 2001). Aquaculture in pond systems where water exchange and effluents are typically low has received less attention until recently. For all semi-intensive and intensive aquaculture production systems, feeds are the source of nutrients added to the systems (inputs), and metabolic and fecal wastes are the outputs (aside from uneaten feed). Thus, all strategies aimed at reducing

the environmental effects of aquaculture production involve analysis of inputs and outputs, and developing methods or approaches to reduce outputs through improving inputs.

Phosphorus and nitrogen of dietary origin are two primary excretory products of aquacultured organisms which are of concern because they may contribute to eutrophication of aquatic systems (Wiesmann et al. 1988; Cowey and Cho 1991; Dosdat et al. 1995). Dietary phosphorus that is not absorbed and deposited in body tissues of cultured fish is released into the water where it may influence primary productivity (Ketola 1985; Ketola and Harland 1993). Dietary protein that is not digested or metabolized for energy rather than for protein accretion by fish results in increased nitrogen excretion, primarily in the form of ammonia. Growth and health of fish can be adversely affected by relatively low concentrations of ammonia or its nitrification product nitrite. Thus, nitrogenous waste products not only may contribute to eutrophication (Handy and Poxton 1993), but may have detrimental effects on cultured fish. Additionally, fish produce organic wastes from undigested components of the diet, which contribute to the biochemical oxygen demand of aquaculture systems (R. I. Johnsen et al. 1993; Kelly and Karpinski 1994) as well as settleable solids in effluent waters (Pillay 1992; Midlen and Redding 1998).

Because waste products associated with aquaculture systems are primarily of dietary origin, manipulation of diet formulations and feeding strategies to increase the assimilation and retention of nutrients by the cultured organism has provided substantial opportunities to enhance production efficiency. This paper will review advancements that have been made in reducing nutrient excretions from fish diets by means of improving diet formulations, ingredient processing, feed manufacturing, and feeding strategies. These advancements will be considered separately for phosphorus, nitrogen, and organic matter, the primary metabolic products of dietary origin.

Discussion

Phosphorus

Phosphorus is a mineral that is generally the most limiting nutrient in freshwater aquatic systems and thus influences primary productivity. It is required at relatively high concentrations in the diets of aquatic organisms to satisfy various metabolic needs (Lall 1991). Phosphorus serves as an inorganic structural component of hard tissues such as bone, teeth, and scales, as well as a constituent of various organic biochemicals including coenzymes, phospholipids, and nucleic acids. Dietary requirements for phosphorus have been reported to range rather widely between 0.3 and 0.9% of diet for various fish species and between 0.3 and 2% of diet for crustaceans such as penaeid shrimp (Gatlin 2000). There is some evidence that fish having scales have a higher dietary phosphorus requirement than do scaleless fish, such as channel catfish *Ictalurus punctatus*.

Reevaluation of minimum dietary phosphorus requirements of aquatic organisms has received considerable attention in recent years because dietary phosphorus in excess of metabolic requirements is readily excreted by the organism, principally in the urine, and thus can directly contribute to eutrophication. In many of these studies, detailed information on phosphorus metabolism and excretion has been obtained to assist in defining minimal requirements (Skonberg et al. 1997; Sugiura et al. 2000a, 2000b).

Satisfying minimal metabolic requirements of aquatic organisms for phosphorus is complicated in that the availability of phosphorus in diet ingredients to fish can vary considerably (NRC 1993; Sugiura and Hardy 2000). Phosphorus that is not absorbed from the diet into the gastrointestinal tract is excreted in the feces and thus unavailable to the organism. Such factors as chemical form of phosphorus, its interactions with other elements such as calcium, and pH of the gastrointestinal tract are known to influence phosphorus availability (Lall 1991; Sugiura et al. 1998a). Phytin phosphorus or phytate is the principal organic form of phosphorus in feedstuffs, and comprises approximately 67% of the phosphorus in plant ingredients. This form of phosphorus is not readily available to monogastric animals including fish and crustaceans because of their lack of phytase, the enzyme required to liberate phosphorus from phytate. Supplementation of microbial phytase to the diet has been shown to increase the availability of phosphorus to several fish species (Cain and Garling 1995; Eya and Lovell 1997; Li and Robinson 1997; Sugiura et al., in press). However, the instability of microbial phytase to the heat normally encountered in extrusion processing of diets has restricted its use in diet formulations for fish and crustaceans. Nevertheless, development of low-temperature manufacturing techniques and application of phytase after extrusion has provided increased opportunities for its use in the diets of aquatic species. The availability of phosphorus in feedstuffs of animal origin as well as inorganic supplements such as dicalcium phosphate or defluorinated rock phosphorus also can vary considerably among fish (NRC 1993) and crustaceans (Davis and Arnold 1994). Information on phosphorus availability of various feedstuffs to fish and crustaceans has increased in recent years to allow more precise formulation of diets to meet minimal metabolic requirements of fish while limiting excess and unavailable phosphorus (Li and Robinson 1996; Sugiura et al. 1998b; Sugiura and Hardy 2000; Weerasinghe et al. 2001). Increasing the incorporation of ingredients containing highly available or low concentrations of phosphorus in diet formulations has become an appealing means of limiting phosphorus excretion in aquaculture.

In most instances, the availability of phosphorus in diets can be calculated from data on the ingredients used to produce the feeds, but in some instances, phosphorus apparent availability coefficients are not additive. This is thought to be due to antagonistic interactions among feed ingredients that lower the availability of dietary phosphorus to fish and other animals. One example is the interaction between dietary calcium level and phosphorus availability, whereby the addition of high-ash (calcium) fish meals in feeds lowers the availability of phosphorus (Sugiura et al. 2000c). Calcium in combination with phytate also reduces the availability of phosphorus in some diet combinations. This antagonistic interaction can be overcome by adding citric acid to the diet to lower the pH and to provide a chelate to prevent the formation of calcium-phytate complexes in the gut (Sugiura et al. 1998a). Understanding and avoiding (or

compensating for) antagonistic interactions among feed ingredients is critical to efforts to improve the availability of phosphorus in fish diets.

Some investigations have shown that the metabolic requirement for phosphorus is greatest in young, rapidly growing organisms (Rodehutscord 1996; Eya and Lovell 1997; Sugiura et al. 2000b). Thus, altering diet formulations and feeding strategies at later life stages also may provide opportunities to limit phosphorus excretion. For example, salmonids maintain a reserve of phosphorus in hard tissues that can be depleted over short periods to compensate for inadequate dietary intake (Hardy et al. 1993). No signs of clinical deficiency or reduced fish performance are evident until body phosphorus reserves are reduced to critical levels. In fish for which there are well-defined production cycles, such as rainbow trout *Oncorhynchus mykiss*, dietary phosphorus levels can be purposefully reduced below required levels during the last stages of production before harvest without reducing fish growth or efficiency, or lowering product quality (Lellis et al., in press). Fish draw upon their tissue phosphorus reserves to meet metabolic needs. As long as fish do not reach critical body phosphorus levels during this period of depletion, no effect on production is noted. This strategy has been demonstrated to reduce the total amount of excreted phosphorus during a production cycle by over 35%, but care must be taken to avoid excessive tissue phosphorus depletion.

Nitrogen

Nitrogen derived from dietary protein represents another enriching nutrient of concern in aquaculture. Nitrogenous wastes are largely dietary in origin with estimates of up to 52–95% of feed nitrogen ending up as waste in the culture environment (Wu 1995). Adequate concentrations of high-quality protein are required in the diet to support rapid growth of fish and crustaceans. Quantitative requirements for dietary protein have been shown to vary widely among aquatic species, but are generally much higher than those of terrestrial animals (NRC 1993). A general relationship between natural feeding habits and dietary protein requirements has been observed among aquatic species such that the more carnivorous species tend to have higher protein requirements compared to omnivorous and herbivorous species (NRC 1993). This is largely attributed to differences in the ability of using dietary carbohydrate for energy. For example, salmonids are typically fed diets containing protein at 40 to 50% of dry weight; whereas, diets for channel catfish typically contain protein at 28 to 36% of dry weight. Catfish are much more able to utilize cooked starch for energy than are salmonids.

The quantity and quality of dietary protein are factors that influence nitrogen excretion. Protein quality is largely determined by a feedstuff's amino acid composition and digestibility. Dietary protein that is not digested into constituent amino acids and absorbed in the gastointestinal tract is readily excreted in the feces. However, the major (75–90%) nitrogenous waste produced in fish is ammonia which is primarily excreted via the gills (Kaushik and Cowey 1991). Ammonia is produced by fish during the catabolism of amino acids. Elevated ammonia production by fish has been readily observed when the concentration of dietary protein is excessive relative to non-protein energy, such that a portion of dietary protein not used for protein accretion is broken down and used for energy. For example, Ballestrazzi et al. (1994) found that ammonia excretion of European sea bass *Dicentrarchus labrax* increased linearly with protein level of the diet. Médale et al. (1995) also found that higher digestible protein relative to digestible energy resulted in increased ammonia excretion in rainbow trout.

Increasing the energy density of diets, particularly by increasing the lipid concentration, has been shown to be an effective strategy to spare dietary protein and limit ammonia production of fish (Kaushik and Cowey 1991). Such responses have been observed for several fish including rainbow trout (Kaushik and de Oliva Teles 1985), carp species (Steffens 1996), Atlantic salmon *Salmo salar* (F. Johnsen et al. 1993), turbot *Scopthalmus maximus* (Andersen and Alsted 1993), lake trout *Salvelinus namaycush* (Jayaram and Beamish 1992), gilthead sea bream *Sparus aurata* (Vergara et al. 1996), and red drum *Sciaenops ocellatus* (McGoogan and Gatlin 1999, 2000). Including higher dietary lipid levels relative to protein generally increases protein sparing but with a concomitant increase in body lipid levels. One of the best examples of increasing energy density of diets by lipid supplementation to enhance growth and protein retention has been seen in the production of Atlantic salmon (Hardy 1999a). Diet formulations for this species have evolved in recent years to include up to 35% lipid; they contained <20% lipid in the late 1980s. Protein retention values over this period have increased from 20–25% to 45% or more as a result of higher dietary lipid levels and the use of higher quality protein sources.

Refinement of diet formulations by including ingredients with high protein digestibility has been used to increase protein retention of fish and thus limit nitrogen excretion. Such manipulations of diet formulations have been augmented by studies of the digestibility and nutritional value of various feed ingredients (Hardy 1999b). However, some highly digestible protein sources, e.g., fish hydrolysates, do not result in higher protein retention due to asynchronous absorption of amino acids from the hydrolysates and intact proteins in diets that lead to higher rates of amino acid catabolism (Stone and Hardy 1989). Cho et al. (1994)

explained that the most beneficial approach to increasing nutrient density of diets generally involves excluding ingredients with low protein and energy contents that are often poorly digested. This strategy was evaluated with red drum and observed to have positive effects on water quality of a closed recirculating system (Jirsa et al. 1997).

In addition to ingredient composition of diets, manufacturing procedures also may influence nutrient digestibility of diets. For example, the use of extrusion processing in the manufacture of fish diets generally has resulted in increased digestibility of diets and reduced ammonia excretion (NRC 1993). This is principally because the heat and pressure associated with extrusion processing improves the digestibility of the protein and carbohydrate fractions and thus increases the digestible energy of the diet. Further advancements in processing techniques for feedstuffs as well as diet formulations are currently being explored to enhance protein digestibility and thus reduce nitrogen excretion.

Another important factor that may influence the utilization of various diet formulations is feeding strategies. As fish typically regulate feed intake to meet an energy need (NRC 1993), the feeding of high-energy diets may reduce intake, which may contribute to a decrease in ammonia production. If feed intake is severely restricted, protein intake may be too low to support rapid growth rates. Therefore, increasing the percentage of dietary protein in conjunction with an increase in energy may help to provide amino acids at levels needed for maximal growth even if feeding activity is reduced (Cho and Bureau 1997). Such interactions between diet formulations and feeding practices have been demonstrated with a number of species including salmonids (Cho and Bureau 1997) and red drum (McGoogan and Gatlin 2000). For other

species such as the channel catfish, there has been a trend of reducing dietary protein concentrations, which does not adversely affect growth as long as feed intake is not limited (Robinson and Li 1997).

Organic Matter

Other components of the diet besides phosphorus and protein may be of concern with regard to waste production in aquaculture. Various organic constituents of the diet besides protein and lipid, such as soluble and fibrous carbohydrates, become fecal waste products when not digested by fish. These organic wastes will contribute to the biochemical oxygen demand of aquaculture systems (R. I. Johnsen et al. 1993; Kelly and Karpinski 1994) as well as settleable solids in effluent waters (Pillay 1992; Midlen and Redding 1998). The primary means of limiting these organic wastes in aquaculture has been to formulate diets using ingredients with high nutrient digestibility. Another approach with some terrestrial animals had been the supplementation of diets with microbial enzymes such as amylases and proteases to facilitate the digestion of carbohydrate and protein, respectively, and hence improve nutrient utilization (Lobo 1999). However, at this time such enzyme supplements have been evaluated only to a limited extent with fish.

*Accomplishments to Date and Outlook
for the Future*

Impressive reductions in the amount of phosphorus excreted by fish have been achieved over the past decade by reducing the level of total phosphorus in the diet, and by formulating the diet so that the amount of available phosphorus closely matches the metabolic requirements of the fish. For example, rainbow trout diets used in the USA in 1990 typically contained approximately 2% total phosphorus, of which about 65% was available (~1.4% available phosphorus) (Table 1). The emphasis in trout feed formulation at that time was to ensure that sufficient levels of available phosphorus were present in the diet. Diets contained high levels of available phosphorus because the dietary requirement was not precisely understood, available phosphorus levels in feed ingredients were not known, and there was no information or even awareness of antagonistic interactions among feed ingredients in diets. It is now known that trout require between 0.55 and 0.7% available phosphorus, depending upon their size. Trout diets in 1990 typically contained twice as much available phosphorus as the fish actually required. Most of this excess available phosphorus was absorbed and excreted via the urine as soluble phosphates, which are virtually impossible to remove except by plants. Fecal phosphorus excretion by trout accounted for the balance (about 0.8% of total dietary phosphorus) (Fig. 1). Today, trout feeds are formulated to contain 1.1–1.2% total phosphorus, of which 0.7–0.9% is available (Table 1). Hence, urinary losses have been reduced by 70%, and fecal losses by 50% (Fig. 1). Further reduction of phosphorus excretion must be achieved by increasing the availability of dietary phosphorus, and simultaneously reducing the total phosphorus levels in diets. The trend toward higher use of plant protein sources in fish diets to replace fish meal will increase the level of phytate in diets, making it necessary to utilize phytase, or possibly use low-phytate grains (Sugiura et al. 1999).

Reducing nitrogen losses from aquaculture

TABLE 1. *Comparison of typical rainbow trout feed formulations from 1990 to low-phosphorus formulation used in 2000.*

	1990 formulation	2000 formulation
	Percent of diet	
Fish meal, menhaden	38.0	0.0
Fish meal, herring, low ash	0.0	40.0
Blood meal	2.0	5.0
Feather meal	0.0	4.0
FAQ poultry by-product meal	4.0	0.0
Low-ash poultry by-product meal	0.0	5.0
Soybean meal	5.0	5.0
Meat and bone meal	4.0	0.0
Wheat middlings	30.0	0.0
Ground whole wheat	8.0	21.5
Fish oil	6.5	17.0
Vitamin/mineral supplements	2.5	2.5
TOTAL	100.0	100.0
Proximate/Chemical Composition (%, as-fed basis)		
Crude protein	38.0	44.0
Crude lipid	12.0	22.0
Ash	13.0	8.5
Phosphorus, total	2.0	1.2

Phosphorus balance, 1990 trout feed

A

Phosphorus balance, 2000 trout feed

B

FIGURE 1. *Phosphorus balance of rainbow trout as influenced by typical diets produced in 1990 (A) and 2000 (B).*

is more problematic than reducing phosphorus because fish, like all animals, have an upper limit on the amount of dietary proteins (amino acids) that are incorporated into tissue proteins, rather than catabolized. Tissue proteins are in a constant state of flux, being continually degraded and resynthesized. Nitrogen is excreted as a result of this protein turnover. Anabolic hormones lower the rate of turnover, thus lowering the rate of nitrogen excretion, suggesting that it is possible to achieve approximately 70% protein retention, but administration of anabolic hormones for this purpose is not anticipated in aquacultural production of food fish. Given the fact that protein retention has doubled in Atlantic salmon farming (from ca. 22% to 45%), it is likely that protein retention values can be increased in other species of farmed fish.

Relatively large improvements were made from 1990 to 2000 in protein retention of rainbow trout (Fig. 2) primarily by altering diet formulations (Table 1). Other strategies for increasing protein retention include balancing dietary amino acid levels, perhaps in a factorial manner that changes with life history stage or size of fish, optimizing dietary protein (and amino acids) and energy levels, and using enzyme supplements to facilitate digestion of less available nutrients in various feedstuffs.

Reducing organic material excreted by fish basically revolves around increasing the digestibility of feed ingredients, especially those of plant origin. As mentioned earlier, feed processing technology will be a key in this effort, as will the use of enzyme supplements to facilitate digestion of starches and cellulose components of plant-derived feed ingredients.

Nitrogen balance, 1990 trout feed

A

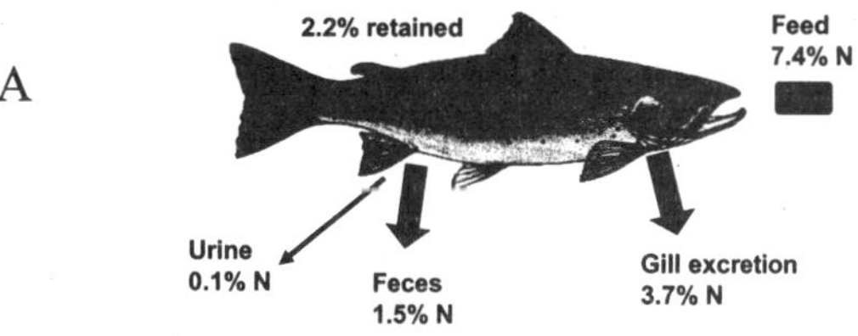

Nitrogen balance, 2000 trout feed

B

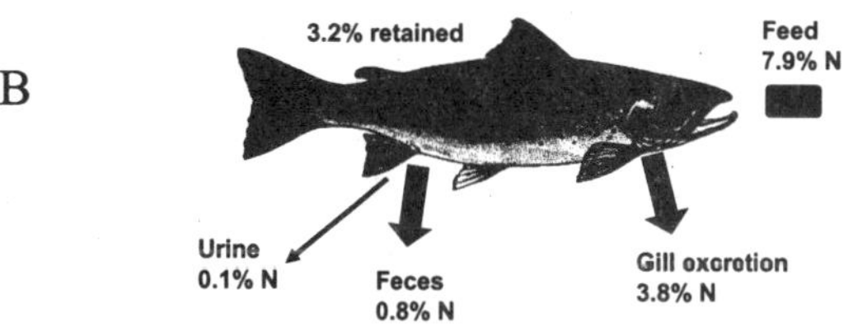

FIGURE 2. *Nitrogen balance of rainbow trout as influenced by typical diets produced in 1990 (A) and 2000 (B).*

Literature Cited

Andersen, N. G. and N. S. Alsted. 1993. Growth and body composition of turbot (*Scopthalmus maximus* (L.)) in relation to different lipid/protein ratios in the diet. Pages 479–491 *in* S. J. Kaushik and P. Luquet, editors. Fish nutrition in practice. INRA, Paris, France.

Ballestrazzi, R., D. Lanari, E. D'Agaro, and A. Mion. 1994. The effect of dietary protein level and source on growth, body composition, total ammonia and reactive phosphate excretion of growing sea bass (*Dicentrarchus labrax*). Aquaculture 127:197–206.

Black, K. D. 2001. Environmental impacts of aquaculture. CRC Press, Boca Raton, Florida, USA.

Cain, K. D. and D. L. Garling. 1995. Pretreatment of soybean meal with phytase for salmonid diets to reduce phosphorus concentrations in hatchery effluents. The Progressive Fish-Culturist 57:114–119.

Cho, C. Y. and D. P. Bureau. 1997. Reduction of waste output from salmonid aquaculture through feeds and feeding. The Progressive Fish-Culturist 59:155–160.

Cho, C. Y., J. D. Hynes, K. R. Wood, and H. K. Yoshida. 1994. Development of high nutrient-dense, low pollution diets and prediction of aquaculture wastes using biological approaches. Aquaculture 124:293–305.

Cowey, C. B. and C. Y. Cho, editors. 1991. Nutritional strategies and aquaculture waste. Proceedings of the First International Symposium on Nutritional Strategies in Management of Aquaculture Wastes. University of Guelph, Guelph, Ontario, Canada.

Davis, D. A. and C. R. Arnold. 1994. Estimation of apparent phosphorus availability from inorganic phosphorus sources for *Penaeus vannamei*. Aquaculture 127:245–254.

Dosdat, A., F. Gaumet, and H. Chartois. 1995. Marine aquaculture effluent monitoring: methodological approach to the evaluation of nitrogen and phosphorus excretion by fish. Aquaculture Engineering 14:59–84.

Eya, J. C. and R. T. Lovell. 1997. Net absorption of dietary phosphorus from various inorganic sources and effect of fungal phytase on net absorption of plant phosphorus by channel catfish *Ictalurus punctatus*. Journal of the World Aquaculture Society 28:386–391.

Gatlin, D. M. III. 2000. Minerals. Pages 532–540 *in* R. R. Stickney, editor-in-chief. Encyclopedia of aquaculture. John Wiley & Sons, Inc., New York, New York, USA.

Goddard, S. 1996. Feed management in intensive aquaculture. Chapman & Hall, London, UK.

Handy, R. D. and M. G. Poxton. 1993. Nitrogen pollution in mariculture: toxicity and excretion of nitrogenous compounds by marine fish. Review in Fish Biology and Fisheries 3:205–241.

Hardy, R. W. 1999a. Problems and opportunities in fish feed formulation. Aquaculture Magazine 25:56–60.

Hardy, R. W. 1999b. Alternative protein sources. Feed Management 50:25–28.

Hardy, R. W., W. T. Fairgrieve, and T. M. Scott. 1993. Periodic feeding of low-phosphorus diet and phosphorus retention in rainbow trout (*Oncorhynchus mykis*). Pages 403–412 *in* S. J. Kaushik and P. Luquet, editors. Fish nutrition in practice. INRA, Paris, France.

Jayaram, M. G. and F. W. H. Beamish. 1992. Influence of dietary protein and lipid on nitrogen and energy losses in lake trout, *Salvelinus namaycush*. Canadian Journal of Fisheries and Aquatic Science 49:2267–2272.

Jirsa, D. O., D. A. Davis, and C. R. Arnold. 1997. Effects of dietary nutrient density on water quality and growth of red drum *Sciaenops ocellatus* in closed systems. Journal of the World Aquaculture Society 28:68–78.

Johnsen, F., M. Hillestad, and E. Austreng. 1993. High energy diets for Atlantic salmon, Effects on pollution. Pages 391–401 *in* S. J. Kaushik and P. Luquet, editors. Fish nutrition in practice. INRA, Paris, France.

Johnsen, R. I., O. Grahl-Nielsen, and B. T. Lunestad. 1993. Environmental distribution of organic waste from a marine fish farm. Aquaculture 118:229–244.

Kaushik, S. J. and C. B. Cowey. 1991. Dietary factors affecting nitrogen excretion by fish. Pages 3–19 *in* C. B. Cowey and C. Y. Cho, editors. Nutritional strategies and aquaculture waste. Proceedings of the First International Symposium on Nutritional Strategies in Management of Aquaculture Wastes. University of Guelph, Guelph, Ontario, Canada.

Kaushik, S. J. and A. de Oliva Teles. 1985. Effect of digestible energy on nitrogen and energy balance in rainbow trout. Aquaculture 50:89–101.

Kelly, L. A. and A. W. Karpinski. 1994. Monitoring BOD outputs from land-base fish farms. Journal of Applied Ichthyology 10:368–372.

Ketola, H. G. 1985. Mineral nutrition: Effects of phosphorus in trout and salmon feeds in water pollution. Pages 465–473 *in* C. B. Cowey, A. M. Mackey, and J. G. Bell, editors. Nutrition and feeding in fish. Academic Press, New York, New York, USA.

Ketola, H. G. and B. F. Harland. 1993. Influence of phosphorus in rainbow trout diets on phosphorus discharges in effluent water. Transactions of the American Fisheries Society 122:1120–1126.

Lall, S. P. 1991. Digestibility, metabolism, and excretion of dietary phosphorus in fish. Pages 21–36 *in* C. B. Cowey and C. Y. Cho, editors. Nutritional strategies and aquaculture waste. Proceedings of the First International Symposium on Nutritional Strategies in Management of Aquaculture Wastes. University of Guelph, Guelph, Ontario, Canada.

Lellis, W. A., F. T. Barrows, F. M. Dong, and R. W. Hardy. In press. Development of low-phosphorus finishing diets for rainbow trout (*Oncorhynchus mykiss*). Aquaculture Research.

Li, M. H. and E. H. Robinson. 1996. Phosphorus availability of common feedstuffs to channel catfish *Ictalurus punctatus* as measures by weight gain and bone mineralization. Journal of the World Aquaculture Society 27:297–302.

Li, M. H. and E. H. Robinson. 1997. Microbial phytase can replace inorganic phosphorus supplements in channel catfish *Ictalurus punctatus* diets. Journal of the World Aquaculture Society 28:402–406.

Lobo, P. 1999. Using enzymes to enhance corn and soy-based broiler diets. Feed Management 50:17–20.

McGoogan, B. B. and D. M. Gatlin III. 1999. Dietary manipulations affecting growth and nitrogenous waste production of red drum, *Sciaenops ocellatus*. I. Effects of dietary protein and energy levels. Aquaculture 178:333–348.

McGoogan, B. B. and D. M. Gatlin III. 2000. Dietary manipulations affecting growth and nitrogenous waste production of red drum, *Sciaenops ocellatus*. II. Effects of energy level and nutrient density at various feeding rates. Aquaculture 182:271–285.

Médale, F., C. Brauge, F. Vallée, and S. J. Kaushik. 1995. Effects of dietary protein/energy ratio, ration size, dietary energy source and water temperature on nitrogen excretion in rainbow trout. Water Science and Technology 31:185–194.

Midlen, A. and T. Redding. 1998. Environmental management for aquaculture. Chapman & Hall, London, UK.

NRC (National Research Council). 1993. Nutrient requirements of fish. National Academy Press, Washington, D.C., USA.

Pillay, T. V. R. 1992. Aquaculture and the environment. John Wiley and Sons, Inc., New York, New York, USA.

Robinson, E. H. and M. H. Li. 1997. Low protein diets for channel catfish *Ictalurus punctatus* raised in earthen ponds at high density. Journal of the World Aquaculture Society 28:224–229.

Rodehutscord, M. 1996. Response of rainbow trout (*Oncorhynchus mykiss*) growing from 50 to 200 g to supplements of dibasic sodium phosphate in a semipurified diet. Journal of Nutrition 126:324–331.

Skonberg, D. I., L. Yogev, R. W. Hardy, and F. M. Dong. 1997. Metabolic response to dietary phosphorus intake in rainbow trout (*Oncorhynchus mykiss*). Aquaculture 157:11–24.

Steffens, W. 1996. Importance and benefit of using lipids in fish nutrition. Fett/Lipid 9:292–299.

Stone, F. E. and R. W. Hardy. 1989. Plasma amino acid changes in rainbow trout (*Salmo gairdneri*) fed freeze-dried fish silage, liquefied fish, and fish meal. Pages 419–426 *in* Proceedings of the 1988 Aquaculture International Congress. Vancouver, British Columbia, Canada.

Sugiura, S. H. and R. W. Hardy. 2000. Environmentally friendly feeds. Pages 299–310 *in* R. R. Stickney, editor-in-chief. Encyclopedia of aquaculture. John Wiley & Sons, Inc., New York, New York, USA.

Sugiura, S. H., J. K. Babbitt, F. M. Dong, and R. W. Hardy. 2000a. Utilization of fish and animal by-product meals in low-pollution feeds for rainbow trout, *Oncorhynchus mykiss* (Walbaum). Aquaculture Research 31:585–593.

Sugiura, S. H., F. M. Dong, and R. W. Hardy. 2000b. A new approach to estimating the minimum dietary requirement of phosphorus for large rainbow trout based on nonfecal excretions of phosphorus and nitrogen. Journal of Nutrition 130:865–872.

Sugiura, S. H., F. M. Dong, and R. W. Hardy. 2000c. Primary responses of rainbow trout to dietary phosphorus concentrations. Aquaculture Nutrition 6:235–245.

Sugiura, S. H., F. M. Dong, and R. W. Hardy. 1999. Availability of phosphorus and trace elements in low-phytate varieties of barley and corn for rainbow trout (*Oncorhynchus mykiss*). Aquaculture 170:285–296.

Sugiura, S. H., F. M. Dong, and R. W. Hardy. 1998a. Effects of dietary supplements on the availability of minerals in fish meal; preliminary observations. Aquaculture 160:283–303.

Sugiura, S. H., F. M. Dong, C. K. Rathbone, and R. W. Hardy. 1998b. Apparent protein digestibility and mineral availabilities in various feed ingredients for salmonid feeds. Aquaculture 159:177–202.

Sugiura, S. H., J. Gabaudan, F. M. Dong, and R. W. Hardy. 2001. Dietary microbial phytase supplementation and the utilization of phosphorus, trace minerals and protein by rainbow trout [*Oncorhynchus mykiss (Walbaum)*] fed soybean meal-based diets. Aquaculture Research 32:583-592.

Vergara, J. M., L. Robainá, M. Izquierdo, and M. de la Higuera. 1996. Protein sparing effect of lipids in diets for fingerlings of gilthead sea bream. Fisheries Science 62:624–628.

Weerasinghe, V., R. W. Hardy, and N. F. Haard. 2001. An *in vitro* method to determine phosphorus digestibility of rainbow trout *Oncorhynchus mykiss* (Walbaum) feed ingredients. Aquaculture Nutrition 7:1–9.

Wiesmann, D., H. Scheid, and E. Pfeffer. 1988. Water pollution with phosphorus of dietary origin by intensively fed rainbow trout (*Salmo gairdneri* Rich.). Aquaculture 69:263–270.

Wu, R. S. S. 1995. The environmental impact of marine fish culture: towards a sustainable future. Marine Pollution Bulletin 3:159–166.

Genetic Implications of Escaped and Intentionally-Stocked Cultured Fishes

REGINAL M. HARRELL

University of Maryland, Wye Research and Education Center, Post Office Box 169,

Queenstown, Maryland 21658 USA

ABSTRACT

There exists a paradox in the way we view genetic variation between fish produced for food or ornamental purposes and the natural genetic diversity that exists with wild populations or those fish produced in hatcheries for stock enhancement or restoration. For more than two and a half decades there have been increasing concerns about genetic contamination of native fishes by escaped farm-reared fish, stock enhancement programs, restoration efforts, and stock transfer programs. Much of the information that drives these concerns come primarily from population genetics and conservation theory, both which are evolving. In many cases where empirical fisheries data exist, the information is focused on salmonids with their strong natal homing instincts and to endangered species where any action may be construed as detrimental to the continued viability of the population and may not be applicable across all species. Other supporting information about genetic contamination is derived from species and situations other than fish. I review some of the more germane and pronounced issues from the perspective of an aquaculturist with a genetics background as opposed to a geneticist with an aquaculture background. I present the information from a logical perspective that allows managers, producers, scientists, environmentalists, fishermen, and politicians to have an open dialogue as to the implications and ramifications of altering the gene-flow mechanisms of fish populations regardless of whether it is from an environmental, management, or a biological perspective.

"Paradox (*n*): 1. A seemingly contradictory statement that may nonetheless be true… 3. An assertion that is essentially self-contradictory, though based on a valid deduction from acceptable premises (AHDEL 1992)."

Why is this concept of "genetic pollution" considered a paradox, especially between aquaculture and fisheries management? It is a paradox because of conflicting goals and objectives. For instance, from a food-fish production perspective, selective breeding and lowering genetic diversity are often the goals, while in wild fisheries management or stock enhancement programs, maximizing diversity is one of the major goals. Genetic variation is the "energy" that drives evolutionary opportunities. It is also the basis for selection for a particular type. Yet, from a genetics perspective, genetic load slows the evolutionary process because of the variation of fitness.

Genetic load is a concept that encompasses the relative decrease in fitness of a population due to the presence of genotypes that have less than the highest fitness. It can also be expressed as the average number of lethal mutations per individual in a population (Doncaster 1995). Regardless of the specific goals and objectives, there is always the "stumbling block" of genetic loading that makes realization of goals difficult. This is true whether it is for food-fish production or for stocking a system with native fish for enhancement or restoration.

In aquaculture for food-fish production, a manager may actually inbreed his stock to eliminate or reduce genetic load, thereby getting rid of those "less than fit" genotypes not suited for the goals of the facility and thus increasing economic production and improving "farm" management. However selecting for one trait (i.e., fast growth) usually ends up being negatively correlated to another equally important trait (i.e., disease resistance, or temperament). This elimination or decrease in variation conversely may be poorly suited for management of wild populations.

The concept of "genetic contamination" can also involve conflicts between emotionalism, theory, the availability of the amount or type of corroborating empirical information, and what can be done about the situation. My purpose here is to address some of the issues in conservation genetics from the perspective of an aquaculturist with a genetics background as opposed to a geneticist with an aquacultural background. I will address this question of genetic pollution or contamination being a paradox throughout the context of this paper by focusing on conservation genetics in general and some of the more germane fisheries issues in particular.

When the biodiversity movement started in the late 20th century, "saving our planet" became a "war cry" of many well-intentioned individuals who had little real comprehension of what biodiversity (genetic diversity in this case) encompassed and who immediately became self-professed experts in environ-mental and population genetic issues per se. This is not to discount the fact that some very good science and scientists were, and are, the true driving forces of this discipline. However when emotionalism becomes involved it can, and has, obfuscated the truth, which can result in individuals with personal agendas inveigling the public with misinformation, misguided loyalties, and often, plain and simple ignorance of the facts. These problems led Daley (1993) to conclude that fisheries scientists are polarized on the issue with little to no middle ground.

It is true that biodiversity of the planet is rapidly being diminished (Kohm 1991; Soulé 1991); however, it is not due to one specific cause. In fact, the most commonly mentioned reasons for the decline include loss of habitat and overexploitation (Frankham 1999), not genetic pollution. As much as 34% of fish species worldwide are listed as threatened, vulnerable, or endangered (Primack 1998). Among all the causes of biodiversity loss, habitat destruction appears to be the most significant, albeit global climate change may supplant habitat loss in the future (Frankham 1999). Genetic factors (in our case genetic pollution) such as inbreeding and genetic drift are believed to increase the possibility of extinction in small populations (Gilpin and Soulé 1986) and dramatically accelerate biodiversity loss thereby providing a negative feedback (e.g., "the extinction vortex" Frankham 1999). However, Vrijenhoek (1994) felt that "Despite the potential significance of this hypothesis for conservation management, we lack evidence that inbreeding and genetic drift have caused an 'extinction vortex' in nature."

The Link To Conservation Genetics Theory

To understand why aquaculture is being targeted as a contributing cause of declining natural populations through genetic contamination of wild stocks by accidental or intentional releases of aquaculture products, it is important to discuss the concepts of conservation genetics. Thus at the heart of the genetic pollution issue for aquaculture is conservation genetics theory. The central problem is the loss of genetic variation resulting in erosion of evolutionary flexibility, which potentially leads to a poorer match of organism to environment, increasing the probability of extinction (Meffe 1986).

Basically, aquacultured fish and shellfish are touted as having a lower overall genetic diversity than conspecific wild populations due to the intentional or unintentional selection and/or domestication that occurs in a hatchery environment. Because they have a lowered overall genetic diversity, they are not as well suited for the natural environment as naturally adapted organisms (less fit).

Interestingly genetic variability is a double-edged sword. When considering genetic variability from a loading perspective, even in wild populations, fitness is always lower than maximum because of the presence of less fit individuals in the entire population. This is very difficult to track because part of the overall decline in fitness is due to the number of lethal mutations. Populations with a high number of lethal mutations normally do not follow Hardy-Weinberg equilibrium because many of these genes are expressed in the early developmental stages and therefore are never seen in the population. An example of genetic load that most people are familiar with is sickle-cell anemia. In this case those individuals heterozygous for the gene have a significant advantage in areas where malaria exits. Those homozygous recessive individuals die at an early age, while those homozygous dominant are susceptible to malaria. We have little knowledge of what similar types of scenarios exist in fisheries, but for lack of a better example, consider the case for albinism. The genes for albinism are essentially recessive (Tave 1993), and you rarely see albino fish in the wild because they are obvious to predators and become prime targets for a meal. Yet those genes are part of the overall genetic variability of a population. Thus from a wild fisheries perspective, a population with heterozygous individuals present with genes for albinism would be, in theory, less fit than another population without the genes present for albinism. Anecdotally, albino fish often bring a "value-added" price to an aquaculture market, and the ability to produce large numbers of pure albinos may be a target goal.

In addition to genetic variability, conservation genetics includes the concepts of fitness and domestication (Frankel 1974; Soulé and Wilcox 1980; Frankel and Soulé 1981; Schonewald-Cox et al. 1983; Soulé 1986; Hedrick and Miller 1992; Frankham 1995, 1999). While the first two concepts (variability and fitness) are directly applicable to wild populations, the latter is most commonly associated with animal or plant breeding (e.g., aquaculture). Until recently the latter also applied only to working with defined breeds and varieties. Interestingly, the area of genetics conservation from an animal breeding perspective actually predates conservation genetics as a discipline within genetics (Barker 1994). In fact, one of the earliest recognitions of the importance of conserving genetic diversity came from animal breeding researchers and was first addressed in 1946 at a FAO meeting in Copenhagen (Barker 1994). However, genetic pollution from aquaculture for food-fish and variety production has been brought to the forefront in two recent publications (Goldberg and Triplet 1997; and Goldberg et al. 2001). In both publications the authors include genetic concerns in the broad form of biological pollution and use Atlantic salmon *Salmo salar* as their model for discussion. There can be no deniability that escapes from aquaculture operations occur and that escaped fish can, and have, interbred with native fish. What is debatable is the genetic impact these escapees have on wild conspecific populations (some of this debate will be discussed below in the section on outbreeding depression). We have a lot of theory, some good information on salmonids and species groups other than fish, and laboratory studies all that support the issue of genetic impacts of captive or domesticated fish on wild populations. What we do not have, other than information from Atlantic and Pacific salmonids, is information on the long-term impacts of hatchery fish on wild populations of other species groups.

In general, while conserving genetic diversity is important for food-fish production and even ornamental fish production because it provides the raw material for future selection efforts, aquaculture (with the exception of aquaculture for stocking purposes) intentionally conflicts with maximizing genetic diversity. In fact, aquaculture for food-fish and variety production strives for directed or convergent selection (Table 1). As previously stated for these purposes, aquaculturists intentionally select for traits such as a specific body configuration with high fillet to carcass yields, certain temperaments for rearing in crowded environments, high food conversion efficiencies, and/or specific colorations and fin shapes. Much of this breeding does lead to domestication, domestication effects, and/or selection (discussed below), and it does have the potential for negative genetic implications if such fish escape into the wild as suggested by Goldberg and Triplet (1997) and Goldberg et al. (2001). However, the ultimate genetic consequences of escaped aquaculture animals interbreeding with wild animals is not clearly established.

I will not address genetic pollution from the context of breeding for production or variety improvement, as there are many aquaculture production texts on these subjects. Rather I will address the potential impact if these or even less stringently maintained animals were to be placed intentionally or unintentionally into natural and wild environments. I also will not discuss the issue of exotic or non-indigenous species intro-ductions and their potential impact from a genetic pollution perspective. For discus-sion of the latter, the reader is referred to Courtenay (1995).

Neutralist versus Selectionist

Genetic theory as it related to conser-vation biology is constantly evolving primarily due to our increasing understanding of genetic variability at the molecular level (Hedrick and Miller 1992). Yet, much of what one believes about the genetic aspects of conservation biology depends on whether you classify yourself as a Neutralist or a Selectionist. While considering the differing perspectives is really peripheral to genetic pollution from aquaculture operations, it is important to understand it from the perspective of how conservation-genetics theory fits together.

A Neutralist view of genetic variability is that genetic variation is being maintained by a balance between mutation and genetic drift (Loeschcke et al. 1994). Stated another way, new mutations are diluted by the random loss of genes due to a finite population size (genetic drift, Kimura 1983, as stated by Loeschcke et al. 1994). Genetic loading factors heavily into this viewpoint. A Selectionist perspective assumes most genetic variability is maintained by natural selection, even at the molecular level (Ayala 1984, as stated by Loeschcke et al. 1994). However, proof for either perspective is still missing (Loeschcke et al. 1994).

Conservation genetic theory obviously has its roots in population and evolutionary biology, and has best been applied to small, closed populations such as those characteristic of endangered or threatened species (Meffe 1986; Frankham 1999). While small is a relative term, the reality is that the theory can be applied to any finite population and all fisheries populations are finite. There are even formulas to calculate population sizes needed to maximize genetic diversity (Tave 1993). Yet, how this fits within the concept of metapopulations remains to be determined (Young 1999). One must always keep in mind the perspective that most of this theory is based on an "idealized population" concept which is rarely, if ever, realized in nature. An idealized population is one in which all individuals are monoecious or hermaphroditic, self-compatible, have a Poisson distribution of offspring, discrete generations, and exhibit random mating (Wright 1977).

How does this relate to the issue of aquaculture and fisheries management? Since the mid to late 1800s, interest and active participation in fisheries and fisheries management have grown in the United States. This is a result of increasing pressures on the natural resources by recreational and commercial fishermen; habitat alterations such as dams, reservoirs, and channelization; and, chemical, industrial, and biological degradation of our aquatic environments. All these factors have resulted in fluctuating fish population structures, even to the extent of complete population restructuring or extirpation of natural fish populations.

Two of the tools used extensively in treating the symptoms of these pressures placed on natural populations is farm-rearing fish for human consumption and the stocking of hatchery-reared fish into natural waters for stock augmentation. During the 1960s and into this new century interests in aquaculture for food-fish purposes has grown tremendously with global production exceeding 39 million tons in 1998 (Goldberg et al. 2001). In the U.S. this equated to roughly 445,000 tons valued at just under $1 billion or about 1/3 of the economic worth of the wild harvest (Goldberg et al. 2001). Hatchery-produced fish have been and still are being stocked for mitigation purposes, stock enhancement, restoration, or even the creation of a new population or fishery. But it is clear that use of these two tools is not going to eliminate the pressures on fisheries. Waples (1999) expressed this concern by stating that for more than a century hatcheries have been viewed as a treatment of the symptom of declines in abundance rather than addressing the causes of these declines. As such, there is an ongoing debate (Selectionist perspective) that hatcheries and hatchery fish can negatively affect the genetic make-up and integrity of wild populations (e.g., Allendorf and Ryman 1987; Hindar et al. 1991; Waples 1991). Campton (1995) felt some of the factors that led to this belief are related to the

general observations that: 1) the abundance of wild fish decline after the introduction of hatchery fish (Hilborn 1992; Washington and Koziol 1993), and 2) hatchery fish are adapted to hatchery environments which may reduce the fitness of a natural population (Taylor 1991). The question of the validity and use of hatchery fish in natural situations is still being lively debated.

In the past 10 to 20 years there have been an abundance of peer-reviewed fisheries articles that present opposing perspectives, at least one series of national debates (Philipp et al. 1993), and a major symposium devoted to the appropriate uses of fish and their effects in aquatic systems. The latter resulted in an excellent synoptic book edited by Schramm and Piper (1995). Included in this book are several solid reviews about the issue of genetic pollution. At a minimum the reader is encouraged to examine the following chapters: Busack and Currens (1995), Campton (1995), Carmichael et al. (1995), and Leary et al. (1995). Other noteworthy refereed papers include Krueger et al. (1981), Allendorf and Ryman (1987), Kapuscinski and Philipp (1988), Hindar et al. (1991), Utter et al. (1993), Currens and Busack (1995), and Waples (1999). Kapuscinski and Jacobson (1987) also provide an excellent layman's review of the issues and principles.

Clearly at the heart of the issue of genetic pollution effects is the disagreement among conservation and fisheries biologists themselves, the majority of whom would classify themselves as Selectionists. Some scientists argue quite well that sufficient empirical information clearly exists (i.e., Hindar et al. 1991; Busack and Currens 1995; Waples 1990, 1991, 1999; Clifford et al. 1998). Others indicate the evidence is not available, sparse, or not directly applicable (i.e., Utter et al. 1993; Campton 1995; Incerpi 1996). Campton (1995) expanded the issue by indicating there was insufficient evidence to separate direct biological genetic effects from biologically independent factors.

He focused on three general areas mentioned by Waples (1991) and Krueger and May (1991): 1) genetic effects of hatcheries and artificial propagation on hatchery fish; 2) direct genetic effects of hatchery fish on wild populations which are due to natural spawning and the potential of interbreeding; and 3) the indirect genetic effects of hatchery fish on wild populations due to ecological interactions (i.e., competition, predation, disease transfer) or other ecological factors not associated with naturally spawning hatchery fish. Waples (1999) added a fourth area to the list: temporary relaxation during the culture phase of selection that would otherwise occur in the wild. Independently, Campton and Waples both make strong arguments for their perspective, which requires the reader to synthesize the information to determine which, if either, argument is more valid.

Another rarely mentioned issue of the debate is that most of the empirical information about intentionally stocking hatchery fish sympatrically with wild conspecifics is derived from the study of salmonids. While there are a few papers considering nonsalmonids (e.g., Vrijenhoek et al. 1985; Philipp et al. 1985; Philipp 1991; Philipp and Whitt 1991; Harrell et al. 1993; Vrijenhoek 1994, 1996; Bartley et al. 1995; Bulak et al. 1995; Forshage and Fries 1995; Philipp and Claussen 1995; Fiumera et al. 2000), the overwhelming majority of the evidence to date comes from salmon, and in particular Pacific salmonids *Oncorhynchus spp.* Most of the impact of food-fish aquaculture escapees on wild populations comes from Atlantic salmon. While this is not surprising due to the issues associated with the biology and management of salmon, it remains to be seen if it is appropriate to universally apply genetic conservation strategies developed around salmonid biology and behavior across all species. Salmonids are undeniably a complex group of fishes with their strong natal homing and semelparity nature (Allendorf and

Phelps 1980; Hjort and Schreck 1982; Allendorf and Ryman 1987; Waples et al. 1990; Taylor 1991; Waples 1991; Hilborn 1992; Gharrett and Smoker 1993a; Washington and Koziol 1993; Busack and Currens 1995; Hard 1995; Maynard et al. 1995; Reisenbichler 1997). In the several key papers dealing with Atlantic salmon, including Cross and King (1983), Hindar et al. (1991), Volpe et al. (2000), King et al. (2001), and Spidle et al. (2001), the same question of stereotyping applies. But, is it appropriate to use them as a model for all species?

Consider the striped bass *Morone saxatilis* for example. This is a species in which major stocking efforts have been proclaimed as a great success story (Field 1997). This author (unpublished data) was directly involved in a similar situation with stock enhancement of striped bass into a coastal river in South Carolina. Through stocking and mark-recovery efforts I was able to determine that as high as 80% of a given year-class strength was attributed to hatchery–produced fish from a paired mating of four females to eight males. Unlike salmon striped bass males mature at 2 yr where it takes females 4 or more years to mature, both sexes are iteroparous and do not exhibit year-class spawning fidelity. Therefore, consanguineous matings are going to be more limited than if it were the same situation as with semelparous salmonids that mostly all spawn at the same age and may be primarily of hatchery origin. So, is it appropriate to use salmon as a model for all species conservation management?

For simplification of the issue of genetic pollution, Campton (1995), using terminology such as "negative genetic effect," stated it could be any factor (be it biological or anthropogenic) that reduces abundance or the effective population size of a natural population. Likewise, any factor that reduces viability or reproductive capability of an individual in a population can be interpreted as a negative genetic effect.

Table 1 presents some of the factors, biological, biologically independent, and abiotic, that can have a negative genetic effect on fish populations. Some of these will be discussed below. Personally, I liked the way Busack and Currens (1995) presented their case for discussing genetic diversity issues and will use their general outline to present my perspective. I will focus on three major areas of concern: loss of genetic variability within populations, loss of genetic variability among

TABLE 1. *Factors affecting genetic diversity in stock enhancement and restoration programs. The realistic column is the author's opinion that the action in the solution column can be achieved: (0) = neutral probability, (+) = good probability, (-) = poor probability.*

Issue	Problem	Solution	Realistic
Small Ne	Highly fecund fish, space limitations	Use more broodstock, equal sex ratio; equal family size	0
Temporal spawning dates	Space limitations, manpower	Prioritize needs, collect entire spawning season	+
Disproportionate contribution to progeny	Presence of dominant males or differential survival	Stock equal numbers of progeny and discard surplus	0
Artificial selection	Bias in broodstock collection, hatchery practices, elimination of poor performance progeny	Random choice of broodstock, maintain equal treatment of all progeny and captive fish	0
Convergent selection	Physical and chemical factors result in differential survival (pond to pond or tank to tank variability)	Maintain as equal treatments and conditions as possible	-
Exotic species	Non-native strains, stocks, or species are used for stocking, interspecific and intraspecific introgression	Sterilization, avoidance	+
Directed selection	Intentional or unintentional selection for desirable trait	Eliminate	+
Outbreeding depression	Loss of coadapted gene complexes	Use broodstock from system of stocking or as similar to system as possible	0
Inbreeding depression	Loss of genetic variability and fitness resulting from related breedings	Rotate captive broodstock, cross year classes, designed breeding scheme	+
Domestication	Loss of genetic variability towards hatchery adapted broodstock	Same as inbreeding	+
Undefined goals and objectives	Lack of background information or stock structure or management needs	Gather baseline data before stocking efforts begin, monitor stocks after stocking, define long-range goals and objectives	+
Evaluation of hatchery performance	Supervisors and management biologists not aware of implications of genetic stock program	Educations, acceptance, rewards	+

populations, and domestication. Within-population variation is the variation upon which natural selection acts. If this variation is reduced, there are fewer bases for future selective change (adaptation) within populations (Meffe 1986). Loss of variation among populations results in convergence of populations toward one "type" and a narrower range of "options" for the species (Meffe 1986). Both types of variation should be maximized to maintain full potential for evolutionary change within a species. My intent is to provide a basis from which the reader can make an informed decision about where our knowledge exists and where the gaps are, with the hope of furthering productive open debate thereby stimulating some of the needed research to answer germane questions.

Loss of Genetic Variation Within Populations

This category is best equated with the general concept of diversity and was given the following definition by Busack and Currens (1995): "Loss of within-population variability (diversity) is the reduction in quantity, variety, and combinations of alleles in a population. Quantity is the proportion of an allele in the population. Variety is the number of different kinds of alleles." While there are many different perspectives of this area, the two main genetic principles that affect this variation are random genetic drift and inbreeding (Busack and Currens 1995).

Random Genetic Drift

Random genetic drift is simply the alteration of gene frequencies through chance processes alone. Looked at another way, it is a prolonged bottleneck leading to repeated loss of variance where ultimately all loci become fixed with complete absence of genetic variance (Meffe 1986). Even with moderate selection pressures, drift can be a potent force in small populations.

In fish populations, because fecundity is usually so high, partners produce more gametes than actually fuse to become zygotes. Hence, the next generation is only a sample of the number and variety of alleles from the parent generation. Likewise, they are not exact copies of the parents (Busack and Currens 1995). In addition, in the real world, sampling of gametes is not random. This is due to issues such as fitness that affect genetic diversity (Busack and Currens 1995). As a result in a closed population, over time, and depending on population size, variation can and will be lost, especially in small populations.

Loss of variation has its greatest influence on rare alleles, which are considered important for the long-term response to selection and survival of the species (Allendorf 1986). Hedrick and Miller (1994) caution that rare alleles may be rare due to the simple fact that there has been some negative selection against them. In fact, Hedrick et al. (1986) warned that breeding programs designed to maintain rare alleles are operationally impossible and counterproductive for maintenance of genetic variation in general. They suggested selection for specific alleles could result in faster loss of variation at most other loci. Further, it is difficult to determine which rare alleles are advantageous, neutral, or detrimental. Thus, genetic variation at some genes may be treated differently because alleles at these genes may indicate a unique founder contribution or they may have important selective affects (Hedrick and Miller 1994).

One of the factors that drives this principle is effective population size (N_e). Genetic drift can be minimized by increasing N_e (Allendorf 1993). An effective population size is the size of an ideal population that has the same rate of increase in homozygosity or gene-frequency drift as the actual population being studied (Crow and Denniston 1988 as cited by Allendorf 1993). Factors affecting N_e include unequal sex ratios, disproportionate progeny distribution (family size), varying population sizes, bottlenecks, founder effects, inbreeding, and other factors. Meffe (1986) discussed random genetic drift and put it in a perspective understandable to most individuals. He stated that the longer the period of drift and the smaller the population, the greater will be the loss of variance. For example, only 60% of original variance will remain in a population of 10 after 10 generations. A population of 100 will retain 95% in that time, but if followed for 100 generations (100 years in an annual fish), variance is reduced to 60.6%. For 1000 years this value drops to 0.67% (Meffe 1986).

Sex Ratio. With genetic theory, any time the sex ratio of spawning adults departs from 1:1, N_e and genetic variation are reduced. Effective population size with respect to sex ratio is determined as:

$$N_e = \frac{4NmNf}{Nm + Nf}$$

where Nm and Nf are the number of breeding males and females respectively (Meffe 1986). One can easily see that skewed sex ratios will present a lowering of N_e. For example, with a population census of 500 fish, we can compare N_e under the condition of equal sex ratio and a skewed ratio of 50 males and 450 females. In the equal sex ratio example, $N_e = N = 500$ fish. However, in the skewed ratio where N still equaled 500, N_e equaled 180. Thus, the equal sex ratio is nearly 2.8 times larger, in a genetic sense, than the skewed sex ratio even though the census sizes were equal.

In hatcheries where artificial fertilization occurs, this becomes problematic. Hatcheries are large, relatively expensive facilities that normally have to operate within a short biological window for any given species. Therefore expediency and spatial concerns have historically been crucial to the operational procedure of the facility. Before the issues of conservation genetics came to the forefront, quite often for the sake of time and to "hedge

your bets on having viable sperm", the sperm of multiple males was often pooled to fertilize the eggs of multiple females or a single male was used more than once to fertilize multiple females. This contributed to having unequal sex ratios providing the progeny for the next generation. In addition, Gharrett and Shirley (1985) and Withler (1988) found that in pooling the milt of Chinook salmon *O. tshawytscha* there was a disproportion contribution the males represented. In other words, although an equal number of males and females could be used to fertilize the eggs, quite often a single or alpha male would fertilize the majority of the eggs. This would mean that even if the hatchery manager had been trying to maximize N_e by using equal sex ratios he could end up with a highly skewed sex ratio. Most hatchery managers today who are aware of conservation genetic issues make every possible effort to balance sex ratios as close to equal as possible.

Interestingly, most of the arguments for this mechanism of mate pairing are associated with small populations and/or endangered species. By default and because the best information available to us at this time comes from studying salmonids, most of the fish genetic conservation mantra accepts this fact. However, contrary to dogma, a growing body of evidence is showing that salmonids are not monogamous as originally thought, but they are naturally polygamous (Bentzen et al. 2001; Garant et al. 2001; Taggart et al. 2001). How this information will impact future consideration of equal sex ratios with larger populations remains to be seen.

Disproportionate family size. In an idealized population, the number of offspring per family is distributed in a Poisson fashion. Deviations from this distribution, with some matings producing more offspring, will bias the representation of contributed gametes in the next generation and thereby lower N_e (Meffe 1986). A biased progeny distribution will affect N_e as approximately:

$$N_e = \frac{4N - f}{2 + V_k}$$

where N_e is the variance effective population size, N is the census population size, and V_k is variance in progeny number (Crow and Denniston 1988 as cited by Allendorf 1993). By controlling for reproduction under captive conditions, it is possible to have N_e actually surpass N. This is effected by using equal full-sib families and mating each male and female to create a single family. Under this scenario V_k is zero and N_e is nearly double N (Allendorf 1993). Reducing differences in family size also lowers the effect of selection under captivity (domestication), as there are no reproductive differences between individuals if all matings produce the same number of progeny (Allendorf 1993).

Waples (1999) addressed the issue of equal family size, particularly dealing with minimizing risks with better management. While his idea that better management helps avoid risks is grounded in some truth, he correctly argues that genetic changes in cultured populations cannot be avoided entirely and that many of the risks are negatively correlated. For instance, while equalizing sex ratio and/or progeny distribution may help reduce impacts of inbreeding and loss of genetic variation, in studies with houseflies and *Tribolium,* it was demonstrated that such relaxing of the intensity of natural (or domestication) selection could erode fitness (Bryant and Read 1998; Lomnicki and Jasienski 2000). Recently however, Fernández and Caballero (2001) pointed out, that with computer simulations, equalization of family sizes does not greatly increase loss of fitness. Waples (1999) clarifies the issue by pointing out that while equalizing the sex ratios and progeny distribution is theoretically important, the real key is equalizing reproductive output in the next generation once fish are put into a natural environment.

It is unusual that the effort is made to equalize family size in most hatcheries unless the species in question is listed as endangered or possibly threatened, for a variety of spatial, economic, and manpower reasons. There is an innate tendency among hatchery personnel to keep all strong and healthy progeny from one spawn over that of a weaker group from another spawn. Right or wrong, usually all surviving progeny from an individual mating are stocked regardless of family size. Thus, from a hatchery perspective and based on Waple's (1999) argument of the key being the equal family contribution in the next generation it is hard to justify culling large numbers of progeny. If one were to proceed in this manner, it would be essential to appropriately address the question as to when this is to be done. Would it be in the larval, juvenile, or subadult stage? What additional selection pressures are added by culling (i.e., domestication effects)? Allendorf (1993) and Waples (1999) both discuss some of the implications to these questions.

Allendorf (1993) further suggested, for practical reasons, equalizing family size was best used only when population size was so small that genetic drift is a major concern. Others (Allendorf and Ryman 1987; Kapuscinski and Jacobson 1987) suggest the additional expense is unwarranted in situations where the population is relatively large that genetic drift is minimized. Thus maintaining "equal family size," if not almost impossible, is challenging.

A Selectionist, however, could argue that the conclusion of not making the effort to equalize family size and sex ratios in large populations is based on assuming selective neutrality and therefore is worthy of consideration (Allendorf 1993). This is because of the differential survival of hatchery-produced fish over that of their wild conspecifics, and how natural selection differentially affects hatchery-produced fish over that of wild progeny.

One reason for having hatchery production in the first place is to circumvent the mortalities when a species is most vulnerable: the larval to juvenile stages. The survival rate of a given spawn is sometimes several orders of magnitude greater in a hatchery than in the wild. This is reputed to lead to concerns about domestication (see below). Yet, Geiger et al. (1997) defined the greatest mortality in salmonids as occurring after they are released as juveniles (up to 90%). Therefore, efforts to equalize family size (and equal sex ratios) in captivity can easily be nullified by events that occur later in the life cycle (Geiger et al. 1997).

From a Neutralist perspective, however one could argue that due to 90% survival to stocking, the total genetic variability available from a given cross is significantly greater than if the parents had spawned in the wild and the mortality had occurred in the larval stage or even before hatch. One could even make an argument for the case that those surviving wild progeny are better adapted to their environment than hatchery fish. However, this is not the issue. The issue ultimately deals with fitness. It goes back to the point made earlier by Waples (1999) that it is not the genetic variability of the year-class strength that matters, it involves the reproductive output to the next generation. While Waples (1999) focused on equal reproductive output, which has empirically been demonstrated as important to N_e, one could argue further that even equal reproductive output is secondary to total genetic variability of the next generation. Which then begs the question: so why worry about it at all? It is clear to me the concern is for small, almost extinct populations where genetic drift plays a major role in overall genetic diversity and not in populations where there may be thousands to tens of thousands of a given population. Therefore, from a genetic drift perspective it is a moot point whether the mortality occurs in the larval to juvenile stage as is normally found in the wild or after fish are stocked, as is the case with hatchery fish. It is the total variability that is transferred to the next generation that matters.

Vrijenhoek (1994) discussed the importance of this variability with *Poeciliopsis monacha* and *P. occidentalis* as it relates to fitness. He stated that the remnant variation of depressed populations of these two species was crucial to survival, growth, fecundity, and developmental stability. With *P. monacha* loss of variability to the next generation after extinction and recolonization resulted in loss of developmental stability, tolerance to physical extremes, competitive ability, and susceptibility to increased parasite loads.

Inbreeding

Wright (1977) discusses that genetic theory predicts that inbreeding in small populations will lead to expression of deleterious recessive alleles, which can be manifested in lowering fecundity, high infant mortality, and reduced growth rates, all eventually leading to the possibility of population extinction. Yet, inbreeding in itself does not lead to population genetic variety or frequency changes (Falconer 1981). Instead it increases the number of homozygotes in a population, which, in turn, leads to changes in the frequency of phenotypes and, if selection acts on these phenotypes, then gene frequencies change (Busack and Currens 1995). Inbreeding is rarely an issue in the wild, unless it concerns remnant or endangered species, simply due to the size of the populations found in the wild and the lowered probability of mating with an individual related by descent. Inbreeding can be problematic in endangered fish populations and restoration efforts or in aquaculture operations where the numbers of broodstock are relatively small.

The effects of inbreeding on reproductive fitness have been known since Darwin (Frankham 1999) and have continued to be demonstrated in many studies with naturally outbreeding populations (Falconer and Mackay 1996). Tave (1993) discussed multiple effects of inbreeding depression (poor phenotypic performance) on qualitative and quantitative characteristics of captive fish populations. Ralls and Ballou (1986) showed that 41 of 44 captive populations of mammals experienced higher mortality in inbred lines as opposed to outbred offspring. Inbreeding depression can result from phenotypic expression of homozygous geneotypes for rare, harmful alleles that are normally suppressed in heterozygotes (Busack and Currens 1995). Remember the concept of genetic loading where not all organisms in the population are at its highest fitness level and the more the lethal mutations, the greater the genetic load.

If heterozygotes do perform better than homozygotes, then a decrease in heterozygosity will lead to a deceased performance (Waldman and McKinnon 1993). Loss of heterozygosity may reduce a population's ability to respond to future environmental change, such that opportunities for evolution are limited or the possibility for extinction increases (Franklin 1980). Yet, Caro and Laurenson (1994) caution that not all declines in populations exhibiting clinical signs of inbreeding are directly attributable to genetics. They found that classic cases of signs of inbreeding in cheetahs were due to ecological factors (predation, disease, fire, etc.) and not necessarily genetics.

Templeton and Read (1994), in an excellent discussion and follow up to Jacquard's (1975) paper on the confusion of the term inbreeding, outline several of the often contradictory meanings of inbreeding and how it can lead to misunderstandings on the role of inbreeding in genetic management. They discuss three biological meanings: 1) as a measure of shared ancestry in paternal and maternal lineages; 2) as a measure of genetic drift in a finite population and; 3) as a measure of a system of mating in a reproducing population. They emphasize that the different meanings of the word inbreeding must be kept separate to avoid incorrect management recommendations and evaluations.

Co-Ancestry. The most commonly accepted idea of inbreeding relates to co-ancestry, or the mating of two individuals sharing ancestors, otherwise known as identity by descent (IDB) or pedigree inbreeding (Templeton and Read 1994). The key and probably most misunderstood component of this concept is that pedigree inbreeding F_p is applied to an individual and not a population. Because the values for F_p range from 0 to 1, corresponding to no shared ancestry (0) or shared ancestry (>0) it is impossible to measure an "avoidance of inbreeding" (Templeton and Read 1994). An individual is either inbred (>0) or not (0). Thus, you cannot measure an individual as being produced by a mating that avoided or minimized inbreeding relative to the mating options available to the population (Templeton and Read 1994). This is an important distinction for managing any population, especially if it is a small population because of the possibility that some shared ancestry exists. This is rarely discussed or examined in hatchery fish.

Templeton and Read (1994) explain that because F_p is an individual-level measure, it tells nothing about the population-level attributes such as the population's system of mating or the level of genetic diversity. By using the F_p values from several individuals in a common analysis you can obtain some idea of the population-level phenomena such as inbreeding depression. Thus, it is essential to know the mating system once fish are released and interbreeding with natural stocks (i.e., fitness effects). Avoidance of inbreeding depression requires avoidance of pedigree inbreeding.

Because inbreeding depression has a genetic basis, in theory it can be eliminated or reduced by evolutionary changes at contributing loci (Templeton and Read 1994) or by outcrossing or immigration (Keller et al.

2001). However, it is not as simple as eliminating deleterious recessive alleles through IDB matings as inferred by Lande (1988). This is because inbreeding depression can be related to the system of mating, size of the population, the length of time breeding programs and supplementation of stocks occurring, and synergistic and epistatic interactions among loci (Templeton and Read 1994; Waples and Do 1994; López-Fanjul et al. 2000).

Inbreeding as it relates to genetic drift. As previously stated, random genetic drift is associated with random sampling of the genes passing from one generation to the next in a finite population, which requires random mating. However, does random mating indicate no inbreeding? No! Templeton and Read (1994) explain that it means mating regardless of ancestry. So in small or any finite population, the possibility of pedigree mating or IDB can always occur. Of course the probability and magnitude depends on the population size. As population size decreases, the probability of an individual mating with a relative (or itself in a monoecious population) concomitantly increases. This relationship allows calculation of heterozygosity (H) as a means to express a population's genetic variability (Templeton and Read 1994). Templeton and Read (1994) discuss how conservation biologists and managers misunderstand this concept and immediately equate heterozygosity to genetic variation, which may not be the case. Inbreeding does not always decrease genetic variability (Templeton and Read 1994). What is true is that genetic drift can increase the probability of IDB of a population and genetic drift can cause loss of genetic variation in a population; thus, an increase in inbreeding and a decrease in variation can be correlated because of a common factor, genetic drift (Templeton and Read 1994).

To summarize, Templeton and Read (1994) state that many management programs avoid inbreeding to preserve genetic variability. However, it is genetic drift and selection that are the real problems. "An effective management program for genetic diversity must be based on accurate identification of the factors that actually erode genetic diversity in managed populations. Otherwise, the program faces the possibility of utilizing management options that have the opposite of the desired effect."

Because most of the information to date comes from information associated with captive populations and not natural populations, or deals with computer simulations, Busack and Currens (1995), Carmichael (1999), and Frankham (1999) pose a series of questions that need to be addressed to understand the impact of loss of within-population variability on various populations and how that relates to overall genetic stability. Many of these questions are difficult to answer yet are crucial to managing a population and supporting the issues of drift and genetic variation. What are the critical thresholds for loss of within-population variability? Can it be contained? What level of heterozygosity is desirable? How large of an effective population should there be to maintain genetic variability in the long run? How do we judge genetic impairment? At what point do we consider a population genetically damaged? What course of action should we take? Definitive answers for these questions still elude us. However, as suggested in Table 1, managing loss of within-populations by manipulating broodstock numbers, sex ratios, and even family size is possible if not logistically difficult. In spite of this, Campton (1995) still maintains that there is no strong evidence that drift effects have reduced variation or affected fitness.

Loss of Genetic Variation Among Populations

Busack and Currens (1995) offer the

following working definition of this component of genetic variation: "Loss of among-population variability is the reduction in differences in quantity, variety, and combinations of alleles among populations." This is the variation lost during convergent selection (Table 1) that drives individual populations toward a common type (Stahl 1983; Meffe 1986; Quinn et al. 2000) and includes issues such as timing of spawning, fitness, loss of adaptation, and loss of co-adapted gene complexes. Among-population variation can be at its greatest risk when isolated populations are mixed, which results in a genetic admixture of the two previously distinct entities (Campton 1987).

In order to maintain overall genetic diversity, gene flow from natural sources is essential. However, minimizing gene flow between populations is important to maintaining variation among populations because the mechanism for reducing this form of genetic variation is gene flow at excessive levels or from non-native sources (Busack and Currens 1995). Excessive gene flow may reduce performance of individual populations by disrupting their genetic organization (Shields 1982). At the multipopulational level this means a reduced ability to respond to evolutionary opportunities presented by environmental change. At the individual population level it represents a loss of genetic uniqueness and associated reduction in performance (Busack and Currens 1995). To agree with these statements one has to assume that populations are reproductively isolated (Busack and Currens 1995), which would obviate this loss of genetic variation from applying to one of the more prevailing ecological concepts found in today's literature, that of metapopulations (Hanski 1997). In the latter situation, isolated discrete patches of suitable habitat are connected by individuals of sufficient number migrating between the populations to have a significant effect on the demography or genetic structure of each component population (Stacey et al. 1997; Young 1999).

Selection for a Type

One of the usually accepted facts that hatcheries or perturbated environments can do is place selective pressure on a given species. Three types of selective pressures that can result in a given "type" of population include artificial, convergent, and directed (Table 1). Artificial selection will be discussed below under domestication.

Convergent selection is usually associated with the physical attributes of a hatchery or modified natural environment and is an artificial selective force. Meffe (1986) discusses convergent selection from the context of fishes that have been reared in physically distinct habitats being suddenly thrust into a common environment where the same physical and chemical regimes are applied uniformly across all populations. He states that if this pressure continues for several generations, or if the forces are sufficiently strong, the practice can select for a common stock, thereby eliminating among-stock variability. An example of this effect was with Atlantic salmon (Stahl 1983). It is difficult to manage or eliminate convergent selective forces.

Directed or directional selection (Table 1) is usually confined to a hatchery environment and can be controlled and eliminated if attention is given to the problem from the outset. Some of the impacts of directed selection can be felt in fitness traits, coadapted gene complexes, and populations expressing local adaptation. All of these factors limit or reduce overall genetic variability among populations. It may be intentional or unintentional depending on the characteristic being expressed and the purpose of the rearing.

Timing of Spawning

Over the course of a natural spawning cycle

wild fish make take several weeks to months to complete spawning of an entire population. Within this given population, different cohorts may spawn at different times of the season thus having different modal peaks (Crawford 1979 as cited by Waples 1999; Taylor 1980; Gharrett and Smoker 1993a). There is good evidence that such life-history traits (Hard 1995; Secor 2000; Quinn et al. 2000) have a genetic basis (Gharrett and Smoker 1993a, 1993b) thereby demonstrating genotypic diversity to represent this phenotypic temporal spawning difference. One way in which hatcheries can affect this among-population diversity is to select individual broodstock for spawning purposes from a single segment of that spawning season and not across the season as a whole. More often than not this is a management decision made either by the hatchery personnel or by the management biologists in charge of the fishery. Such is the case in many fisheries management efforts to minimize opportunities for interbreeding with wild fish and allowing for selective harvest of hatchery fish (Waples 1999). From a hatchery perspective, it is a matter of expediency. Many hatcheries are designed for multi-species production and they operate in a "first come, first spawned" scenario. In other words, during a given season, the first species to spawn is collected from the wild first or is spawned from captive fish first regardless of the length of the spawning season. Once production quotas are met, the second species is sought after and spawned, and so forth until all the targeted species are spawned or the biological seasons ends. What this does not take into account and where it causes a genetic pollution concern is that the hatchery personnel are performing a de facto directed selection for early spawning fish, or type selection. In theory, over time by adding the progeny of these fish to the wild you may disproportionately skew the "natural population" spawning time period toward the early spawners, hence, the "excessive gene

flow" mentioned by Busack and Currens (1995). Indeed, this has been done for many captive populations of fish, and in some cases intentionally. However, from an ecological perspective, this may not be the best option because of competition (Hayes 1987), survival effects (Taylor 1980; Secor 2000), temperature effects (Secor and Houde 1995), and availability of food resources for larval fish that are linked to timing of spawning (Warlen 1994; Rutherford et al. 1997; Limburg et al. 1999). Hard (1995) discussed the theoretical issues of life-history patterns and the potential impacts supplementation programs may have on such traits with Pacific salmon. From these discussions, one could assume if given sufficient time and selection intensity such directed efforts could not only reduce overall diversity of the population, but also push the entire population toward a "type" that has a high expression of the given trait being selected. Whether this selection pressure is transferred to wild populations has still yet to be demonstrated. Although circumstantial, evidence exists that does tend to support that conclusion (Campton 1995). Absence of baseline data for most wild populations and pedigree data for hatchery populations makes it difficult to make definitive statements concerning this matter (Campton 1995). Obviously this can be avoided by sampling the spawning population across the entire temporal season.

Fitness

How does fitness relate to the concept of genetic variability? First we must separate the concept of fitness from an environmental perspective and fitness from a genetic perspective. We often equate fitness to the strongest, smartest, most competitive organism of a given population. Or we equate it to the environment and an individual's competitive advantage to survive within an environment. Although these factors can be characteristics of an individual that exhibits fitness, it does not capture the true meaning of genetic fitness. From an evolutionary and genetic perspective, fitness is simply the ability to propagate their genes into the next generation. Therefore, the fittest individuals in a population are those that leave the greatest number of descendents in relation to the number of descendents left by other less fit individuals (Doncaster 1995). Different fitness components include first breeding age, fecundity, survival, and mate selection (i.e., non-random mating, Burt 1995). Fitness fluctuates between generations because of increases in natural selection and decreases due to mutations and changes in the environment (Burt 1995). The fluctuation generally falls between 1 and 10% per generation.

Because fitness is directly tied to reproductive output, it is logical to examine a selective advantage of mate choice. Hatcheries, through artificial propagation, remove the option of mate choice by artificially pairing males to females. Burt (1995) discusses the importance of mate choice by addressing the cumulative effect of choice on a population adaptation. He states that removing mate choice for many generations (which can occur in a hatchery where captive stocks are used as progenitors of a given year-class, or where wild fish are taken on an annual basis and mate pairings are either predetermined or selected at random, both of which negate mate choice) would lead to a halving of survival probabilities. This is even true if the selection pressure maintaining choice is a few percentage points or less. However, he goes on to state that fitness need not be inherited from parent to offspring for there to be a genetic advantage to mate choice. Using an example of annual birds that hatch, live a year to reproduce, then die due to a virus that affects different juvenile genotypes each year but only affects adult male plumage, he suggests that a female rejecting dull-colored males would increase the probability of her offspring's survival. Similarly, in iteroparous species, by rejecting males showing the symptoms of "this year's virus," avoidance of disastrous loses of this year's brood may be circumvented, even if different genotypes are susceptible in different years and there is little variance in lifetime fitness (Burt 1995). Surprisingly, we know little about the anatomy and physiology of female choice with the exception that female sticklebacks are found to be more sensitive to the color red during breeding season (Cronly-Dillon and Sharma 1968).

There are many studies that address fitness issues with fish. Fleming and Gross (1992, 1993) in a series of experiments found that hatchery males were less aggressive and active and more submissive than wild males. Differences between females were not as pronounced, but there were temporal differences, which were putatively the cause of differential mortality of eggs because of environmental effects. Similarly, Berejikian et al. (2001) observed in controlled laboratory studies that captive male coho salmon were subservient to wild males and were not successful in contributing to the genetic makeup of the resulting progeny. They equated some of the competitive inferiority to be due to environmental conditions that the captive males were held as not being conducive to promoting appropriate spawning body color and shape, thereby altering natural spawning behavior development. Waples (1999) addresses the issue of fitness effects by hatchery fish on wild populations and cites several examples: Chilcote (1997), Bugert et al. (1992), and a synoptic paper by Reisenbichler (1997). Overall, Waples (1999) felt that these were solid examples where the fitness of hatchery fish in the wild could be reduced compared with natural fish. However, in each case by his own admission, the evidence was mostly circumstantial and definitive proof remains to be seen, as many other factors could be the causative agent (e.g., environmental, stock transfers as opposed to culture effects, etc). Similarly, Taylor (1991) and Campton (1995) felt the information available on the validity of local adaptation is too circumstantial from which to draw any firm conclusions. Definite proof is difficult to detect because natural selection is so difficult to study (Busack and Currens 1995). Waples (1999) quotes Busack and Currens' (1995) comments that rigorous research designed to detect such effects has yet to prove the impact of hatchery-reared fish on natural populations. He rightly concludes his argument by stating that we lack a consensus on what constitutes a reasonable approach to this issue given the uncertainty involved and the potential consequences for actions taken or not taken.

Coadapted Gene Complexes and Local Adaptation

Tied directly to fitness are the concepts of adaptation to a local environment and coadapted gene complexes (Dobzhansky 1970; Templeton 1986). When two genetically divergent or reproductively isolated populations interbreed in which the resultant progeny are less adapted to their environment, the potential for decreases in fitness exists (Campton 1995). This loss of fitness is commonly known as outbreeding depression (Templeton 1986).

Although it has not been definitively shown to have a genetic basis, an example of local adaptation may be with Chesapeake Bay striped bass. The eggs from Chesapeake Bay striped bass have a natural proclivity to float as opposed to conspecifics from other East Coast drainage systems in which the eggs are slightly heavier than the specific gravity of water and require moving water to keep the eggs in suspension (Harrell 1987; Rees and Harrell 1990). The Chesapeake Bay tributaries in which the fish annually spawn are relatively short and tidal. At each tidal change, flow stops

and changes directions. Without water movement, eggs without this positive buoyancy would sink to the bottom where they possibly would die from micro-environmental hypoxic conditions. Thus the genetic advantage to floating eggs would be that they would stay in suspension in the water column during slack tides, and such is the case. The concern is that if it is genetically controlled and non-Chesapeake Bay strain striped bass interbred with Chesapeake Bay striped bass females this trait could be lost and significantly affect recruitment (Harrell et al. 1993). This however is still speculative and whether this trait is genetically or nutritionally controlled (Chesapeake Bay striped bass females having a higher egg lipid content than other strains and captive broodstock (Harrell and Woods 1995)) remains to be confirmed.

The genomic coadapted fitness premise is that over time, natural selection acts on localized populations and favors those individuals best suited to localized environmental conditions or that epistatic interactions are better suited to fitness traits and are carried on to the next generation. Thus, when excessive gene flow or disruptive gene flow from exogenous sources, such as stocking hatchery fish, enters into the population (Busack and Currens 1995), these adaptive combinations are lost and the adaptive significance is decreased. Genomic coadaptation fitness decreases result from epistatic or chromosomal interactions among alleles at different loci having an overall negative effect on the organism. As new immigrating alleles from a previously isolated population replace existing alleles in another now interbreeding population new, less favorable allelic combinations may be formed, reducing performance (Busack and Currens 1995). However, this may not show up until the second generation (Gharrett and Smoker 1991). Campton (1995) explains that when populations interbreed, the positive allelic combinations can be broken up in the F_2 generation due to meiotic recombinations in the F_1 generation following interpopulation matings.

Genomic coadaptation is at best a controversial concept that is not accepted by all conservation and fisheries biologists (Purdom 1993; Tave 1993). The problem arises because the concept was developed from theory and empirical information with *Drosophila,* which has a very limited number of chromosomes for these interactions to occur. However, there are other examples that do have data to point to the veracity of this concept (Templeton 1986; Lynch 1991; Mitton 1993). In fact, there is a growing body of literature that supports the concept of outbreeding depression in general in fisheries (i.e., Stahl 1981, 1983; Emlen 1991; Waples 1991; Gharrett and Smoker 1991, 1993a with salmonids; Philipp 1991; Philipp and Claussen 1995; Philipp and Whitt 1991 with largemouth bass *Micropterus salmoides*). Philipp et al. (1993) stated past and present hatchery practices have been conducive to supporting increased gene flow and resulting in this loss of adaptive significance. Some of these practices include releasing fish outside their original distribution, transferring eggs and fish to meet production needs, and using hatchery stocks of mixed ancestry (Busack and Currens 1995). Yet, despite this claim and the concerns of hybridization (inter and intraspecific) causing loss of among-population variation, others claim that in some environments native genotypes persist (Wishard et al. 1984; Currens et al. 1990). I will not discuss the impacts of hybridization in this paper as such because most of the principles associated with this complex issue deal directly with loss of variation both within and among populations. I suggest the papers by Campton (1987), Kerby and Harrell (1990), Hindar et al. (1991), Busack and Currens (1995), Campton (1995), Leary et al. (1995), and Waples (1999).

Philipp and Claussen (1995), using largemouth bass as an example, present an important perspective of fish genetics and genetic pollution in that they examine outbreeding depression from a reductionist viewpoint by discussing the impact of stock transfers and their impact on fitness. The basis of their argument is founded on the Stock Concept (Kutkuhn 1981; Philipp 1991), which states that because species are composed of multiple distinctive units (e.g., stocks, populations, demes), for truly effective conservation it is these units, not the species as a whole, that must be considered the operational unit of concern for management purposes (Philipp and Claussen 1995). Their basic treatise follows much of the arguments found for salmonids and supports the claim that stock transfers negatively affect the genetic resources of a species by diminishing genetic diversity among populations and likely decrease the fitness trait differences among stocks. Their long-term studies document genetic and physiological differences among stocks of largemouth bass and demonstrated that nonnative stocks exhibit poorer fitness and performance traits than native stocks. Originally the basis for their argument was between the two subspecies of largemouth bass, the northern strain, *M. s. salmoides,* and the Florida strain *M. s. floridanus.* Now however, they demonstrated significant fitness differences between the same strains from two different river systems, which argue for their point of the Stock Concept. Hence, geographical boundaries forcing reproductive isolation are more important than previously expected with fish populations. Interestingly, Bulak et al. (1995) examining largemouth bass strain dynamics in South Carolina, where a true hybrid zone for the two strains exists, found a distinct trend. While intergrades performed well throughout most of the state, there was indeed a proclivity for the Florida strain to be more focused in the warmer, southern coastal part of the state. The northern strain tended toward the inland, cooler, northern part of the state.

Whether it is a function of local adaptation or genomic coadaptation, the reality of outbreeding depression is still debated. Ironically, genetic admixture and resultant gene flow that is contributory to outbreeding depression does have an end result in the increase in total genetic diversity. While this overall increase in diversity disrupts these putative adaptive complexes and breaks down local adaptation, it should theoretically be better suited to adjust to major catastrophic changes or perturbations in localized environmental conditions: a trait one might think would be a better long-term option, given today's anthropogenic effects.

Domestication

"Domestication is the changes in quantity, variety, or combination of alleles within a captive population or between a captive population and its source population in the wild as a result of selection in an artificial environment" (Busack and Currens 1995). Although this sounds like the same definition as "within-population variance" mentioned above there are two major differences: 1) changes in diversity by genetic drift are random while changes in diversity due to domestication are directly related to specific traits and selection of those traits; and, 2) the rate of diversity loss through random genetic drift is inversely proportional to population size while in domestication the rate is dependent on the genetic nature of the trait and the selection pressure being applied (Busack and Currens 1995). Campton (1995) coined the phrase as "simply natural selection in an artificial or domestic environment." The end result generally focuses on adaptation to a human-controlled environment and favors captive hatchery stocks with increased fitness and lowers fitness of wild or natural stocks (Kohane

and Parsons 1988). Thus, at the heart of domestication is artificial selection in which particular phenotypic traits that have a genetic basis (especially behavioral and physiological) are usually (though not always) intentionally selected for breeding, husbandry, and management needs. Problems arise when these traits are maladaptive to natural environments and these fish are released into the wild and compete and interbreed with native stocks. It is important to recognize that domestication selection works primarily on quantitative and not qualitative traits and therefore is difficult to measure because of environmental effects on genotypic expression (Hard 1995).

Waples (1999) focused his paper on the concept of domestication (particularly domestication selection), an area in which hatcheries can have an impact on populations. He even added a fourth area to Campton's (1995) list of factors that lead to genetic change in cultured populations: temporary relaxation of selection that would otherwise occur in the wild (see argument above in section on disproportionate family size about fitness erosion under natural selection relaxation). He also differentiated between domestication as a "state or condition of a population" while domestication selection is the "process that leads to domestication." I agree with Waples on this point in that it is important to remember that domestication is indeed a process, not an end point. It is very difficult, because of the dynamics of the definition to unequivocally state that at any given point or generation an individual fish or population is "domesticated."

The forces at work here are obviously a combination of selection pressures, artificial and domestication. Hatcheries have a tendency to relieve certain natural selection pressures (i.e., food competition, predation effects, physical and chemical environmental fluctuations) because they provide a stable environment, artificial diets, and size segregation. However, they simultaneously introduce artificial selection pressures that may not be conducive to survival in a natural environment (i.e., crowding, prophylactic treatment for diseases, artificial diets, handling, etc.). In other words, hatcheries artificially select for those individuals who can adapt to an artificial, captive environment. Those genotypes that cannot adjust to such an environment are lost. In theory, the longer you keep a given cohort confined to a hatchery environment (be it days, weeks, months, of even generations) the more "domesticated" and selected for survival in an artificial environment they become.

Obviously domestication efforts have been in place ever since man has been trying to culture fish for consumptive purposes. However, concerns about the impact domestication may have on native fishes did not come about until the 1950s and 1960s (Greene 1952; Miller 1954; Vincent 1960; Flick and Webster 1962, 1964). Many studies (essentially all with salmonids) indicate that hatchery-reared fish do not perform well in natural environments (i.e., Miller 1954; Reisenbichler and McIntyre 1977; Hynes et al. 1981; Bachman 1984; Hindar et al. 1991; Maynard et al. 1995). While others indicate that performance of hatchery fish is superior (Mason et al. 1967; Nickelson et al. 1986; Lepage et al. 2000) or similar (Symons 1969; Rhodes and Quinn 1999). One must be careful in considering any of these studies, however, because it is often difficult to separate phenotypic differences in such studies with actual genetic links. For example, Waples (1999), in his *Corollary 1* under *Myth 2* about domestication selection being avoided if there is no mortality, argues that size is a genetic effect that transcends the culture period. I would argue that while size is indeed genetically linked, environmental and even maternal conditions can have a dramatic impact. It is well known that female size and nutritional competency at spawning will significantly affect egg and larval size (Zastrow et al. 1989; Harrell and Woods 1995). Once a size differential is expressed under culture conditions it becomes magnified as time proceeds, even with all conditions being equal. Therefore, while some studies comparing performance of hatchery-produced fish to that of wild stocks do indeed identify a clear genetic basis, others do not. Busack and Currens (1995), Campton (1995), Hard (1995), and Waples (1999) all have some excellent discussion and appropriately stress the significance of domestication and how it relates to the issue of genetic pollution.

Summary and Conclusions

Without doubt there have been some genetic impacts of aquacultured fishes on natural stocks, with the majority coming from intentional stocking programs designed to enhance or restore natural stocks. Is this a form of genetic pollution? I believe you would find little argument to the contrary. However, is the impact any different than the short and long-term genetic effects on fish populations brought about naturally by genetic load or anthropogenically by changes to natural environments? Several individuals have made a strong case that the latter is actually more insidious and far-reaching than accidental or intentional hatchery introductions have ever been. Others argue for the former. Like other points so closely tied to this issue, it may well be a personal perspective.

I originally referred to this issue as being a paradox and I still believe that to be the case. We perceive we know the truth, but do we really know the answers with our current level of understanding of genetic theory and diversity. Almost everyone who has previously reviewed this issue has recognized that theory drives most of our action, or better yet, reaction. The one thing that kept coming to me as I reviewed the literature for this paper is how much geneticists agree on the concept and principles of conservation genetics, but how often they disagreed that the support for an opposing perspective really exists. Much is founded on how much one wants to read into a scientific article. As such, there is still an open debate about whether sufficient collaborative empirical information exists as there are obviously opposing viewpoints that one could use to argue whichever perspective you chose to side on. There are some that even argue that because an "idealized population," the underlying assumption upon which most conservation genetic theory is based, does not even exist in nature and therefore the theory is obviated. However, one reviewer of this paper commented that denying the empirical evidence exists for most of the opposing viewpoints is sort of like a smoker saying that we still have not proven that smoking is bad for you. An excellent point indeed! The reviewer goes on to argue that the evidence is there; you just have to interpret the literature.

Throughout this paper I have continually referenced three papers Busack and Currens (1995), Campton (1995), and Waples (1999). I referenced them because each brought a unique, but somewhat similar, perspective to the issue. I also used these references because they are all geneticists looking at the issue from the other side of the mirror than I do as an aquaculturist. While I do not agree with all they present, I do support most. What is clear to me is that ideologically conservation genetics and their arguments are sound. Thus, even from a practical perspective, conservation genetics in fisheries management has considerable merit. But does it have limitations? Absolutely! I believe it was Einstein that said if we knew the answers we would not call it research. I have reviewed the literature and tried to make the point that the issues are germane to aquaculture and managing our fisheries resources, but the information and science associated with conservation genetics is far from exact and, like

our knowledge of aquaculture, it is obviously dynamic.

Campton (1995) called for delineation between biotic and biologically independent factors as they affect genetic diversity and addressed each appropriately. Others emphasized the impact of stocking on natural populations or looked at ways to increase within- and among-population genetic variation in those fish being released into the wild. I agree with Busack and Currens (1995) that it ultimately is the variation that is transferred to the next generation that is essential. This is especially relevant since 90% of mortality in a salmon hatchery occurs after juveniles are released (Geiger et al. 1997). Genetic variation in the hatchery from a natural population perspective therefore is irrelevant and will remain so unless, and until, it is transferred to the next generation. However, because it is the raw material for survival, it is essential to have the resources available in the current generation for it to be expressed in the future. The point is that we need to focus on the long-term instead of the short-term, and as Waples (1999) emphasized we need to agree on what constitutes a reasonable approach to answering the questions germane to the issue.

Last, I brought forth the point that most of the empirical information we use to formulate our decisions and action plans are based on salmonid biology and raised the question of the validity of uniformly applying this across all species. I even pointed out that new information about salmon not being monogamous strikes at the heart of what was once considered dogma for managing salmon populations. How will such new information change the way we think about these issues in the future? How will it affect the way we manage our hatcheries and our fishery resources? When I was a younger hatchery biologist, I remember the "old-timers" telling me that if fish ever got sick or started dying on the station to "put wheels under them." At the time it made a lot of sense because our hatchery performance evaluation was based on numbers not quality. Much of what was done in the 1950s, 1960s, and even 1970s was not that it was wrong at the time; it was just that our knowledge is improving, and so is the way we go about our work and stewardship. Meffe (1986) stated it best: "The point here is that, although our current knowledge of conservation genetics is incomplete and perhaps inaccurate in some cases, it is the best information presently available."

Literature Cited

AHDEL, 1992. The American heritage dictionary of the English language, third edition. Houghton Mifflin Company, Boston, Massachusetts, USA.

Allendorf, F. W. 1986. Genetic drift and the loss of rare alleles versus heterozygosity. Zoo Biology 5:181–190.

Allendorf, F. W. 1993. Delay of adaptation to captive breeding by equalizing family size. Conservation Biology 7:416–419.

Allendorf, F. W. and S. R. Phelps. 1980. Loss of genetic variation in a hatchery stock of cutthroat trout. Transactions of the American Fisheries Society 109:537–543.

Allendorf, F. W. and N. Ryman. 1987. Genetic management of hatchery stocks. Pages 141–159 *in* N. Ryman and F. Utter, editors. Population genetics and fishery management. University of Washington Press, Seattle, Washington, USA.

Ayala, F. J. 1984. Molecular polymorphism: how much is there and why is there so much? Developmental Genetics 4:379–391.

Bachman, R. A. 1984. Foraging behavior of free-ranging wild and hatchery brown trout in a stream. Transactions of the American Fisheries Society 113:1–32.

Barker, J. S. F. 1994. Animal breeding and conservation genetics. Pages 381–395 *in* V. Loeschcke, J. Tomiuk, and S. K. Jain, editors. Conservation genetics. Birkhauser Verlag, Boston, Massachusetts, USA.

Bartley, D. M., D. B. Kent, and M. A. Drawbridge. 1995. Conservation genetics diversity in a white seabass hatchery enhancement program in Southern California. American Fisheries Society Symposium 15:249–260.

Bentzen, P. B., J. B. Olsen, J. E. McLean, T. R. Seamons, and T. P. Quinn. 2001. Kinship analysis of Pacific salmon: insights into mating, homing, and timing of reproduction. The Journal of Heredity 9:127–136.

Berejikian, B. A., E. P. Tezak, L. Park, E. LaHood, S. L. Schroder, and E. Beall. 2001. Male competition and breeding success in captively reared and wild coho salmon (*Oncorhynchus kisutch*). Canadian Journal of Fisheries and Aquatic Sciences 58:804–810.

Bryant, E. H. and D. H. Reed. 1998. Fitness decline under relaxed selection in captive populations. Conservation Biology 13:665–669.

Bugert, R., K. Petersen, G. Mendel, L. Ross, D. Milks, J. Dedloff, and M. Alexandersdottir. 1992. Lower Snake River compensation plan; Tucannon River spring Chinook salmon hatchery evaluation plan. U.S. Fish and Wildlife Service, Boise, Idaho, USA.

Bulak, J., J. Leitner, T. Hilbish, and R. A. Dunham. 1995. Distribution of largemouth bass genotypes in South Carolina: initial implications. American Fisheries Society Symposium 15:226–235.

Burt, A. 1995. The evolution of fitness. Evolution 49:1–8

Busack, C. A. and K. P. Currens. 1995. Genetic risks and hazards in hatchery operations: fundamental concepts and issues. American Fisheries Society Symposium 15:71–80.

Campton, D. E. 1987. Natural hybridization and introgression in fishes: methods of detection and interpretation. Pages 333–344 *in* N. Ryman and F. Utter, editors. Population genetics and fishery management. University of Washington Press, Seattle, Washington, USA.

Campton, D. E. 1995. Genetic effects of hatchery fish on wild population of Pacific salmon and steelhead: what do we really know? American Fisheries Society Symposium 15:337–353.

Carmichael, G. 1999. How about other myths being addressed, too? An open letter to the membership of the American Fisheries Society. Fisheries (Bethesda) 24:36–37.

Carmichael, G. J., J. N. Hanson, J. R. Novy. K. J. Meyer, and D. C. Morizot. 1995. Apache trout management: cultured fish, genetics, habitat improvements, and regulations. American Fisheries Society Symposium 15:112–121.

Caro, T. M. and M. K. Laurenson. 1994. Ecological and genetic factors in conservation: a cautionary tale. Science 263:485–486.

Chilcote, M. 1997. Conservation status of steelhead in Oregon. Oregon Department of Fish and Wildlife, Portland, Oregon, USA.

Clifford, S. L., P. McGinnity, and A. Ferguson. 1998. Genetic changes in Atlantic salmon (*Salmo salar*) populations of Northwest Irish rivers resulting from escapes of adult farm salmon. Canadian Journal of Fisheries and Aquatic Sciences 55:358–363.

Courtenay, Jr., W. R. 1995. The case for caution with fish introductions. American Fisheries Society Symposium 15:413–424.

Crawford, B. A. 1979. The origin and history of the trout brood stocks of the Washington Department of Game. Washington State Game Department Fishery Research Report, Olympia, Washington, USA.

Cronly-Dillon, J. and S. C. Sharma. 1968. Effect of season and sex on the photopic spectral sensitivity of the three-spined stickleback. Journal of Experimental Biology 49:679–687.

Cross, T. F., and J. King. 1983. Genetic effects of hatchery rearing in Atlantic salmon. Aquaculture 33:33–40.

Crow, J. F. and C. Denniston. 1988. Inbreeding and variance effective population numbers. Evolution 42:482–495.

Currens, K. P., and C. A. Busack. 1995. A framework for assessing genetic vulnerability. Fisheries 20(12):24–31.

Currens, K. P., C. B. Schreck, and H. W. Li. 1990. Allozyme and morphological divergence of rainbow trout (*Oncorhynchus mykiss*) above and below waterfalls in the Deschutes River, Oregon. Copeia 1990:730–746.

Daley, W. J. 1993. The use of fish hatcheries—polarizing the issue. Fisheries 18(3):4–5.

Dobzhansky, T. 1970. Genetics of the evolutionary process. Columbia University Press, New York, New York, USA.

Doncaster, C. P. 1995. Lexicon of evolutionary genetics and related topics. Page 10 at <http://www.geodata.soton.ac.uk/biolgy/gene/html>.

Emlen, J. M. 1991. Heterosis and outbreeding depression: a multilocus model and an application to salmon production. Fisheries Research 12:187–212.

Falconer, D. S. 1981. Introduction to quantitative genetics, 2nd edition. Longman, New York, New York, USA.

Falconer, D. S. and T. F. C. Mackay. 1996. Introduction to quantitative genetics, 4th edition, Harlow. Essex: Longman.

Fernandez, J. and A. Caballero. 2001. Accumulation of deleterious mutations and equalization of parental contributions in the conservation of genetic resources. Heredity 86:480–488.

Field, J. D. 1997. Atlantic striped bass management: where did we go right. Fisheries 22(7):6–8.

Fiumera, A. C., P. G. Parker, and P. A. Fuerst. 2000. Effective population size and maintenance of genetic diversity in captive-bred populations of a Lake Victoria cichlid. Conservation Biology 14:886–892.

Fleming, I. A. and M. R. Gross. 1992. Reproductive behavior of hatchery and wild coho salmon (*Oncorhynchus kisutch*) – does it differ? Aquaculture 103:101–121.

Fleming, I. A. and M. R. Gross. 1993. Breeding success of hatchery and wild coho salmon (*Oncorhynchus kisutch*) in competition. Ecological Applications 3:230–245.

Flick, W. A. and D. A. Webster. 1962. Problems in sampling wild and domestic stocks of brook trout (*Salvelinus fontinalis*). Transactions of the American Fisheries Society 91:140–144.

Flick, W. A. and D. A. Webster. 1964. Comparative first year survival and production in wild and domestic strains of brook trout, *Salvelinus fontinalis*. Transactions of the American Fisheries Society 93:58–69.

Forshage, A. A. and L. T. Fries. 1995. Evaluation of the Florida largemouth bass in Texas, 1972–1993. American Fisheries Society Symposium 15:484–491.

Frankel, O. H. 1974. Genetic conservation: our evolutionary responsibility. Genetics 78:53–65.

Frankel, O. H. and M. E. Soulé. 1981. Conservation and evolution. Cambridge University Press, Cambridge, England.

Frankham, R. 1995. Conservation genetics. Annual Review of Genetics 29:305–327.

Frankham, R. 1999. Quantitative genetics in conservation biology. Genetical Research 74:237–244.

Franklin, I. R. 1980. Pages 135–149 *in* M. E. Soulé and B. A. Wilcox, editors. Conservation biology: an evolutionary approach. Sinauer, Sunderland, Massachusetts, USA.

Garant, D., J. J. Dodson, and L. Bernatchez. 2001. A genetic evaluation of mating system and determinants of individual reproductive success in Atlantic salmon (*Salmo salar* L.). Journal of Heredity 92:13–145.

Geiger, H. J., W. W. Smoker, L. A. Zhivotovsky, and A. J. Gharrett. 1997. Variability of family size and marine survival in pink salmon (*Oncorhynchus gorbuscha*) have implications for conservation biology and human use. Canadian Journal of Fisheries and Aquatic Sciences 54:2,684–2,690.

Gharrett, A. J., and S. M. Shirley. 1985. A genetic examination of spawning methodology in a salmon hatchery. Aquaculture 47:245–256.

Gharrett, A. J. and W. W. Smoker. 1991. Two generations of hybrids between even- and odd-year pink salmon (*Oncorhynchus gorbuscha*): a test for outbreeding depression? Canadian Journal of Fisheries and Aquatic Sciences 48:1,744–1,749.

Gharrett, A. J. and W. W. Smoker. 1993a. A perspective on the adaptive importance of genetic infrastructure in salmon populations to ocean ranching in Alaska. Fisheries Research 18:45–58.

Gharrett, A. J. and W. W. Smoker. 1993b. Genetic components in life history traits contribute to population structure. Pages

197–202 *in* J. Cloud, editor. Genetic conservation of salmonids fishes. NATO Advanced Study Institute Series A: Life Sciences. Plenum Press, New York, New York, USA.

Gilpin, M. E. and M. E. Soulé. 1986. Minimum viable populations: processes of species extinction. Pages 19–34 *in* M. E. Soulé, editor. Conservation biology: the science of scarcity and diversity. Sinauer, Sunderland, Massachusetts, USA.

Goldberg, R. J., and T. Triplett. 1997. Murky waters: environmental effects of aquaculture in the United States. Environmental Defense Fund Publications, Washington, D.C., USA.

Goldberg, R. J., M. S. Elliot, and R. L. Naylor. 2001. Marine aquaculture in the United States. Pew Oceans Commission, Arlington, Virginia, USA.

Greene, C. W. 1952. Results from stocking brook trout of wild and hatchery strains at Stillwater Pond. Transactions of the American Fisheries Society 81:43–52.

Hanski, I. 1997. Metapopulation dynamics: from concepts and observations to predictive models. Pages 69–92 *in* I. A. Hanski and M. E. Gilpin, editors. Metapopulation biology: ecology, genetics, evolution. Academic Press, New York, New York, USA.

Hard, J. J. 1995. Genetic monitoring of life-history characters in salmon supplementation: problems and opportunities. American Fisheries Society Symposium 15:212-225.

Harrell, R. M. 1987. Factors affecting and effecting hatchery production and survival of Chesapeake Bay striped bass. Completion report of the University of Maryland, Horn Point Environmental Laboratories to U.S. Fish and Wildlife Service (Contract 14-16-0009-85-012)

Washington, D.C., USA.

Harrell, R. M., and L. C. Woods III. 1995. Comparative fatty acid composition of eggs from domestic and wild striped bass (Morone saxatilis). Aquaculture 133:225–233.

Harrell, R. M., X. L. Xu, and B. Ely. 1993. Evidence of introgressive hybridization in Chesapeake Bay *Morone.* Molecular Marine Biology and Biotechnology 2:291–299.

Hayes, J. W. 1987. Competition for spawning space between brown trout (*Salmo trutta*) and rainbow trout (*S. gairdneri*) in a lake inlet tributary, New Zealand. Canadian Journal of Fisheries and Aquatic Sciences 44:40–47.

Hedrick, P. W. and P. S. Miller. 1992. Conservation genetics: Techniques and fundamentals. Ecological Applications 2:30–46.

Hedrick, P. W. and P. S. Miller. 1994. Rare alleles, MHC and captive breeding. Pages 187–204 *in* V. Loeschcke, J. Tomiuk, and S. K. Jain, editors. Birkhauser Verlag, Boston, Massachusetts, USA.

Hedrick, P. W., P. F. Broussard, F. W. Allendorf, J. A. Beardmore, and S. Orzack. 1986. Protein variation, fitness and captive propagation. Zoo Biology 5:91–99.

Hilborn, R. 1992. Hatcheries and the future of salmon in the northwest. Fisheries 17(1):5–8.

Hindar, K. N. Ryman, and F. Utter. 1991. Genetic effects of culture fish on natural fish populations. Canadian Journal of Fisheries and Aquatic Sciences 48:945–957.

Hjort, R. C. and C. B. Schreck. 1982. Phenotypic differences among stocks of hatchery and wild coho salmon, *Oncorhynchus kisutch*, in Oregon, Washington, and California. U.S. National Marine Fisheries Service Fishery Bulletin 80:105–119.

Hynes, J. D., E. H. Brown, Jr., J. H. Helle, N. Ryman, and D. A. Webster. 1981. Guidelines for the culture of fish stocks for resource management. Canadian Journal of Fisheries and Aquatic Sciences 38:1867–1876.

Incerpi, A. 1996. Hatchery-bashing: a useless pastime. Fisheries 21(5):28.

Jacquard, A. 1975. Inbreeding: one word, several meanings. Theoretical Population Biology 7:338–363.

Kapuscinski, A. R. and L. D. Jacobson. 1987. Genetic guide for fisheries management. Minnesota Sea Grant Research Report 17. Department of Fisheries and Wildlife, University of Minnesota, St. Paul, Minnesota, USA.

Kapuscinski, A. R. and D. P. Philipp. 1988. Fisheries genetics: issues and priorities for research and policy development. Fisheries 13(6):4–10.

Keller, L. F., K. J. Jeffery, P. Arcese, M. A. Beaumont, W. M. Hochachka, J. N. M. Smith, and M. W. Bruford. 2001. Immigration and the ephermerality of a natural population bottleneck: evidence from molecular markers. Proceedings of the Royal Society (Ser B.) 268 (1474):1387–1394.

Kerby, J. H. and R. M. Harrell. 1990. Hybridization, genetic manipulation, and gene pool conservation of striped bass. Pages 159–190 *in* R. M. Harrell, J. H. Kerby, and R. V. Minton, editors. Culture and propagation of striped bass and its hybrids. Striped Bass Committee, Southern Division, American Fisheries Society, Bethesda, Maryland, USA.

Kimura, M. 1983. The neutral theory of molecular evolution. Cambridge University Press, Cambridge, England.

King, T. L., S. T. Kalinowski, W. B. Schill, A. P. Spidle, and B. A. Lubinski. 2001. Population structure of Atlantic salmon (*Salmo salar* L.): a range-wide perspective from microsatellite DNA variation. Molecular Ecology 10:807–821.

Kohane, M. J., and P. A. Parsons. 1988. Domestication. Evolutionary change under stress. Evolutionary Biology 23:31–38.

Kohm, K. A., editor. 1991. Balancing on the brink of extinction. Island Press, Washington D.C., USA.

Krueger, C.C., and B. May. 1991. Ecological and genetic effects of salmonid introductions in North America. Canadian Journal of Fisheries and Aquatic Sciences 48(Supplement 1):66–77.

Krueger, C. C., A. J. Gharrett, R. R. Dehring, and F. W. Allendorf. 1981. Genetic aspects of fisheries rehabilitation programs. Canadian Journal of Fisheries and Aquatic Sciences 38:1877–1881.

Kutkuhn, J. H. 1981. Stock definition as a necessary basis for cooperative management of Great Lakes fish resources. Canadian Journal of Fisheries and Aquatic Sciences 38:1476–1478.

Lande, R. 1988. Genetics and demography in biological conservation. Science 241:1455–1460.

Leary, R. F., F. W. Allendorf, and G. K. Sage. 1995. Hybridization and introgression between introduced and native fish. American Fisheries Society Symposium 15:91–101.

Lepage, O., O. Overli, E. Petersson, T. Jarvi, and S. Winberg. 2000. Differential stress coping in wild and domesticated sea trout. Brain Behavioral Evolution 56:259–268.

Limburg, K. E., M. L. Pace, and K. K. Arend. 1999. Growth, mortality, and recruitment of larval *Morone* spp. in relation to food availability and temperature in the Hudson River. Fishery Bulletin 97: 80–91.

Loeschcke, V., J. Tomiuk, and S. K. Jain. 1994. Conservation genetics. Birkhauser Verlag, Boston, Massachusetts, USA.

Lomnicki, A., and M. Jasienski. 2000. Does fitness erode in the absence of selection? An experimental test with Tribolium. Journal of Heredity 91:407–411.

López-Fanjul, C., A. Fernández, and M. A. Toro. 2000. Epistatsis and the conversion of non-additive to additive genetic variance at population bottlenecks. Theoretical Population Biology 58:48–59.

Lynch, M. 1991. The genetic interpretation of inbreeding depression and outbreeding depression. Evolution 45:622–629.

Mason, J. W., O. M. Brynidson, and P. R. Degurse. 1967. Comparative survival of wild and domestic strains of brook trout in streams. Transactions of the American Fisheries Society 96:313–319.

Maynard, D. J., T. A. Flagg, and C. V. W. Mahnken. 1995. A review of seminatural culture strategies for enhancing the postrelease survival of anadromous salmonids. American Fisheries Society Symposium 15:307–316.

Meffe, G. K. 1986. Conservation genetics and the management of endangered fishes. Fisheries 11(1):14–23.

Miller, R. B. 1954. Comparative survival of wild and hatchery-reared cutthroat trout in a stream. Transactions of the American Fisheries Society 83:120–130.

Mitton, J. B. 1993. Theory and data pertinent to the relationship between heterozygosity and fitness. Pages 17–41 *in* N. W. Thorhill, editor. The natural history of inbreeding and outbreeding. University of Chicago Press, Chicago, Illinois, USA.

Nickelson, T. E., M. F. Solazzi, and S. L. Johnson. 1986. Use of hatchery coho salmon (*Oncorhynchus kisutch*) presmolts to rebuild wild populations in Oregon coastal streams. Canadian Journal of Fisheries and Aquatic Sciences 43:2443–2449.

Philipp, D. P. 1991. Genetic implications of introducing Florida largemouth bass, *Micropterus salmoides floridanus.* Canadian Journal of Fisheries and Aquatic Sciences 48(Supplement 1):58–65.

Philipp, D. P. and J. E. Claussen. 1995. Fitness and performance differences between two stocks of largemouth bass from different river drainages within Illinois. American Fisheries Society Symposium 15:236–243.

Philipp, D. P. and G. S. Whitt. 1991. Survival and growth of northern, Florida, and reciprocal F_1 hybrid largemouth bass in central Illinois. Transactions of the American Fisheries Society 119:2–15.

Philipp, D. P., W. F. Childers, and G. S. Whitt. 1985. Correlations of allele frequencies with physical and environmental variables for populations of largemouth bass, *Micropterus salmoides* (Lacepede). Journal of Fish Biology 27:347–365.

Philipp, D. P., J. M. Epifanio, and M. J. Jennings. 1993. Point/counterpoint: conservation genetics and current stocking practices—are they compatible? Fisheries 18(12):14–16.

Primack, R. B. 1998. Essentials of conservation biology, 2nd edition. Sinauer, Sunderland, Massachusetts, USA.

Purdom, C. E. 1993. Genetics and fish breeding. Chapman and Hall, New York, New York, USA.

Quinn, T. P., M. J. Unwin, and M. T. Kinnison. 2000. Evolution of temporal isolation in the wild: genetic divergence in timing of migration and breeding by introduced chinook salmon populations. Evolution 54:1372–1385.

Ralls, K. and J. D. Ballou. 1986. Proceedings of the workshop on genetic management of captive populations. Zoo Biology 5:81–238.

Rees, R. A., and R. M. Harrell. 1990. Artificial spawning and fry production of striped bass and hybrids. Pages 43–72. *in* R. M. Harrell, J. H. Kerby, and R. V. Minton, editors. Culture and propagation of striped bass and its hybrids. Striped Bass Committee, Southern Division, American Fisheries Society, Bethesda, Maryland, USA.

Reisenbichler, R. R. 1997. Genetic factors contributing to declines of anadromous salmonids in the Pacific Northwest. Pages 223–244 *in* D. J. Stouder, P. A. Bisso, and R. J. Naiman, editors. Pacific salmon and their ecosystems: status and future opportunities. Chapman and Hall, New York, New York, USA.

Reisenbichler, R. R. and J. D. McIntyre. 1977. Genetic differences in growth and survival of hatchery and wild steelhead trout (*Salmo gairdneri*). Journal of the Fisheries Research Board of Canada 34:123–128.

Rhodes, J. S. and T. P. Quinn. 1999. Comparative performance of genetically similar hatchery and naturally reared juvenile coho salmon in streams. North American Journal of Fisheries Management 19:670–677.

Rutherford, E. S., E. D. Houde, and R. M. Nyman. 1997. Relationship of larval-stage growth and mortality to recruitment of striped bass, *Morone saxatilis*, in Chesapeake Bay. Estuaries 20:74–198.

Schonewald-Cox, C. M., S. M. Chambers, B. MacBryde, and L. Thomas, editors. 1983. Genetics and conservation: a reference for managing wild animal and plant populations. Benjamin/Cummings, Menlo Park, California, USA.

Schramm, H. L. and R. G. Piper, editors. 1995. Uses and effects of cultured fishes in aquatic ecosystems. American Fisheries Society Symposium 15. American Fisheries Society, Bethesda, Maryland, USA.

Secor, D. H. 2000. Spawning in the nick of time? Effect of adult demographics on spawning behavior and recruitment of Chesapeake Bay striped bass. ICES Journal of Marine Science 57: 403–411.

Secor, D. H. and E. D. Houde. 1995. Temperature effects on the timing of striped bass egg production, larval viability, and recruitment potential in the Patuxent River (Chesapeake Bay). Estuaries 18:527–533.

Shields, W. M. 1982. Philopatry, inbreeding, and evolution of sex. State University of New York Press, Albany, New York, USA.

Soulé, M. D., editor. 1986. Conservation biology: the science of scarcity and diversity. Sinauer Associates, Sunderland, Massachusetts, USA.

Soulé, M. D. 1991. Conservation: tactics for a constant crisis. Science 253:744–750.

Soulé, M. D. and B. A. Wilcox. 1980. Conservation biology: an evolutionary-ecological perspective. Sinauer, Sunderland, Massachusetts, USA.

Spidle, A. P., W. B. Schill, B. A. Lubinski, and T. L. King. 2001. Fine-scale population structure in Atlantic salmon from Maine's Penobscot River drainage. Conservation Genetics 2:11–24.

Stacey, P. B., V. A. Johnson, and M. L. Taper. 1997. Migration within metapopulations:

the impact upon local population dynamics. Pages 267–291 *in* I. A. Hanski and M. E. Gilpin, editors. Metapopulation biology: ecology, genetics, evolution. Academic Press, New York, New York, USA.

Stahl, G. 1981. Genetic differentiation among natural populations of Atlantic salmon (*Salmo salar*) in northern Sweden. Pages 95–105 *in* N. Ryman, editor. Fish gene pools. Ecological Bulletin Number 34. Stockholm, Sweden.

Stahl, G. 1983. Differences in the amount and distribution of genetic variation between natural populations and hatchery stocks of Atlantic salmon. Aquaculture 33:23–32.

Symons, P. E. K. 1969. Greater dispersal of wild compared with hatchery-reared juvenile Atlantic salmon released in streams. Journal of the Fisheries Research Board of Canada 26:1867–1876.

Taggart, J. B., I. S. McLaren, D. W. Hay, J. H. Webb, and A. F. Youngson. 2001. Spawning success in Atlantic salmon (*Salmo salar* L.): a long-term DNA profiling-based study conducted in a natural stream. Molecular Ecology 10:1047–1060.

Tave, D. 1993. Genetics for fish hatchery managers, 2nd edition. AVI, Westport, Connecticut, USA.

Taylor, E. B. 1991. A review of local adaptation in Salmonidae, with particular reference to Pacific and Atlantic salmon. Aquaculture 98:185–207.

Taylor, S. G. 1980. Marine survival of pink salmon fry from early and late spawners. Transactions of the American Fisheries Society 109:79–82.

Templeton, A. R. 1986. Coadaptation and outbreeding depression. Pages 105–116 *in* M.E. Soulé, editor. Conservation

biology: the science of scarcity and diversity. Sinauer Associates, Sunderland, Massachusetts, USA.

Templeton, A. R. and B. Read. 1994. Inbreeding: one word, several meanings, much confusion. Pages 91–105 *in* V. Loeschcke, J. Tomiuk, and S. K. Jain editors. Conservation genetics. Birkhauser Verlag, Boston, Massachusetts, USA.

Utter, F., K. Hindar, and N. Ryman. 1993. Genetic effects of aquaculture on natural salmonids populations. Pages 144–165 *in* K. Heen, R. L. Monahan, and F. M. Utter, editors. Salmon aquaculture. Wiley, New York, New York, USA.

Vincent, R. E. 1960. Some influences of domestication upon three stocks of brood trout (*Salvelinus fontinalis* Mitchell). Transactions of the American Fisheries Society 89:35–52.

Volpe, J. P., E. B. Taylor, D. W. Rimmer, and B. W. Glickman. 2000. Evidence of natural reproduction of aquaculture-escaped Atlantic salmon in a coastal British Columbia River. Conservation Biology 14:899–903.

Vrijenhoek, R. C. 1994. Genetic diversity and fitness in small populations. Pages 33–53 *in* V. Loeschcke, J. Tomiuk, and S. K. Jain editors. Conservation genetics. Birkhauser Verlag, Boston, Massachusetts, USA.

Vrijenhoek, R. C. 1996. Conservation genetics of North American desert fishes. Pages 367–397 *in* J. C. Avise and J. L. Hamricks editors. Conservation genetics: case histories from nature. Chapman Hall, New York, New York, USA.

Vrijenhoek, R. C, M. E. Douglass, and G. K. Meffe. 1985. Conservation genetics of endangered fish populations in Arizona. Science 229:400–402.

Waldman, B. and J. S. McKinnon. 1993. Inbreeding and outbreeding in fishes, amphibians, and reptiles. Pages 250–282 *in* N. W. Thorhill, editor. The natural history of inbreeding and outbreeding. University of Chicago Press, Chicago, Illinois, USA.

Waples, R. S. 1990. Conservation genetics of Pacific salmon, part 2. Estimating effective population size and the rate of loss of genetic variability. Journal of Heredity 81:267–276.

Waples, R. S. 1991. Genetic interaction between hatchery and wild salmonids: lessons from the Pacific Northwest. Canadian Journal of Fisheries and Aquatic Sciences 48:124–133.

Waples, R. S. 1999. Dispelling some myths about hatcheries. Fisheries 24(2):12–21.

Waples, R. S. and C. Do. 1994. Genetic risk associated with supplementation of Pacific salmonids: captive broodstock programs. Canadian Journal of Fisheries and Aquatic Sciences 51(Supplement 1):310–329.

Waples, R. S., G. A. Winans, F. M. Utter, and C. Mahnken. 1990. Genetic approaches to the management of Pacific salmon. Fisheries (5):15:9–25

Warlen, S. M. 1994. Spawning time and recruitment dynamics of larval Atlantic menhaden, *Brevoortia tyrannus*, into a North Carolina estuary. Fishery Bulletin 92:420–433.

Washington, O. M. and A. M. Koziol. 1993. Overview of the interaction and environmental impacts of hatchery practices on natural and artificial stocks of salmonids. Fisheries Research 18:105–122.

Wishard, L., J. Seeb, F. M. Utter, and D. Stefan. 1984. A genetic investigation of suspected redband trout populations. Copeia 1984:120–132.

Withler, R. E. 1988. Genetic consequences of fertilizing Chinook salmon (*Oncorhynchus tshawytscha*) eggs with pooled milt. Aquaculture 68:15–25.

Wright, S. 1977. Evolution and the genetics of populations, volume 3. University of Chicago Press, Chicago, Illinois, USA,

Young, K. A. 1999. Managing the decline of Pacific salmon: metapopulation theory and artificial recolonization as ecological mitigation. Canadian Journal of Fisheries and Aquatic Sciences 56:1700–1706.

Zastrow, C. E., E. D. Houde, and E. H. Saunders. 1989. Quality of striped bass (Morone saxatilis) eggs in relation to river source and female weight. Rapports et procès-verbaux des réunions Commission internationale pour \-exploration scientifique de la Mer Méditerranée 191:34–42.

Pathogens of Cultured Fishes: Potential Risks to Wild Fish Populations

VICKI S. BLAZER

U.S. Geological Survey, National Fish Health Research Laboratory, Leetown Science Center,

1700 Leetown Road, Kearneysville, West Virginia, 25430 USA

SCOTT E. LAPATRA

Clear Springs Food, Inc., Research Division, P.O. Box 712, Buhl, Idaho, 83316 USA

ABSTRACT

Infectious diseases are important factors of any fish population, both wild and cultured. Infection and the manifestation of clinical disease depend on factors associated with, and the interactions of, host, pathogen, and the environment. Assessing the risk of pathogen transmission between cultured and wild fishes is difficult. Natural and anthropogenic influences, other than infectious organisms, affect the health of wild fishes. In addition, serious deficiencies exist in our knowledge of the current distribution of pathogens, factors affecting pathogen transmission, associated risks of various diseases to wild populations, and management strategies to minimize and avoid pathogen impacts in aquatic ecosystems. Aquaculture can impact wild fish population in terms of disease by the spread or amplification of pathogens as well as other factors involved in disease management, such as the chemical control and treatment of potentially pathogenic organisms. Conversely, pathogen transmission from wild fishes can adversely affect cultured species. The risks associated depend in large part on the type of aquaculture facility (recirculating, raceway, pond, net-pen) and the species cultured. Examples exist in which the importation, movement, and culture of nonindigenous fish species have introduced pathogens that are now impacting indigenous fish populations, including threatened and endangered species. These and other selected examples of pathogens of wild and cultured fishes will be presented to illustrate the difficulties in assessing risk and developing effective management strategies.

Concern has been expressed over potential negative effects of cultured fishes in aquatic ecosystems, and controversy exists about whether the operation of hatcheries and other intensive fish culture facilities is damaging to the environment (Hilborn 1992; Martin et al. 1992). Some of this controversy has centered on fish specifically cultured for release, both for sportfishing and restoration or stock enhancement efforts. It is interesting to note that disease or pathogen dissemination is not mentioned in many of these debates. Four threats to wild fishes of "artificial" production noted by Goodman (1990) and Hilborn (1992) were direct competition for food and other resources, predation of cultured fish, on wild fish, genetic dilution of wild fish stocks by cultured fish, and increased fishing pressure on wild stocks due to artificial production. An international symposium and workshop on the uses and effects of cultured fishes in aquatic ecosystems was held in Albuquerque, New Mexico, in 1994 (Schramm and Piper 1995). Again, very little recognition of pathogen dissemination as a significant area of concern is evident in those proceedings. The Pacific Northwest Fish Health Protection Committee, a coalition of state and federal conservation agencies, Indian tribes, private producers, and the American Fisheries Society Fish Health Section, sponsored a conference in 1997 and a resulting issue of the *Journal of Aquatic Animal*

Health (volume 10, issue 2, 1998) to review information on risks to the environment of pathogens of wild and cultured fish. It was recognized then, and is still the case, that serious deficiencies exist in our knowledge of current distribution of pathogens, potential impacts of disease in aquatic ecosystems, factors affecting pathogen transmission between wild and cultured fish populations, associated risks of various diseases, and management strategies to minimize and avoid pathogen impacts in aquatic ecosystems (Moffitt et al. 1998).

Disease is also a concern to the aquaculture industry. The *Catfish '97* study (USDA 1997), a report of food-size catfish production in Alabama, Arkansas, Louisiana, and Mississippi, representing 95.9% of the total national sales in 1996, contains a number of interesting findings. First, the primary economic loss to the catfish industry was disease and was estimated to account for 45.4% of all losses. Second, the bacterial diseases enteric septicemia *Edwardsiella ictaluri* and columnaris *Flavobacterium columnare* were reported in 78.1% of all operations, winter kill (caused by opportunistic fungal/oomycete organisms) in 35.8% of the operations, and proliferative gill disease (caused by a parasite) in 19.8% of the operations. In general, disease problems increased as operation size increased. In addition to actual losses due to infectious agents, considerable economic resources were spent on disease prevention and management (USDA 1997).

We recognize, although not specifically addressed in this chapter, that water quality parameters associated with prevention or treatment of infectious disease and parasites in intensive culture have the potential to adversely affect wild fish health either directly or indirectly. Chemical treatments can be toxic to nontarget aquatic species (Weston 1991) and can also, in lower concentrations, be immunosuppressive (Zelikoff 1994). The

impacts of antibiotics entering aquatic ecosystems from a variety of sources, primarily human and animal wastes, are just beginning to be recognized. These antibiotics can also be immunosuppressive (Rijkers et al. 1980; Lunden et al. 1998). But perhaps of more significance is the development of antibiotic resistant strains of potential pathogens in aquatic ecosystems. Development of antibiotic resistance in bacterial fish pathogens has been linked to aquaculture facilities (Smith et al. 1994; Schmidt et al. 2000), although the impact on wild fish has not been examined. Chemicals used in aquaculture are approved for use by the U.S. Food and Drug Administration or the Environmental Protection Agency. Rigorous target animal safety and environmental studies are required before any chemical is approved and can legally be used on fish. The number of antibiotics approved for and used in aquaculture is very small, only two (MacMillan 2001), relative to those used in other intensive animal production and by humans (Levy 1998) and in other countries.

Pathogens are important factors in health of both cultured and wild fish populations. Our ability to identify causes of disease outbreaks and mortality, and our understanding of pathogen transmission, predisposing factors, and control of pathogens is far greater for cultured than for wild populations. Certainly diseases occur "naturally" in wild populations, and the prevalence and impact of an individual pathogen depend on factors associated with host, pathogen, and the environment. Understanding the interactions of these three is also generally much more difficult in wild than in cultured fishes. Resource managers are concerned about the effects, including pathogens, aquaculture facilities have on the environment. Conversely, aquaculturists are concerned about the transfer and impact of infectious agents harbored by wild fishes on cultured fish. In this chapter we will discuss disease concepts in general and the factors that influence pathogen transmission. We will provide examples of pathogens, introduced or suspected to be introduced, through cultured fish that have impacted wild populations. We also acknowledge the groups and programs already addressing these issues and discuss research gaps/needs and methodologies being developed to address some of the research needs. We will concentrate on U.S. aquaculture in general, food fish as well as tropical fish importation and culture for the aquarium trade, and culture of fish for release in restoration efforts.

Concepts and Issues of Disease in Wild Populations

Disease has been defined as any impairment that interferes with or modifies the performance of normal functions, including responses to environmental factors such as toxicants and climate, nutrition, infectious agents, inherent or congenital defects, or any combination of these factors (Wobeser 1981). Here we will focus on infectious diseases and the pathogens that cause them. However, even the manifestation of infectious disease in a population depends on complex interactions of variables such as genetics, nutrition, environment, pathogen, and host organisms. Understanding the complex interactions between host, pathogen, and environment is the challenge (Hedrick 1998; Reno 1998).

The list of important infectious disease problems of cultured fishes has expanded greatly during the past two decades. Plumb (1997) lists a variety of reasons that include newly evolved agents, agents that may have previously existed but are now emerging as problems due to intensive culture or environmental perturbations, culture of a greater variety of fish species with varying susceptibility to potential pathogens, and increased numbers of researchers with more diverse backgrounds working in the area of fish health. In some incidences, pathogens are not new but are now found in a new geographic location or species. Although some pathogens appear to have species or group (e.g., salmonid) specificity, others adapt and survive in a variety of hosts. For example, a study of iridovirus-like agents from redfin perch *Perca fluviatilis* and a frog from Australia, sheatfish *Silurus qlanis* in Germany and black bullhead *Ameiurus melas* in France suggests all appear to be similar (Hedrick et al. 1992).

Assessing the risk of pathogen transmission between cultured and wild fishes and potential effects on wild fish populations is difficult. Natural and anthropogenic influences, such as climate change, floods, drought, impoundments, dams, chemical contaminants, and fishing pressure can impact fish populations and increase the difficulty of assessing risk from disease. Studies that document the prevalence and impacts of infections or specific pathogens on free-ranging fish are limited. Most often, surveys of fish health in wild populations document contaminant effects or other environmental stressors and emphasize histopathological or physiological changes rather than pathogen isolation and identification (USEPA 1995; Schmitt and Dethloff 2000). The U.S. Fish and Wildlife has initiated a wild fish survey (http://wildfishsurvey.fws.gov) with the objective of applying a standardized approach to compare watersheds for distribution of parasites and pathogens. This, and surveys like it, will help to fill an information void. We need a better understanding of the issues concerning the presence of a pathogen versus actual disease. Infection is defined as invasion of a host by a pathogenic agent, while clinical disease is a condition that results in morbidity and sometimes mortality, as a consequence of infection. Clinical disease is apparent by various signs of infection, poor feed conversion, reduced growth, and reduced fecundity, and may range from peracute (extremely rapid progression) to chronic (slow progression and long duration). Many infectious organisms can be carried by individuals without overt signs of disease, i.e., asymptomatic infections. Asymptomatic infection may progress to clinical disease or may remain subclinical—a carrier state in which the host may function as a reservoir of infection to other organisms. With the continuing advances and increased sensitivity of DNA-based and other molecular diagnostic tools, controversy has developed over interpreting the significance of positive results. It is critical to understand the difference between the presence of a component (e.g. protein, DNA) of a pathogen or even the pathogen itself, and actual manifestation of disease. Lastly, it is often difficult in the natural environment to find moribund or recently dead fish, unless the pathogen causes massive fish kills or large numbers of fish with external lesions. Sick or abnormally behaving fish will die or be rapidly eaten by predators. Hence, a low number of fish found in a wild fish survey infected with a highly virulent pathogen could be very important while the occurrence of pathogens of lower virulence may be less important unless there is an epidemic (McVicar 1988).

When assessing the risk of a particular pathogen being spread or introduced by cultured fish, a variety of factors need to be considered. These include:

- Can the pathogen survive free in the environment? If so, for how long and under what environmental conditions?

- How is the pathogen transmitted—water, food, parasites, sex products? If vertical (intergenerational) transmission is a concern, can it be spread via eggs, sperm, or both?

- What effect does dose, route, and duration of exposure have on likelihood of infection or disease? Is there a threshold level of exposure required to initiate infection or disease?

- Does the pathogen survive in fresh, frozen, or processed fish products and can it be spread through unprocessed fish offal?

- Are there intermediate hosts required as part of the pathogen's life cycle, and if so, what environmental factors regulate their populations?

- Is the organism an obligate pathogen always associated with disease or is it a facultative pathogen—one that may be commonly present in the environment and only causes disease in stressed hosts or if high numbers of the infectious agent are present?

- What environmental stressors enhance susceptibility of the host to the particular pathogen?

- What is the probability of the pathogen surviving at a threshold concentration needed to initiate infection coincident with a susceptible fish species being present?

Potential Interactions of Cultured and Wild Fish Populations in Terms of Pathogen Transmission

The aquaculture industry has the potential to impact wild fish populations through infectious disease in a number of ways:

- Importation or culture of nonindigenous species can introduce previously unknown or exotic pathogens to an area. This may be importation of the fish themselves, of various intermediate hosts such as snails that may be used in the aquarium trade, worms such as tubificids that may be used as a food source, and contaminated water used to ship fish and other organisms.

- Movement of various strains or families of fish (indigenous or nonindigenous) from other countries/areas/states/watersheds can introduce new pathogens or strains of pathogens into a particular area.

- Intensive fish culture can exacerbate diseases or amplify pathogens that previously existed in wild populations. Many infectious organisms, including the viruses, bacteria, fungi, and parasites that commonly cause disease outbreaks in cultured fish can be part of the natural environment. However, the organisms may not be present in sufficient numbers to initiate infection or overcome the natural defenses of the host. The intensive nature of aquaculture increases the likelihood of disease outbreaks because of crowding, less than optimal water quality and nutritional status, and other stressors associated with captivity that can lower disease resistance. In addition, high nutrient loads in the water, often associated with aquaculture can enhance proliferation of opportunistic pathogens.

The risk of pathogens being discharged from an aquaculture facility certainly depends greatly on the type of facility and its effluent treatment program. In addition, the interest and resources provided for research in the area of pathogen impacts of cultured fish on wild fish populations depend greatly on the type of fish cultured and the types of wild fishes likely to be impacted. For instance, semi-closed system aquaculture probably represents the least risk potential for releasing pathogens into the environment. Waste water can be treated appropriately for the organisms of concern, and access to the system or fish by intermediate hosts, such as snails and worms, or carriers of organisms, such as fish-eating birds, is generally restricted. However, of the five principal fish groups cultured in the U.S. (catfish, trout, salmon, tilapia, and hybrid striped bass), only tilapia is primarily cultured using semi-closed system aqua-culture (USDA 1995).

Catfish, primarily *Ictalurus punctatus*, represents by far the largest U.S. aquaculture industry, principally located in the South where longer growing seasons and warm water are conducive to production (USDA 1995). Catfish are cultured in ponds that are reused year after year and only drain directly into streams or rivers when they are periodically emptied for renovation. Channel catfish are not highly prized as a sportfish and are not endangered. Nor is it necessary to get broodfish or gametes from wild populations. However, these open systems offer free access to intermediate hosts such as birds and snails. Little information is available on infectious diseases of wild populations of channel catfish or other ictalurid fishes because few, if any, surveys have been conducted. Conversely, most salmonid species are cultured in flow-through raceway facilities with water continually discharged into streams and rivers after passing through the aquaculture operation. Although effluent water may be treated to meet effluent permit discharge requirements, the water is generally not disinfected, due to economic constraints. Many trout species, highly regarded as sportfish, are raised and stocked by state and federal hatcheries for stock enhancement, and some species/stocks are threatened or endangered. For these reasons there are considerable resources made available when a disease such as whirling disease is believed to cause declines of natural populations.

On the opposite end of the risk continuum is net-pen culture, particularly of migratory fish species such as salmon. Most grow-out of farmed salmon takes place in net pens or sea cages anchored in nearshore coastal sites where the fish are fed and grown to harvestable size. Use of the natural environment in this way potentially allows for pathogens from farmed species to be transmitted to invertebrates, other fishes, plankton, and birds as well as the water and sediment, which can then serve as vectors of fish pathogens. For instance, marine plankton can carry viable *Aeromonas salmonicida,* an important bacterial pathogen of salmonids (Nese and Enger 1993), while sea lice are known to be a vector of infectious salmon anemia virus, an important viral pathogen of Atlantic salmon *Salmo salar* (Nylund et al. 1994). There has been much debate in recent years concerning the culture of salmon and whether there is a greater risk of wild fish transmitting diseases to cultured fish or vice versa, and a number of reviews of studies conducted in Canada and the European countries are available (Håstein and Lindstad 1991; Kent 1994; McVicar 1997; Stephen and Iwama 1997; Bakke and Harris 1998). Additional factors that complicate this issue are the decline of wild salmon stocks, the commercial and sport fishing industries for salmon, the migratory/anadromous nature of most species, acquisition of eggs and sperm from wild populations, and the recent listing of species such as Atlantic salmon as endangered.

Examples of Pathogens Impacting Wild Populations

In this section we will use a number of illustrations of pathogens impacting wild populations, that are known or suspected, to have been introduced through cultured fishes. The purpose is not to provide a comprehensive list of pathogens or to thoroughly describe individual diseases and causes, but rather to use these examples to illustrate the various routes by which pathogens can be introduced, factors that determine potential risks and the difficulty in controlling pathogens once they are introduced into an area/watershed.

Parasites

Helminth and sporozoan parasites are often well adapted to their definitive host and hence may not cause serious damage or death to that host. However, alternative hosts that become infected may not have the same adaptations and so may be more susceptible. Examples of parasitic diseases that have impacted wild populations provide insights into the different life cycles that must be understood, difficulties in identifying sources and modes of transmission, and the conclusion that removing a parasite once it is established is generally an unrealistic option.

Cestode Bothriocephalus acheilognathi. The Asian tapeworm *Bothriocephaus acheilognathi* (synonyms *B. gowkongenesis, B. opsariichthydis,* and *B. phoxini*) is a cestode in which the procercoid development occurs in a number of genera of copepod intermediate hosts (Paperna 1991). This parasite was apparently first identified in grass carp *Cenopharyngodon idella* in the Amur river in southern Siberia (Yukhimenko 1970) and spread to farmed and native cyprinids and some noncyprinids via introductions of grass carp and common carp to Europe and Asia for aquaculture (Hoffman and Schubert 1984). The tapeworm was identified in North America in 1975 (Hoffman 1980) and became established in fish farms where golden shiners *Notemigonus crysoleucas,* fathead minnows *Pimephales promelas* and grass carp were raised (Scott and Grizzle 1979). More recently, the use of poeciliids (such as mosquitofish *Gambusia affinis*) for mosquito control and possible releases of exotic fishes from aquaria have been suggested as mechanisms for introductions of this parasite into native fish populations in areas such as Hawaii (Font and Tate 1994). A study of the parasites of native Hawaiian fishes indicated native fish from streams where no exotic fishes were found were completely free of adult helminths. Conversely, in two rivers that did contain exotic species, nematodes and Asian tapeworms were found in the exotic species as well as the native fishes (Font and Tate 1994).

The life cycle of this parasite involves shedding of the eggs into the water column from mature worms found in the intestine of host fish. Intermediate stages develop within the eggs in 3–5 d, depending on water temperature (3–4 d at 16–19 C; 2 d at 22–25 C), and are eaten by copepods in which the procercoids develop. Development of infective procercoids is also temperature dependent and may take 4–12 d. The procercoids infect fish when infected copepods are eaten. The tapeworm matures in the fish intestine, again dependent on temperature, and can begin to lay eggs in 20–25 d (Bauer et al. 1969).

Large numbers of tapeworms (30–156) can fill the intestine, causing blockage and perforation of the gastrointestinal tract. The parasite attaches to the intestinal wall and lesions include mild to severe catarrhal enteritis with hemorrhaging and connective tissue proliferation occurring at the point of attachment (Scott and Grizzle 1979; Hoffman 1980). Fish may become emaciated and infestation with this parasite can cause up to 90% mortality in heavily infected carp (Paperna 1991).

The variety of copepod intermediate hosts and fish hosts have aided in the spread of this parasite. It appears that copepod hosts are readily found within the parasite's temperature range (Hoffman 1980). Temperature appears to be the one controlling factor—egg hatching is delayed from 10–28 d at temperatures of 14–15 C and development does not occur below 12 C (Liao and Shih 1956). Eggs are also sensitive to drying and freezing (Bauer et al. 1969).

This tapeworm has had an adverse effect on the baitfish industry, but more recently effects on native fish populations are a concern. Surveys of native cyprinids from the Little Colorado River revealed varying rates of infection with the Asian tapeworm. Mean prevalences of 28% (range 0–78%) for humpback chub *Gila cypha* and 8% (range 0–46%) in speckled dace *Rhinichthys osculus* with individual fish having worm numbers as high as 46 and 28, respectively, have been reported (Clarkson et al. 1997). These authors suggested the rapidity with which the Asian tapeworm has spread to different drainages of the Colorado River Basin means it will eventually have a cosmopolitan basin distribution in areas suitable to the parasite's life history. The humpback chub is an endangered species, and this parasite is considered one of the more serious threats to its conservation and recovery. The parasite was also found in the San Juan River drainage in 1994 and poses a threat to native speckled dace, roundtail chub *Gila robusta* and the endangered Colorado pikeminnow *Ptychocheilus lucius* in this drainage (Landye et al. 1999).

Digenetic Trematode Centrocestus formosanus. Centrocestus formosanus is a heterophyid trematode for which the first intermediate host is the thiard snail *Melanoides tuberculata* (Chen 1942). *M. tuberculata*, an exotic freshwater snail, is now widely distributed in tropical, subtropical, and some temperate regions (Abbott 1973). It has been used as a biological control agent for other freshwater snails, particularly those that serve as intermediate hosts for intestinal schisto-somiasis (Madsen 1990). The definitive hosts are piscivorous birds and mammals (Chen 1942; Scholtz and Salgado-Maldonado 2000). More than 40 species of fish in Mexico, mainly poeciliids and cyprinids, have been reported with infections of *C. formosanus* (Salgado-Maldonado et al. 1995).

The cercariae of the parasite are released by the snail, burrow into the gills of fishes, and take up residence along the cartilaginous rod of the gill filament, where they develop into metacercariae. They elicit an unusual pathological response—that of extensive proliferation of cartilage, leading to distortion of gill filaments, epithelial hyperplasia, and fusion of gill lamellae and even fusion of adjacent gill filaments. Infected fish often have flared and hemorrhagic gills (Blazer and Gratzek 1985; Mitchell et al. 2000). Heavy infestations can result in reduction of respiratory surface and hence, respiratory distress (Blazer and Gratzek 1985; Paperna 1991).

It is not known how *C. formosanus* entered the United States and very little recognition or study has been directed toward this parasite. There are a number of possibilities for the route of entry—introduction of the infected snail host (through the aquarium trade or by extension of it range from other countries), movement of infected aquarium fishes, and spread by migratory, fish-eating birds. The parasite is widespread in Mexico and two possibilities have been suggested for its introduction. It was first recorded parasitizing the gills of the introduced carp *Mylopharingodon pisceus* and other fish from a fish farm (López-Jiménez 1987) and later in natural populations of Poeciliids, highly appreciated as ornamental fishes (Salgado-Maldonado et al. 1995). Hence, introduced fish were suggested as a source. Other reports suggest a more likely source to be the importation of the infected snail host from Asia in 1979 (Scholz and Salgado-Maldonado 2000). However, it was also noted in that report that *C. formosanus* has never been documented from countries where *M. tuberculata* had lived for decades, including regions of the southern United States, which appears not to be the case. The snail was apparently in Texas in 1964 (Murray 1971) and in Florida as early as 1971 (Russo 1973). The parasite was discovered in Texas in 1990 in San Antonio where the snail was originally discovered (H. Murry, personal communication). It is interesting to note that what we now believe to be the same parasite was found in a variety of pond-held aquarium fish in Florida in the early 1980s (Blazer and Gratzek 1985). The parasite was identified at that time only as a heterophyid trematode and the snail host as a cone snail. However, recent comparisons of photographs (V. Blazer) of the snails, cercariae and metacercariae from these early cases to the current findings in the Comal and San Marcos rivers (T. M. Brandt, U.S. Fish and Wildlife Service, San Marcos, Texas, USA and A. J. Mitchell, U.S. Dept. of Agriculture, Stuttgart, Arkansas, USA) suggest the snails found to be infected in the Florida fish farm in the 1980s were *M. tuberculata* and the parasite was *C. formosanus*.

Recently this parasite has been reported to cause gill lesions in a number of threatened or endangered species in Texas, including the fountain darter *Etheostoma fonticola*, Devils River minnow *Dionda diaboli*, Rio Grande darter *Etheostoma grahami*, Proserpine shiner *Cyprinella proserpina*, Comanche Springs pupfish *Cyprinodon elegans* and Pecos gambusia *Gambusia mobilis*, as well as at least 11 other species of non-listed fishes (McDermott 2000; Mitchell et al. 2000). The endangered fountain darter is endemic to only two spring-fed rivers in central Texas, the Comal and San Marcos rivers (U.S. Fish and Wildlife Service 1995). Mitchell et al. (2000) recently examined darters from both rivers and found varying parasite prevalence between rivers, 3% to 100% respectively. Varying intensity of infection in the Comal river darters, from 8–1,524 cysts per fish, were also noted. The few infected darters from the San Marcos river only contained 1–2 cysts per fish, even though the snail host was abundant in places within the river.

A report on the use of bayluscide for snail control in commercial fish farms in Florida suggests *M. tuberculata* is well established and that the *C. formosanus* metacercarial infections of gills can cause significant losses to the industry (Francis-Floyd et al. 1997). The snail's tolerance of salinities between 0 and 30 ppt (Roessler et al. 1977), and hydrogen sulfide up to 3.4 µg/L in oxygen-depleted water (Heller and Ehrlich 1994) mean the potential for spread of this parasite in the southern U.S. is great.

Myxosporidian Parasite Myxobolus cerebralis. Whirling disease, caused by *Myxobolus cerebralis*, is an example of a fish disease that has received much public, political, and research attention in the last decade. *Myxobolus cerebralis* (synonyms *Myxobolus chondrophagus, Lentospora cerebralis,* and *Myxosoma cerebralis*) was first detected in 1883 in Germany in both rainbow trout *Oncorhynchus mykiss* and brook trout

Salvelinus fontinalis (Hofer 1903). Neither of these species is indigenous to Europe and was introduced as imported eggs. Brown trout *Salmo trutta*, a resistant salmonid, is native to Europe. Hoffman (1970) suggested brown trout is the natural host of this parasite and hence clinical signs and mortalities were not noted until more susceptible nonindigenous species were introduced. Throughout the early 1900s the pathogen spread throughout Europe and parts of Asia, adversely impacting cultured trout populations (Hoffman 1970). The impact on wild trout populations, which were primarily brown trout, was reported as negligible (Christensen 1972). In Russia clinical disease was unknown in natural populations where infections were usually light (Bogdanova 1970). Hoffman (1970) attributed the dissemination throughout Europe to unrestricted transfers of wild rainbow trout, development of a frozen table trout market, and brown trout acting as possible carriers.

In the U.S. whirling disease was first detected in 1956 from brook trout in Pennsylvania (Hoffman 1962). Potential causes of the initial outbreaks are imported, frozen European table trout fed to the hatchery trout, viscera from infected imported table trout discarded in the stream, or the importation of infected, live brown trout with no clinical signs by government and private hatcheries in the 1950s (Hoffman 1990, Hnath 1996). The subsequent spread of *M. cerebralis* in the U.S. was attributed to transfers of live fish by both government agencies and commercial trout growers (Hoffman 1970, 1990). Although widely distributed by the 1970s, clinical whirling disease was only reported in fish from aquaculture facilities. However, few surveys of wild fish for low-level infections were conducted during this time. A survey in Michigan did indicate that the parasite had become established in native brook and brown trout found below an aquaculture facility that contained infected fish (Yoder 1972). Despite

the apparent lack of effect on wild fish populations, *M. cerebralis* was considered a resource threat because of the susceptibility of rainbow trout that are indigenous to the western U.S. and widely dispersed throughout most states. In 1968, Title 50 Wildlife and Fisheries Act (Code of Federal Regulations, Title 50, Section 13.7) became effective and required salmonids and salmonid eggs to be certified free from *M. cerebralis* and viral hemorrhagic septicemia virus (VHSV) prior to importation into the United States. Although much was known about susceptibility of various salmonid species, pathology and distribution, the life cycle of this parasite was still unknown.

In the 1980s the complex life cycle of *M. cerebralis* was finally elucidated (Markiw and Wolf 1983; Wolf and Markiw 1984), and independently confirmed (El-Matbouli and Hoffman 1989). It was discovered that the intermediate worm host *Tubifex tubifex* was required to complete the life cycle. Infected fish harbor *Myxobolus* spores in the bone and cartilage of head and backbone. Upon death and decomposition, the spores are released and ingested by the tubificid worm in which the infective stage, referred to as triactinomyxon spores, develop and are released. The triactinomyxon attaches to the epithelium of the fish, inserts the sporoplasms that divide, and migrate along nerves to the head and backbone cartilage (El-Matbouli et al. 1995). Although fish can be lifetime carriers, spores are not released until death and are not transmitted either vertically or horizontally. A variety of mechanisms have been implicated for the distribution of *M. cerebralis* including transfers of subclinically infected live or processed salmonids, migrating anadromous fish, fish-eating birds (Taylor and Lott 1978; El-Matbouli and Hoffman 1991), contaminated boats, boots and other equipment (Modin 1998). Movement of infected tubifex worms could also spread the pathogen.

In 1993, *M. cerebralis* was removed from

Title 50 because it was believed to be primarily a hatchery problem and thus manageable. States and regions managed *M. cerebralis* in different ways. Some states required destruction of all fish on a facility, quarantine of affected facilities, and treatment of ponds and streams with chlorine and quicklime. Other states restricted stocking of infected fish and replaced earthen ponds and raceways with concrete. In California, the problem was first noted in 1965 in fish from a private trout farm (Horsch 1987). Between 1965 and 1984, *M. cerebralis* was detected in five commercial operations and one state hatchery. In each case infected stocks, estimated to include 660,000 kg of catchable and brood fish and about 2.3 million fingerlings, were destroyed. Sampling since 1984 has revealed many positive wild fish populations in California despite eradication and cleanup efforts. Current policy does not require destruction of infected fish, but rather limits distribution of state or commercially raised fish to known enzootic waters, or to terminal waters with no sustaining salmonid fish populations. Despite this change in policy, surveys have revealed little evidence of clinical disease or population declines. In many waters positive for *M. cerebralis,* fish numbers and year-class composition of rainbow and brown trout have remained constant (Modin 1998). A similar lack of effect on wild populations has been reported in other states such as New York and Pennsylvania where *M. cerebralis* is enzootic (Hnath 1996; Schachte and Hulbert 1998).

In 1988, the Colorado River Fisheries and Wildlife Council reclassified *M. cerebralis* from prohibited to notifiable. This re-classification required inspection but did not require destruction of infected fish or disinfection of facilities and gave the states flexibility to regulate disposition of fish from infected facilities. Subsequently, in Colorado, population declines in wild rainbow trout have been related to the presence of *M. cerebralis*

(Walker and Nehring 1995). Severe declines in rainbow trout numbers, beginning in 1991, in the Madison River in Montana, a river that has not been stocked since the late 1970s have also been attributed to whirling disease (Vincent 1996).

Why *M. cerebralis* causes population effects in wild fish in some areas but not others is the subject of much current research. On the west coast of the U.S. combined conditions of climate, topography, and the oligotrophic nature of areas such as the West Sierra drainage basins may contribute to the lack of significant impacts on wild salmonid populations (Modin 1998). Fish in drainages most severely affected by *M. cerebralis* may be those in which substantial habitat alteration has occurred. Habitat degradation can increase stress in fish, making them more susceptible to a variety of diseases and may also increase the abundance of the intermediate host *T. tubifex* (Allendorf et al. 2001). Also, some wild fish may avoid the effects of *M. cerebralis* due to life history characterisitics. Salmonids exhibit variation in spawning and early life history as a result of local adaptation to environmental conditions (Metcalfe 1993; Healey and Prince 1995). There is a relatively narrow time period (2–3 mo after hatch) in which fish are likely to become infected and experience mortality from *M. cerebralis* (Markiw 1991, 1992). Hence, individuals that spawn early, during colder temperatures, may avoid exposure of the susceptible young fish to high concentrations of the triactinomyxon stage (McMahon et al. 1999).

Viruses

To assess the risk for introduction and establishment of a fish virus in an aquatic environment a number of factors must be considered. These include:

- host specificity and susceptibility
- ability of the virus to replicate in hosts that do not show clinical signs of disease

- ability of the virus to exist in a carrier state and be shed by infected individuals
- virus liability to pH, cold, heat, and other environmental parameters
- the ability to survive free in the environment long enough to reach and infect another host and
- the virulence or strain of virus.

The examples illustrated below represent viral infections well-established in the United States and one emerging problem, and demonstrate the various parameters associated with different families and strains of viral pathogens.

Infectious hematopoietic necrosis virus. Infectious hematopoietic necrosis virus (IHNV) is a Rhabdovirus that has caused epizootics among wild (Williams and Amend 1976), hatchery-reared freshwater salmonid fishes (Groberg and Fryer 1983) and farmed fish in saltwater (Traxler et al. 1997). This virus affects cultured salmon and trout worldwide and is the most important viral pathogen affecting salmonid fishes in North America (Wolf 1988). The IHN virus is considered endemic in sockeye stocks from Alaska to California, nevertheless, the sockeye salmon populations remain viable in British Columbia and Alaska, where IHN epizootics have occurred for a number of years (Traxler et al. 1997). The virus is also endemic inland to Idaho and has been reported in other states such as Minnesota (Plumb 1972), South Dakota and West Virginia (Wolf et al. 1973), Montana (Holway and Smith 1973), New York (Carlisle et al. 1979), and Colorado (Janeke 1984); however, these outbreaks have usually been associated with the movement of infected eggs or fry, and IHNV did not become established in these areas. High mortalities in marine net-pen culture of Atlantic salmon in British Columbia (Armstrong et al. 1993) and finding the virus in adult sockeye salmon in seawater suggest the possibility of a marine reservoir

(Traxler et al. 1997). A survey of diseases of wild ocean fishes captured near net-pen farms and from open ocean waters of British Columbia found IHN virus in a Pacific herring *Clupea pallasi* captured at a site not associated with salmon farms and in tube-snout *Aulorhynchus flavidus* and shiner perch *Cymatogaster aggregata* from the area near a net-pen farm experiencing an IHN outbreak (Kent et al. 1998). This further suggests that nonsalmonid marine fishes may serve as a reservoir of the virus.

A recent review described factors affecting the pathogenicity of IHNV for salmonid fish and how these factors are involved in the occurrence of disease (LaPatra 1998). Factors that affect the virulence of a pathogen include the contagiousness or invasiveness of the pathogen, the viral strain, and resistance of the virus to host defenses. Different strains of IHNV have been shown to be highly pathogenic to a specific host species. For example, a type 1 electropherotype has been shown experimentally to maintain a high degree of pathogenicity in kokanee salmon *Oncorhynchus nerka* independent of fish size. All IHN epizootics that have been documented in natural populations have occurred in sockeye salmon or kokanee. Furthermore, these natural epizootics were either due to a strain of IHNV that was a type 1 electropherotype or the epizootic occurred in an area where type 1 strains are endemic. Additionally, the majority of these epizootics occurred in 1 or 2-year-old fish, which contradicts the general notion that host susceptibility decreases with increasing size and age. This may suggest that sockeye salmon and kokanee are "natural" hosts of IHNV and that the range of IHNV expanded due to straying of anadromous salmonids, fish transfers, and with subsequent adaptation of the virus to alternative hosts, but with reduced pathogenicity of the viral strain.

Some information is known about IHNV's lability to pH and temperature and survival in different water systems, but the data is limited and sometimes inconsistent. Infectivity studies have indicated the virus is able to survive for some time in the environment, and horizontal transmission may occur via water (fresh and salt), feces, urine, mucus, and feed (Mulcahy et al. 1983; Wolf 1988; Traxler et al. 1993). Recently, LaPatra et al. (2001) evaluated the survival of three IHNV isolates that exhibited antigenic differences. Virus suspended in spring water survived longer than virus incubated in water obtained from a fish farm or the river. Virus suspended in river water exhibited a 99% reduction in virus concentration in 24 h. Survival of IHNV at different temperatures also varied with the IHN isolate tested. This study demonstrated strain differences in sensitivity to environmental conditions and also that very little virus survived after 24 h in river water.

A study by Anderson et al. (2000) evaluated virus trafficking and risk of IHNV to wild populations using molecular methods. The IHNV genetic heterogeneity and viral traffic was monitored at a study site in the Deschutes River watershed in Oregon where epidemics occurred in wild kokanee between 1991 and 1995. Forty-two IHNV isolates collected from this area between 1975 and 1995 were characterized on a genetic basis. Analysis suggested that both virus evolution and introduction of new IHNV strains contributed to the genetic diversity observed. The results indicated that the 1991–1995 epidemics in kokanee from Lake Billy Chinook were due to a newly introduced IHNV type that was first detected in spawning adult kokanee in 1988 and that this virus type was transmitted from the wild kokanee to hatchery fish downstream in 1991.

Foott et al. (2000) described the risk of disease associated with the release of IHNV infected chinook smolts from a hatchery on natural chinook. Historically, when epizootics are detected in the hatchery, juvenile fish are released into the Sacramento River, a practice that has raised concerns over potential impacts to the "natural" or wild chinook salmon juveniles. A survey of hatchery smolts captured down river had a prevalence of IHNV infection ranging from 9–12% over a 2-wk period following release; however, viral infection has not been detected in natural chinook juveniles ($\geq$ 500 fish examined over 4 yr). When uninfected, wild chinook juveniles were co-habitated at different ratios with infected hatchery chinook; virus was not detected in the wild fish from any exposure group. Foote et al. (2000) concluded that the ecological risk to natural stocks from the release of IHNV-infected hatchery chinook smolts was low and suggests that neither effluents nor infected hatchery fish pose a risk to the wild population in this particular system.

Infectious pancreatic necrosis virus. Infectious pancreatic necrosis virus (IPNV) is an aquatic Birnavirus that has a broad geographic distribution and has been detected in a wide range of freshwater and marine, salmonid and nonsalmonid fishes, as well invertebrates and homeotherms (McAllister 1993). Fish can develop an inapparent infection that can persist as a lifelong carrier state, or they can develop overt disease with high mortalities (Wolf 1988). Virus-contaminated eggs, fry, and fingerlings (Wolf 1988; McAllister 1993) as well as birds (Peters and Neukirch 1986; McAllister and Owens 1992) can spread the virus. Isolations from various fish species, invertebrates, and homeotherms have in part been attributed to contact with discharges or products of contaminated fish culture facilities (Sonstegard and McDermott 1972; Bucke et al. 1979).

Yamamoto and Kilistoff (1979) evaluated the persistence of IPNV in brook trout in an isolated natural lake. Subsequent to the stocking of brook trout from a hatchery, at least 90% were determined to be carriers. This particular lake had few resident salmonids so the stocked fish made up a majority of the population. In the first year 88% of the trout examined were positive for IPNV, but this decreased to approximately 50% within 6 yr. Progeny of these stocked fish were also studied. The overall virus isolation from progeny fish over a 2-yr period was 0.2%. Other salmonids (disease-free) were stocked during these 6 yr and none were found to be infected. Other lakes, stocked with the same population of infected fish the first year but with disease-free fish in subsequent years, showed the number of carrier fish decreasing to undetectable within 5–6 yr. However, in these lakes the stocked population represented a much smaller proportion of the population (Yamamoto and Kilistoff 1979).

McAllister and Bebak (1997) monitored effluents from three fish hatcheries (known to contain fish infected with IPNV) for discharge and downstream distribution of IPNV. They found no virus upstream of the hatcheries or in the hatchery spring water supplies. However, virus could be detected 19.3 km (the furthest distance tested) below the hatchery discharge. Virus titers in the streams ranged from 1.3 X 10^4 to 1.9 X 10^1 plaque forming units (PFU) per liter and, unlike the findings with IHNV, little IPNV infectivity was lost as the virus passed downstream. Virus burden downstream was affected by stream dilution parameters. Previous work had shown survival of IPNV in natural waters for at least 8 wk (Wedemeyer et al. 1978). McAllister and Bebak (1997) also sampled 106 resident fish downstream of the hatcheries, but found no IPNV in the 61 nonsalmonid fishes. IPNV was detected in three (all brook trout) salmonid fish. Two of these were adults and believed to be hatchery escapees. A fingerling brook trout, captured about 8 km downstream was believed to be the consequence of instream infection. No clinical signs were found in any of the positive fish. Further work used chronic, low-level exposures, similar to the levels found in stream

waters, to estimate the effects on early life stages of rainbow trout. Infection, IPNV mortality, and carrier prevalence were affected by virus concentration and population density. At the lowest exposure levels, similar to those taken downstream, zero mortality and no carrier fish were detected at low population density. However, at the higher concentrations of virus, similar to near hatchery water samples, mortalities and prevalence of carriers increased. Ambient virus concentration, population density, and excretion augmentation of virus all interact to initiate and sustain infection in a population (McAllister and Bebak 1999).

Infectious salmon anemia virus (ISAV). Infectious salmon anemia (ISA) provides an example of an emerging disease problem that has recently been found in salmon net pen facilities in Maine (Bouchard et al. 2001), despite regulations and restrictions in place in an attempt to keep it out of U.S. aquaculture. It also demonstrates the difficulty in managing fish health in marine net pen culture. The disease, also called hemorrhagic kidney syndrome, was first diagnosed from cultured fish in Norway in 1984 (Thorud and Djupvik 1988) and within a few years became a disease of significant importance to the Norwegian fish farming industry. It was not known outside of Norway until 1997 and has since been detected in Canada (New Brunswick and Nova Scotia) and Scotland (Mullins et al. 1998; Rodger et al. 1998). The disease is caused by an orthomyxovirus (Falk et al. 1997). Clinical signs generally appear 2–4 wk after exposure and include lethargy, anemia, swelling and hemorrhage of the kidney and other organs, ascites, exopthalmia, pale gills, and darkening of the posterior gut (Thorud and Djupvik 1988; Evensen et al. 1991). Mortalities vary greatly and can range from 2 to 80% (Jarp and Karlsen 1997). The virus can be transmitted through feces, blood, mucus, contact with contaminated equipment, and sea lice (Nylund et al. 1994;

Totland et al. 1996). The virus has also been found in the sex products of Atlantic salmon; however, vertical transmission was not observed (Melville and Griffiths 1999). Although primarily a disease of Atlantic salmon in saltwater, ISAV will replicate in a variety of salmonid fishes including rainbow trout, brown trout, and arctic char *Salvelinus alpinus* (Nylund et al. 1997; Snow et al. 2001) and can cause disease in freshwater. Jarp and Karlsen (1997) studied the risk factors associated with ISA outbreaks and found proximity of salmonid net pens to fish slaughterhouses, processing plants, or ISA-positive sites were crucial factors in the spread of disease. A minimum of 5 km between aquaculture units was recommended to reduce the risk of passive transmission of ISAV through sea water. The risk of ISA increased significantly when the nearest slaughterhouse had no disinfecting system for waste water.

Atlantic salmon production in Maine is a major industry, estimated at a worth of $59.5 million in 1998 (USDA 2001). There was much concern about this disease when ISAV was isolated from net pens in New Brunswick, Canada in 1999, only 4.8 km from U.S. net pen sites. An ISAV Action Plan was developed for the Maine Salmonid Industry Fish Health Committee in September of 1999. This is a reportable disease in Maine, meaning within 24 h of a confirmed positive finding the farm must report the finding to the Maine Department of Marine Resources. In addition to reporting protocols, the plan set forth sampling procedures for ISAV monitoring, site disinfection policies, and biosecurity guidelines. Despite these safeguards, the first identification of ISAV in the United States was made in a commerical aquaculture pen in Cobscook Bay, Maine, in 2001 (Bouchard et al. 2001). Biosecurity increased with isolation of the affected cage and site, increased surveillance of neighboring cages, disinfection of equipment, and daily removal of dead fish

for burial in a landfill were initiated (USDA 2001). It remains to be seen how successful these measures will be at preventing the spread of ISAV. In 1990, the Norwegian Ministry of Agriculture issued regulations aimed at reducing the transmission of fish infections, particularly ISAV. These included compulsory veterinary health control in smolt hatcheries, compulsory health certificate at time of sale, regulations concerning transport of live fish, required disinfection of wastewater from salmonid processing plants, that all ISA-infected fish be slaughtered as soon as possible, and all infected biological material and equipment be disinfected (Jarp and Karlsen 1997; Aspehaug et al. 2001). These regulations resulted in a dramatic reduction in new ISA outbreaks (Aspehaug et al. 2001). Similarly, in Scotland a disease containment and eradication policy was enforced when ISA was detected in farmed Atlantic salmon in 1998 and 1999, and no further cases of the disease have been found since then.

Molecular techniques have provided some insight into ISAV, but there is no definitive information on the origin of the virus in different countries. Blake et al. (1999) compared partial cDNA nucleotides and deduced amino acid sequences and suggested the North American isolate may represent a distinct genomic variant from the Norwegian strains. Further comparisons of eight Norwegian strains isolated between 1987 and 1999 with a Canadian and a Scottish strain, using the polymerase gene, also suggest a genetic difference between strains. This work calculated the mutation rate of the gene and suggested the Norwegian and North American isolates diverged around 1900 (Krossøy et al. 2001). These authors suggest two possible hypotheses for the geographic origin of the ISA virus. Rainbow trout, a species in which ISAV can replicate without causing disease was introduced to Europe from North America in the late 1800s and could have carried the virus.

Conversely, sea or brown trout, in which the virus can also replicate without signs of disease, were introduced into North America from Europe in the late 1800s and could have carried the virus to this continent. A third possibility is a marine reservoir; the virus has been reported in wild salmon (Whorisky 2000), however, to date, the virus has not been shown to successfully replicate in the non-salmonid species tested (Aspehaug et al. 2001).

Bacteria

Bacterial diseases are certainly some of the most devastating infectious diseases of cultured fishes. Intensive culture of fishes can increase the risk of bacterial disease by increasing the pathogen load as bacteria are excreted or released from infected fish and by increasing the organic load allowing for proliferation of potential water-borne pathogens.

However, the risk of bacterial pathogens of cultured fishes to wild populations may be less than previously discussed pathogens for a number of reasons.

- Many of the bacteria (*Flavobacterium sp., Aeromonas hydrophila, Vibrio sp., Edwardsiella tarda*) that can cause serious economic loss in intensive culture situations are common inhabitants of the aquatic environment. Often disease caused by these organisms is stress-mediated. For this reason, some of the more devastating bacterial pathogens were not recognized until certain fish species began to be intensively cultured. Disease outbreaks in wild populations also are often associated with degraded ecosystems and environmental stressors.

- There has been more success in terms of chemotherapeutics and vaccine development for bacterial diseases compared to those caused by parasites and viruses.

- In general, bacterial pathogens are easier to culture and identify.

- Bacteria do not have the complex life cycle and intermediate hosts of many parasites.

Edwardsiella ictaluri. E. ictaluri, a gram-negative, enteric bacterial pathogen, the cause of enteric septicemia of catfish, is considered the most important infectious disease of catfish industry. The disease, enteric septicemia of catfish (ESC), and the causative agent, a new bacterial species, were first recognized in 1976 among populations of pond-reared fingerling and yearling channel catfish (Hawke 1979). However, preserved channel catfish samples from 1970, stored at the Stuttgart National Aquaculture Research Center, tested with Gram stains, histology, and immuno-histochemistry, were found to be positive for *E. ictaluri.* This work demonstrated that *E. ictaluri* was causing disease in channel catfish in Arkansas by 1970. Losses due to ESC relate well to the intensification of catfish culture in the Southeast, and Mitchell and Goodwin (1999) suggested increased losses from ESC have been caused by reduced water quality, other stressors, or the increased ease of fish-to-fish transmission associated with intensified culture practices. To date, in the U.S., it has been primarily a disease of channel catfish, although occasional isolations from diseased aquarium fish have been reported (Blazer et al. 1985; Waltman et al. 1985) and a number of species of salmonids have been shown to be susceptible in laboratory exposures (Baxa et al. 1990). Despite the facts that channel catfish survivors of an infection can become carriers even after chemotherapeutic treatment (Klesius 1992), viable *E. ictaluri* have been cultured from rectal samples of fish-eating birds (Taylor 1992), and the bacteria can survive in bottom muds of ponds (although not water) for more than 95 d (Plumb and Quinlan 1986), disease outbreaks in wild populations have not been reported. Chen et al. (1994) reviewed the history of ESC in California and surveyed wild and cultured populations of channel catfish for the bacteria and serum antibody to *E. ictaluri.* They found serum antibody titers in wild fish from a number of areas but never observed clinical disease or were able to isolate bacteria from wild fish. Cultured populations also had antibody titers and in two incidences bacteria could be isolated from asymptomatic carriers.

Renibacterium salmoninarum. Bacterial kidney disease (BKD), caused by the gram positive organism *Renibacterium salmoninarum*, is a significant disease of both cultured and wild salmonids. It was initially observed in Atlantic salmon in Scotland in 1930 (Mackie et al. 1933). Diagnosis of these early cases was based on observing the organism within the characteristic diffuse granulomatous lesions in tissue sections. The bacterium, fastidious and slow-growing, was not successfully cultured until the 1950s (Earp 1950). The disease was first reported in the U.S. from hatchery-reared brown trout, rainbow trout, and brook trout in Massachusetts in 1935 (Belding and Merrill 1935) and in 1936 on the Pacific Coast (Earp et al. 1953). It is now widespread throughout the U.S., but the greatest problems occur in Pacific salmon in the Pacific Northwest and the Great Lakes region (Fryer and Lannan 1993). This particular bacterium is unusual in that it can be spread vertically through the egg (Bullock et al. 1978), as well as horizontally. It causes a chronic disease, often without external signs, and infected fish can be carriers, shed bacteria in the feces and hence the fecal-oral route seems to be an important route for horizontal transmission (Balfry et al. 1996).

It has been suggested that *R. salmoninarum* is a highly successful and unique bacterium with a complex host-pathogen relationship. It has been detected in populations of feral fish with no history of human intervention (Fryer and Lannan 1993) and possibly survives quite well, sequestered within macrophages, in otherwise healthy fish. It has even been hypothesized to be a normal resident of salmonid fishes (Austin and Austin 1987). Certainly transfer and translocation can spread the bacteria, particularly since it can be spread

by eggs and does not require the movement of fish themselves. Shedding by subclinically infected fish may amplify the bacteria in natural systems. Disease most often occurs when the fish are stressed or nutritionally compromised (Bell et al. 1984), which could explain why disease is more common in intensive culture situations than in wild fish populations (Post 1987). However, stressors such as those induced by hydroelectric dams or other anthropogenic influences may activate the bacteria and lead to disease in wild salmonids (Raymond 1988).

Mitchum et al. (1979) diagnosed *R. salmoninarum* in a wild, naturally reproducing population of brook trout. The source of the bacteria was believed to have been infected hatchery fish, however the suspect fish had been stocked 13 yr or more before the initiation of a study to determine the prevalence and effects of BKD in this drainage. Fish were collected upstream and downstream of a lake. BKD was found in all age groups, from 0–4, and difference in prevalence and severity of infection between upstream and downstream sites was noted. A combination of ecological conditions, including lower volumes and velocities of water, occurrence of pools where fish tended to congregate (perhaps allowing for easier horizontal transmission), softer water upstream, and stress due to annual water fluctuations appeared to influence the distribution of BKD in this drainage (Mitchum et al. 1979). In a follow-up study, natural horizontal transmission of *R. salmoninarum* from infected brook trout in this drainage to newly stocked hatchery brook, brown, and rainbow trout was shown. Mortalities of these newly stocked fish due to BKD began within 9 mo (Mitchum and Sherman 1981).

The story of the decline of chinook salmon in Lake Michigan associated with *R. salmoninarum* provides an example of the multiple factors that may contribute to disease and population declines in wild populations

(Holey et al. 1998). Although other salmonids in Lake Michigan have shown clinical signs of BKD, only chinook salmon have experienced epizootics and high mortalities attributed to this disease. It was concluded that the chinook salmon decline in Lake Michigan should be considered the result of ecosystem imbalance, not the result of infection by any one pathogen. Low food availability leading to nutritional stress, high salmon densities, and heavy infestations of an intestinal parasite enabled BKD to manifest and have a population effect.

Conclusions and Recommendations

Disease is an integral part of the existence of all animals, including both cultured and wild fishes. Certainly intensive culture, particularly of non-native species, can and has been involved in the introduction and/or amplification of pathogens and disease in wild populations. In many cases these introductions have been by resource management agencies. Huge gaps exist in our knowledge regarding pathogen distribution, survival, and fate in the environment, and host susceptibility in aquatic ecosystems. Hence, there are many perceptions and misperceptions held by the public and scientific community regarding the spread of infectious agents from cultured to wild aquatic animals. However, studies to examine the prevalence and the impacts of infection or disease on free-ranging fish populations have been limited. The detection of infected fish and measuring the potential impacts of disease in a free-ranging population is both difficult and expensive. Issues associated with sampling free-ranging populations and the possible removal of infected fish by predators are complex.

In contrast to free-ranging populations, artificial propagation of aquatic animals presents a captive, and often intensively monitored population. Captivity coupled with the routine monitoring of the health and performance of fish facilitates the identification of pathogens that have evolved with their hosts in natural environments and may help elucidate various stressors that contribute to disease extent and severity. Conditions that promote or exacerbate infection and disease are generally more prevalent and pronounced in aquaculture facilities than in wild populations. Adverse environmental factors include temperature and oxygen extremes, inadequate water flows, and increased densities. Increased densities under suboptimal conditions are not only conducive to disease but also promote the rapid transmission of pathogens throughout the population. In wild populations an increased incidence of morbidity and mortality could result in extirpation of the host species and its endemic pathogen.

Exposure to potential infectious agents is a continuous process during the life span of any organism. However, exposure to an infectious microorganism does not necessarily result in infection or manifestation of clinical disease. The latter depends on the interaction of several factors including: 1) the health and immunological status of the host, 2) the dose and virulence or contagiousness of the pathogen, and 3) the environmental conditions that affect the host and pathogen interaction. Although clinical disease is easily qualified and quantified, subclinical disease is more difficult to characterize and may only be detected with the assistance of diagnostic tests or aids. However, it must be emphasized that the presence or detection of any infectious agent does not imply the presence of disease. Simply put, *infection*—defined as invasion of a host by a pathogenic agent—is a more common event. In contrast, *disease*—defined as the condition that results in morbidity and, possibly, mortality in the individual host or population as a consequence of infection—is less common. Therefore, asymptomatic infection may be widely distributed throughout wild populations without the clinical manifestation of disease that may subsequently occur due to aquaculture-specific stressors.

State, federal, and tribal pathogen control programs exist and have existed for a long time. However, regulations differ greatly from state to state and regionally. State resource management agencies and/or state agriculture departments generally oversee these programs in public and private aquaculture operations. The U.S. Fish and Wildlife Service also has an importation inspection program (Title 50) to prevent the introduction of foreign animal pathogens, and the National Marine Fisheries Service and the U.S. Department of Agriculture—Animal Plant Health Inspection Service may also be involved under certain circumstances. The goal of all these programs is to prevent the introduction of significant fish pathogens into the U.S., specific states, regions, or facilities. Pathogens are regulated that meet criteria such as: 1) serious pathogens exotic to an area, 2) pathogens known to cause serious problems, 3) pathogens that are highly infectious and easily transmitted, and/or 4) pathogens that regional watershed compacts have agreed are of concern in that region. Additionally, pathogen inspections should be required before fish are brought on to an aquaculture facility and routine disease inspections may be required of fish on the facility. The two primary diagnostic manuals used and accepted internationally that specify analytical methods to qualify and quantify aquatic animal pathogens include the <u>Diagnostic Manual for Aquatic Animal Diseases</u> (Office International Des Epizooties 2000) and the Fish Health Section's <u>Suggested Procedures for the Detection and Identification of Certain Finfish and Shellfish Pathogens</u> (Thoesen 1994). Other manuals have been developed by federal and state agencies, including the U.S. Fish and Wildlife Service and state and tribal resource management agencies, and these are continually being revised as existing techniques are modified

and/or new methods developed. These regulatory control programs are generally directed at cultured food fish or fish in which resource managers have an interest and have sometimes been successful at limiting the introduction of important fish pathogens. However, there is very little control of the ornamental or bait fish industries. Also, control programs exist only for diseases that are known and for which methods of detection have been established. Hence, as more "exotic" species are imported and cultured, the likelihood of new diseases increases.

Risk analysis needs to be applied to aquatic animal health issues, however, gaps in our knowledge exist due, in part, to difficulties in reproducing pathogen life cycles and determining if the agent is in fact active. This is further compounded by the lack of information on pathogen amplification when a host, at various life stages and under different environmental conditions, becomes infected or diseased. Additionally, previously mentioned characteristics of the pathogen that are poorly described and must be considered in any risk assessment include: 1) the ability to multiply and remain viable in water, 2) the survival time outside the host, 3) the number of infectious units required to cause infections and disease, and 4) intermediate hosts or carriers. Studies need to be conducted in order to develop the type of scientific information that could be used to more accurately assess the risk associated with the presence of a pathogen or the movement of a group of fish.

There is very little credible scientific information about the presence or distribution of pathogens in the wild that currently exist or may have existed prior to stocking. There is a need to develop a national survey of free-ranging fish for select pathogens using a standardized approach agreed upon by state, federal, tribal, and commercial entities. A survey of this type will help to identify where pathogens are known to exist. Additionally, this will allow comparisons from state to state or watershed to watershed that may help identify why a pathogen in one area has negative impacts on certain fish stocks while not in others. But most importantly, this information will provide a scientific basis for management decisions regarding stocking and fish transport activities that has been lacking for many years. The U.S. Fish and Wildlife Service has undertaken this for selected fish species and in selected areas, with the development and implementation of the National Wild Fish Health Survey. The real challenge will be to determine how to use this information in establishing new regulation and control programs or modifying existing ones.

Other challenges include the application of advanced technologies for the detection of components of aquatic animal pathogens (e.g., nucleic acids, antigens of, or antibodies to) being developed but sometimes, prematurely utilized in diagnostic and inspection procedures. These tests have the potential to provide valuable scientific information but could also have serious regulatory implications. Additionally, the use of these methods could cause inappropriate devaluation and condemnation of aquatic animal stocks that could have serious legal and political ramifications. There must be a clear process to validate and standardized diagnostic methods and there must be consistent application of diagnostic methods, and requirements for confirmation. We must also strive to understand what the presence of nucleic acids or other components of pathogens means in assessment of potential risks and how this information can be utilized in the development of effective pathogen introduction and disease prevention programs.

For the future it is clear that a coordinated, national strategy on aquatic animal health is needed. This strategy must include the states and the federal Departments of Commerce, Interior, and Agriculture to create a unified program to protect both wild and cultured fish, shellfish, and crustaceans in public and private waters. This unified approach is needed to insure a single national policy that can be implemented and serve aquaculture without jeopardizing natural stocks of aquatic animals.

Acknowledgments

The authors thank Drs. Pete Bullock, Phillip McAllister, James Winton and Thomas Brandt for review of the manuscript and helpful suggestions for improvement.

Literature Cited

Abbott, R. T. 1973. Spread of *Melanoides tuberculata*. Nautilus 87:29.

Allendorf, F. W., P. Spruell and F. M. Utter. 2001. Whirling disease and wild trout: Darwinian fisheries management. Fisheries 26 (5):27–28.

Anderson, E. D., H. M. Engelking, E. J. Emmenegger, and G. Kurath. 2000. Molecular epidemiology reveals emergence of a virulent infectious hematopoietic necrosis (IHN) virus strain in wild salmon and its transmission to hatchery fish. Journal of Aquatic Animal Health 12: 85–99.

Armstrong, R., J. Robinson, C. Rymes, and T. Needham. 1993. Infectious hematopoietic necrosis in Atlantic salmon in British Columbia. Canadian Veterinary Journal 34:312–313.

Aspehaug, V., M. Devold, K. Falk, O. Dale, B. Krossøy, E. Biering, M. Aaset, C. Endresen, and A. Nylund. 2001. Transmission and reservoir of ISAV. Pages 1–3 *in* Proceedings Annual New England Farmed Fish Health Management Workshop, Machias, Maine, USA.

Austin, B. and D.A. Austin. 1987. Aerobic gram-positive rods. Pages 70–87 *in* B. Austin and D. A. Austin, editors. Bacterial fish pathogens: Disease in farmed and wild fish. Ellis Horwood Limited, Chichester,

England.

Bakke, T. A. and P. D. Harris. 1998. Diseases and parasites in wild Atlantic salmon (*Salmo salar*) populations. Canadian Journal of Fisheries and Aquatic Sciences 55 (Suppl. 1):247–266.

Balfry, S. K., L. J. Albright, and T. P. T. Evelyn. 1996. Horizontal transfer of *Renibacterium salmoninarum* among farmed salmonids via the fecal-oral route. Diseases of Aquatic Organisms 25:63–69.

Bauer, O. N., V. A. Musselius and Y. A. Strelkov. 1969. Diseases of pond fishes. Israel Program for Scientific Translations, Jerusaleum, 1973. U.S. Department of Commerce, National Technical Information Service, Springfield, Virginia, USA.

Baxa, D. V., J. M. Groff, A. Wishkovsky, and R. P. Hedrick. 1990. Susceptibility of nonictalurid fishes to experimental infection with *Edwardsiella ictaluri*. Diseases of Aquatic Organisms 8:113–117.

Belding, D. L. and B. Merrill. 1935. A preliminary report upon a hatchery disease of the *Salmonidae*. Transactions of the American Fisheries Society 65:76–84.

Bell, G. R., D. A. Higgs, and G. S. Traxler. 1984. The effect of dietary ascorbate, zinc, and manganese on the development of experimentally induced bacterial kidney disease in sockeye salmon (*Oncorhynchus nerka*). Aquaculture 36:293–311.

Blake, S., D. Bouchard, W. Keleher, M. Opitz, and B.L. Nicholson. 1999. Genomic relationships of the North American isolate of infectious salmon anemia virus (ISAV) to the Norwegian strain of ISAV. Diseases of Aquatic Organisms 35:139–144.

Blazer, V. S. and J. B. Gratzek. 1985. Cartilage proliferation in response to metacercarial infections of fish gills. Journal of Comparative Pathology 95:273–280.

Blazer, V. S., E. B. Shotts, and W. D. Waltman. 1985. Pathology associated with *Edwardsiella ictaluri* in catfish, *Ictalurus punctatus* Rafinesque, and *Danio devario* (Hamilton-Buchanan, 1822). Journal of Fish Biology 27:167–175.

Bogdanova, E. A. 1970. On the occurrence of whirling disease of salmonids in nature in the USSR. Journal of Parasitology 56:399.

Bouchard, D. A., K. Brockway, C. Giray, W. Keleher and P. L. Merrill. 2001. First report of infectious salmon anemia (ISA) in the United States. Bulletin of the European Association of Fish Pathologists 21:86–88.

Bucke, D., J. Finlay, D. McGregor and C. Seagrave. 1979. Infectious pancreatic necrosis (IPN) virus: its occurrence in captive and wild fish in England and Wales. Journal of Fish Diseases 2:549–553.

Bullock, G. L., H. M. Stuckey, and D. Mulcahy. 1978. Corynebacterial kidney disease: egg transmission following iodophore disinfection. Fish Health News 7:51–52.

Carlisle, J. C., K. A. Schat, and R. Elston. 1979. Infectious haematopoietic necrosis in rainbow trout *Salmo gairdneri* Richardson in a semi-closed system. Journal of Fish Diseases 2:511–517.

Chen, H. T. 1942. The metacercariae and adult of *Centrocestus formosanus* (Nishigori, 1924). Lingnan Science Journal 22:93–105.

Chen, M. F., D. Henry-Ford, M. E. Kumlin, M. L. Key, T. S. Light, W. T. Cox and J. C. Modin. 1994. Distribution of *Edwardsiella ictaluri* in California. 1994. Journal of Aquatic Animal Health 6:234–241.

Christensen, N. O. 1972. Panel review on Myxosomiasis (whirling disease in salmonid fishes). Europena Inland Fisheries Commission, Sub-Commission II)Aquaculture) Symposium 8.

Clarkson, R. W., A. T. Robinson and T. L. Huffnagle. 1997. Asian tapeworm (*Bothriocephaus acheilognathi*) in native fishes from the Little Colorado River, Grand Canyon, Arizona. Great Basin Naturalist 57:6–69.

Earp, B. J. 1950. Kidney disease in young salmon. Master's thesis, University of Washington, Pullman, Washington, USA.

Earp, B. J., C. H. Ellis, and E. J. Ordal. 1953. Kidney disease in young salmon. Washington Department of Fisheries, Special Report 1:1–74.

El-Matbouli, M. and R. W. Hoffman. 1989. Experimental transmission of two *Myxobolus* spp. developing bisporogeny via tubificid worms. Parasitology Research 75:461–464.

El-Matbouli, M. and R. W. Hoffman. 1991. Effects of freezing, aging, and passage through the alimentary canal of predatory animals on the viability of *Myxobolus cerebralis* spores. Journal of Aquatic Animal Health 3:260–262.

El-Matbouli, M., R. W. Hoffman, and C. Mandok. 1995. Light and electron microscopic observations on the route of the triactinomyxon-sporoplasm of *Myxoblous cerebralis* from the epidermis into rainbow trout cartilage. Journal of Fish Biology 46:919–935.

Evensen, O., K. E. Thorud, and Y. A. Olsen. 1991. A morphological study of the gross and light microscopic lesions of infectious anaemia in Atlantic salmon (*Salmo salar*). Research in Veterinary Science 51:215–222.

Falk, K., E. Namork, E. Rimstad, S. Mjaaland, and B. H. Dannevig. 1997. Characterization of infectious salmon anemia virus, an orthomyxo-like virus isolated from Atlantic salmon (*Salmo salar* L.). Journal of Virology 71:9016–9023.

Font, W. F. and D. C. Tate. 1994. Helminth parasites of native Hawaiian freshwater fishes: An example of extreme ecological isolation. Journal of Parasitology 80:682–688.

Foott, J. S., R. Harmon, K. Nichols, D. Free, and K. True. 2000. Release of IHNV infected chinook smolts from Coleman NFH: risk assessment of the disease impacts on natural chinook. Page 7 *in* Proceedings of the 41st Annual Western Fish Disease Workshop. Gig Harbor, Washington, USA.

Francis-Floyd, R., J. Gildea, P. Reed and R. Klinger. 1997. Use of Bayluscide (Bayer 73) for snail control in fish ponds. Journal of Aquatic Animal Health 9:41–48.

Fryer, J. L. and C. N. Lannan. 1993. The history and current status of *Renibacterium salmoninarum*, the causative agent of bacterial kidney disease in Pacific salmon. Fisheries Research 17:15–33.

Goodman, M. L. 1990. Preserving the genetic diversity of salmonid stocks: a call for federal regulation of hatchery programs. Environmental Law 20:111–166.

Groberg, W. J. and J. L. Fryer. 1983. Increased occurrences of infectious hematopoietic necrosis virus in fish at Columbia River basin hatcheries: 1980-1982. Oregon State University, Sea Grant College Program, Technical Paper 6620, Corvallis, Oregon, USA.

Håstein, T. and T. Lindstad. 1991. Diseases in wild and cultured salmon: possible interaction. Aquaculture 98:277–288.

Hawke, J. P. 1979. A bacterium associated with disease of pond cultured channel catfish, *Ictalurus punctatus*. Journal of the Fisheries Research Board of Canada 36:1508–1512.

Healy, M. C. and A. Prince. 1995. Scales of variation in life history tactics of Pacific salmon and the conservation of phenotype and genotype. American Fisheries Society Symposium 17:176–184.

Hedrick, R. P. 1998. Relationships of the host, pathogen, and environment: Implications for diseases of cultured and wild fish populations. Journal of Aquatic Animal Health 10:107–111.

Hedrick, R. P., T. S. McDowell, W. Ahne, C. Torhy, and P. de Kinkelin. 1992. Properties of three iridovirus-like agents associated with systemic infections of fish. Diseases of Aquatic Organisms 13:203–209.

Heller, J. and S. Ehrlich. 1994. A freshwater prosobranch, *Melanoides tuberculata* in a hydrogen sulfide stream. Journal of Conchology, London 35:236–241.

Hilborn, R. 1992. Hatcheries and the future of salmon in the Northwest. Fisheries 17(1):5–8.

Hnath, J. G. 1996. Whirling disease in the Midwest (Michigan, Pennsylvania, Ohio, West Virginia, Virginia, Maryland). Pages 18–22 in E. P. Bergersen and B. A. Knopf, editors. Whirling Disease Workshop Proceedings, Denver, Colorado, USA.

Hofer, B. 1903. Uber die Drehkrankheit der Regenbogenforelle. Allgemeine Fischerei Zeitschrift 28:7–8.

Hoffman, G. L. 1962. Whirling disease of trout. Fishery Leaflet 508, United States Department of the Interior, Fish and Wildlife Service, Washington, D.C., USA.

Hoffman, G. L. 1970. Intercontinental and transcontinental dissemination and transfaunation of fish parasites with emphasis on whirling disease (*Myxosoma cerebralis*) and its effect on fish. American Fisheries Society Special Publication 5:69–81.

Hoffman, G. L. 1980. Asian tapeworm *Bothriocephalus acheilognathi* Yamaguti, 1934, in North America. Pages 69–75 *in* H.-H. Reichenbach-Klinke, editor. Contributions to Fish pathology and fish toxicology 8. Gustav Fischer Verlag, New York, New York, USA.

Hoffman, G. L. 1990. *Myxobolus cerebralis*, a worldwide cause of salmonid whirling disease. Journal of Aquatic Animal Health 2:30–37.

Hoffman, G. L. and G. Schubert. 1984. Some parasites of exotic fishes. Pages 233–261 *in* W. R. Courtenay, Jr. and J. R. Stauffer, editors. Distribution, biology and management of exotic fishes. The John Hopkins University Press, Baltimore, Maryland, USA.

Holey, M. E., R. F. Elliott, S. V. Marcquenski, J. G. Hnath, and K. D. Smith. 1998. Chinook salmon epizootics in Lake Michigan: Possible contributing factors and management implications. Journal of Aquatic Animal Health 10:202–210.

Holway, J. E. and C. E. Smith. 1973. Infectious hematopoietic necrosis of rainbow trout in Montana: a case report. Journal of Wildlife Diseases 9:287–290.

Horsch, C. M. 1987. A case history of whirling disease in a drainage system: Battle Creek drainage of the upper Sacramento River basin, California, USA. Journal of Fish Diseases 10:453–460.

Janeke, P. 1984. IHN outbreak in Colorado. American Fisheries Society, Fish Health Section Newsletter 12:6.

Jarp, J. and E. Karlsen. 1997. Infectious salmon anaemia (ISA) risk factors in sea-cultured Atlantic salmon *Salmo salar*. Diseases of Aquatic Organisms 28:79–86.

Kent, M. L. 1994. The impact of diseases of pen-reared salmonids on coastal environments. Pages 85–95 *in* A. Erik, P. K. Hansen and V. Wennevik, editors.

Proceedings of the Canada-Norway Workshop on Environmental Impacts of Aquaculture, Havforskningsinstututtet, Norway.

Kent, M. L., G. S. Traxler, D. Kieser, J. Richard, S. C. Dawe, R. W. Shaw, G. Prosperi-Porta, J. Ketcheson, and T. P. T. Evelyn. 1998. Survey of salmonid pathogens in ocean-caught fishes in British Columbia, Canada. Journal of Aquatic Animal Health 10:211–219.

Klesius, P. H. 1992. Carrier state of channel catfish infected with *Edwardsiella ictaluri*. Journal of Aquatic Animal Health 4:227–230.

Krossøy, B., F. Nilsen, K. Falk, C. Endresen, and A. Nylund. 2001. Phylogenetic analysis of infectious salmon anaemia virus isolates from Norway, Canada and Scotland. Diseases of Aquatic Organisms 44:1–6.

Landye J., B. McCasland, C. Hart, K. Hayden, and J.C. Thoesen. 1999. San Juan River fish health surveys (1992-1999). U.S. Fish and Wildlife Service, Fish Health Report prepared for the San Juan River Biological Committee, Pinetop Fish Health Center, Pinetop, Arizona, USA.

LaPatra, S. E. 1998. Factors affecting pathogenicity of infectious hematopoietic necrosis virus for salmonid fish. Journal of Aquatic Animal Health 10:121–131.

LaPatra, S. E., R. Troyer, W. Shewmaker, G. Jones, and G. Kurath. 2001. Pages 251–258 *in* C. Rogers, editor. Understanding aquatic animal virus survival and trafficking and its role in risk assessment. Proceedings of the International Conference on Risk Analysis in Aquatic Animal Health. Office of International des Epizootics, Paris, France..

Levy, S. B. 1998. The challenge of antibiotic resistance. Scientific American 3:32–39.

Liao, H-h. and L-c. Shih. 1956. On the biology and control of *Bothriocephalus gowkongenesis* Yeh, a cestode parasitizing young grass carp (*Ctenopharyngodon idella*). Acta Hydrobiologica Sinica 2:129–185.

López-Jiménez, S. 1987. Enfermedades más frecuentes de las carpas cultivadas en México. Acuavisión, Revista Mexicana de Acuacultura 2:11–13.

Lunden, T., S. Miettinen, L.-G. Loennstroem, E.-M. Lilius, and G. Bylund. 1998. Influence of oxytetracycline and oxolinic acid on the immune response of rainbow trout (*Oncorhynchus mykiss*). Fish and Shellfish Immunology 8:217–230.

Mackie, T. J., J. A. Arkwright, T. E. Pryce-Tannatt, J. C. Mottram, W. D. Johnson, and W. J. M. Menzies. 1933. The second nterim report of the furunculosis committee. Edinburgh, Scotland.

MacMillan, J. R. 2001. Aquaculture and antibiotic resistance: A negligible public health risk? World Aquaculture 32:49–52.

Madsen, H. 1990. Biological methods for the control of freshwater snails. Parasitology Today 6:237–240.

Markiw, M. E. 1991. Whirling disease:earliest susceptible age of rainbow trout to the triactinomyxid of *Myxobolus cerebralis*. Aquaculture 92:1–6.

Markiw, M. E. 1992. Experimentally induced whirling disease I. Dose response of fry and adults of rainbow trout exposed to the triactinomyxon stage of *Myxobolus cerebralis*. Journal of Aquatic Animal Health 4:40–43.

Markiw, M. E. and K. Wolf. 1983. *Myxosoma cerebralis* (Myxozoa:Myxosporea) etiologic agent of salmonid whirling disease requires a tubificid worm (Annelida: Oligochaeta) in its life cycle. Journal of Protozoology 30:561–564.

Martin, J., J. Webster, and G. Edwards. 1992. Hatcheries and wild stocks: Are they compatible? Fisheries 17(1):4.

McAllister, P. E. 1993. Salmonid fish viruses. Pages 380–400 *in* M. K. Stoskopf, editor. Fish medicine. W. B. Saunders Company, Philadelphia, Pennsylvania, USA.

McAllister, P. E. and J. Bebak. 1997. Infectious pancreatic necrosis virus in the environment: relationship to effluent from aquaculture facilities. Journal of Fish Diseases 20:201–207.

McAllister, P. E. and J. Bebak. 1999. Chronic, low-level exposure to infectious pancreatic necrosis virus (IPNV): An estimation of effects on early life stages of rainbow trout, *Oncorhynchus mykiss*. Page 29 *in* Fifty-fifth Annual Northeast Fish and Wildlife Conference Abstracts, Concord, Connecticut, USA.

McAllister, P. E. and W. J. Owens. 1992. Recovery of infectious pancreatic necrosis virus from the faeces of wild piscivorous birds. Aquaculture 106:227–232.

McDermott, K. 2000. Distribution and infection relationships of an undescribed digentic trematode, its exotic intermediate host, and endangered fishes in springs of west Texas. Master's thesis. Southwest Texas State University, San Marcos, Texas, USA.

McMahon, T. E., A. R. Munro, D. Downing, E. R. Vincent, S. Leathe, G. Grisak, G. Liknes, P. Clancey, B. B. Shepard, and A. V. Zale. 1999. Life history in rainbow trout in relation to whirling disease infection risk. Pages 109–111 *in* Proceedings of the 5th Annual Whirling Disease Symposium, Missoula, Montana, USA.

McVicar, A. H. 1988. Epidemiology/epizootiology: a basis for control of disease in mariculture. Pages 397–405 *in* F. O. Perkins and T. C. Cheng, editors. Pathology in marine science. Academic Press, San Diego, California, USA.

McVicar, A. H. 1997. Disease and parasite implications of coexistence of wild and cultured Atlantic salmon populations. ICES Journal of Marine Science 54:1093–1103.

Melville, K. J. and S. G. Griffiths. 1999. Absence of vertical transmission of infectious salmon anemia virus (ISAV) from individually infected Atlantic salmon *Salmo salar*. Diseases of Aquatic Organisms 38:231–234.

Metcalfe, N. B. 1993. Behavioural causes and consequences of life history variation in fish. Marine Behaviour and Physiology 23:205–217.

Mitchell, A. J. and A. E. Goodwin. 1999. Evidence that enteric septicemia of catfish (ESC) was present in Arkansas by the late 1960's: New insights into the epidemiology of ESC. Journal of Aquatic Animal Health 11:175–178.

Mitchell, A. J., M. J. Salmon, D. G. Huffman, A. E. Goodwin, and T. M. Brandt. 2000. Prevalence and pathogenicity of a heterophyid trematode infecting the gills of an endangered fish, the fountain darter, in two central Texas spring-fed rivers. Journal of Aquatic Animal Health 12:283–289.

Mitchum, D. L. and L. E. Sherman. 1981. Transmission of bacterial kidney disease from wild to stocked hatchery trout. Canadian Journal of Fisheries and Aquatic Sciences 38:547–551.

Mitchum, D. L., L. E. Sherman, and G. T. Baxter. 1979. Bacterial kidney disease in feral populations of brook trout (*Salvelinus fontinalis*), brown trout (*Salmo trutta*) and rainbow trout (*Salmo gairdneri*). Journal of the Fisheries Research Board of Canada 36:1370–1376.

Modin, J. 1998. Whirling disease in California: A review of its history, distribution, and impacts, 1965-1997. Journal of Aquatic Animal Health 10:132–142.

Moffitt, C. M., B. C. Stewart, S. E. LaPatra, R. D. Brunson, J. L. Bartholomew, J. E. Peterson, and K. H. Amos. 1998. Pathogens and diseases of fish in aquatic ecosystems: Implications in fisheries management. Journal of Aquatic Animal Health 10:95–100.

Mulcahy, D., R. Pascho, and C. K. Jenes. 1983. Detection of infectious hematopoietic necrosis virus in river water and demonstration of waterborne transmission. Journal of Fish Diseases 6:321–330.

Mullins, J. E., D. Groman, and D. Wadowska. 1998. Infectious salmon anaemia in salt water Atlantic salmon (*Salmo salar* L.) in New Brunswick, Canada. Bulletin of the European Association of Fish Pathologists 18:110–114.

Murray, H. D. 1971. The introduction and spread of thiarids in the United States. Biologist 53:133–135.

Nese, L. and Ø. Enger. 1993. Isolation of *Aeromonas salmonicida* from salmon lice *Lepeophtheirus salmonis* and marine plankton. Diseases of Aquatic Organisms 16:79–81.

Nylund, A., T. Hovland, K. Hodneland, F. Nilsen, and P. Lovik. 1994. Mechanisms for transmission of infectious salmon anaemia (ISA). Diseases of Aquatic Organisms 19:95–100.

Nylund, A., A. M. Kvenseth, B. Krossøy, and K. Hodneland. 1997. Replication of the infectious salmon anaemia virus (ISAV) in rainbow trout, *Oncorhynchus mykiss* (Walbaum). Journal of Fish Diseases 20:275–279.

Office International Des Epizooties. 2000. Diagnostic manual for aquatic animal diseases, third edition. Office International Des Epizooties Fish Disease Commission, Paris, France.

Paperna, I. 1991. Diseases caused by parasites in the aquaculture of warm water fish. Annual Review of Fish Diseases 1:155–194.

Peters, F. and M. Neukirch. 1986. Transmission of some fish pathogenic viruses by the heron, *Ardea cinerea.* Journal of Fish Diseases 9:539-544.

Plumb, J. A. 1972. A virus-caused epizootic of rainbow trout (*Salmo gairdneri*) in Minnesota. Transactions of the American Fisheries Society 101:121–123.

Plumb, J. A. 1997. Trends in freshwater fish disease research. Pages 33–47 *in* T. W. Flegel and I. H. MacRae, editors. Diseases in Asian aquaculture III. Fish Health Section, Asian Fisheries Society, Manila, Phillipines.

Plumb, J. A. and E. E. Quinlan. 1986. Survival of *Edwardsiella ictaluri* in pond water and bottom mud. Progressive Fish-Culturist 48:212–214.

Post, G. 1987. Textbook of fish health. T. F. H. Publications, Neptune City, New Jersey, USA.

Raymond, H. L. 1988. Effects of hydroelectric development and fisheries enhancement on spring Chinook salmon and steelhead in the Columbia River basin. North American Journal of Fisheries Management 8:1–24.

Reno, P. W. 1998. Factors involved in the dissemination of disease in fish populations. Journal of Aquatic Animal Health 10:160–171.

Rijkers, G. T., A. G. Teunissen, R. van Oosterom, and W. B. van Muiswinkel. 1980. The immune system of cyprinid fish. The immunosuppressive effect of the antibiotic oxytetracycline in carp (*Cyprinus carpio* L.). Aquaculture 19:177–189.

Rodger, H. D., T. Turnbull, F. Muir, S. Millar, and R.H. Richards. 1998. Infectious salmon anaemia (ISA) in the United Kingdom. Bulletin of the European Association of Fish Pathologists 18:115–116.

Roessler, M. A., G. L. Beardsley, and D. C. Tabb. 1977. New records of the introduced snail, *Melanoides tuberculata* (Mollusca:Thiaridae) in south Florida. Florida Scientist 40:87–94.

Russo, T. N. 1973. Discovery of the gastropod snail *Melanoides (Thiara) tuberculata* (Muller) in Florida. Florida Scientist 36:212–213.

Salgado-Maldonado, G., M. I. Rodriguez-Vargas, and J. J. Campos-Perez. 1995. Metacercariae of *Centrocestus formosanus* (Nishigori, 1924) (Trematoda) in freshwater fishes in México and their transmission by the thiarid snail *Melanoides tuberculata*. Studies of Neotropical Fauna and Environment 30:245–250.

Schachte, J. H. and P. J. Hulbert. 1998. A summary of field research and monitoring for *Myxobolus cerebralis* in New York State from 1994 through 1997. Pages 19–21 *in* Proceedings of theWhirling Disease Symposium, Fort Collins, Colorado, USA.

Schmidt, A. S., M. S. Bruun, I. Dalsgaard, K. Pedersen, and J. L. Larsen. 2000. Occurrence of antimicrobial resistance in fish-pathogenic and environmental bacteria associated with four Danish rainbow trout farms. Applied and Environmental Microbiology 66:4908–4915.

Schmitt, C. J. and G. M. Dethloff, editors. 2000. Biomonitoring of Environmental Status and Trends (BEST) Program: Selected methods for monitoring chemical contaminants and their effects in aquatic ecosystems. U.S. Geological Survey, Biological Resources Division, Information and Technology Report USGS/BRD-2000—0005. Columbia, Missouri, USA.

Scholtz, T. and G. Salgado-Maldonado. 2000. The introduction and dispersal of *Centrocestus formosanus* (Nishigori, 1924) (Digenea:Heterophyidae) in Mexico: A review. American Midland Naturalist 143:185–200.

Schramm, H. L., Jr. and R. G. Piper, editors. 1995. Uses and effects of cultured fishes in aquatic ecosystems. American Fisheries Society Symposium 15, American Fisheries Society, Bethesda, Maryland, USA.

Scott, A. L. and J. M. Grizzle. 1979. Pathology of cyprinid fishes by *Bothriocephalus gowkongenesis* (Cestodea: Pseudophyllidae). Journal of Fish Diseases 2:69–73.

Smith, P., M. Hiney, and O. B. Samuelson. 1994. Bacterial resistance to antimicrobial agents used in fish farming: a critical evaluation of method and meaning. Annual Review of Fish Diseases 4:273–313.

Snow, M., R. S. Raynard, and D. W. Bruno. 2001. Comparative susceptibility of Artic char (*Salvelinus alpinus*), rainbow trout (*Oncorhynchus mykiss*) and brown trout (*Salmo trutta*) to the Scottish isolate of infectious salmon anaemia virus. Aquaculture 196:47–54.

Sonstegard, R. A. and L. A. McDermott. 1972. Epidemiological model for passive transfer of IPNV by homeotherms. Nature (London) 237:104–105.

Stephen, G. and G. Iwama. 1997. Salmon aquaculture review. Key Issue B Fish Health. Environmental Assessment Office, Victoria, British Columbia, Canada.

Taylor, P. W. 1992. Fish-eating birds as potential vectors of *Edwardsiella ictaluri*. Journal of Aquatic Animal Health 4:240–243.

Taylor, R. L. and M. Lott. 1978. Transmission of salmonid whirling disease by birds fed trout infected with *Myxosoma cerebralis*. Journal of Protozoology 25:105–106.

Thoesen, J.C., editor. 1994. Suggested procedures for the detection and identification of certain finfish and shellfish pathogens, fourth edition, version 1. Fish Health Section, American Fisheries Society, Bethesda, Maryland, USA.

Thorud, K. and H. O. Djupvik. 1988. Infectious salmon anaemia in Atlantic salmon (*Salmo salar* L.). Bulletin of the European Association of Fish Pathologists 8:109–111.

Totland, G. K., B. K. Hjeltnes, and P. R. Flood. 1996. Transmission of infectious salmon anaemia (ISA) through natural secretions and excretions from infected smolts of Atlantic salmon *Salmo salar* during their presymptomatic phase. Diseases of Aquatic Organisms 26:25–31.

Traxler, G. S., J. R. Roome, and M. L. Kent. 1993. Transmission of infectious haematopoietic necrosis virus in seawater. Diseases of Aquatic Organisms 16:111–114.

Traxler, G. S., J. R. Roome, K. A. Lauda, and S. LaPatra. 1997. Appearance of infectious hematopoietic necrosis virus (IHNV) and neutralizing antibodies in sockeye salmon *Oncorynchus nerka* during their migration and maturation period. Diseases of Aquatic Organisms 28:31–38.

USDA (U.S. Department of Agriculture). 1995. Overview of aquaculture in the United States. Centers for Epidemiology and Animal Health, Animal and Plant Health Inspection Service, Veterinary Services, Fort Collins, Colorado, USA.

USDA (U.S. Department of Agriculture). 1997. Catfish '97, part 1: Reference of 1996 U.S. catfish health & production practices. Report of the National Animal Health Monitoring System, Animal and Plant Health Inspection Service, Veterinary Services N235.597, Fort Collins, Colorado, USA.

USDA (U.S. Department of Agriculture). 2001. Infectious salmon anemia, Maine, USA. Impact worksheet prepared by Center for Emerging Issues, Centers for Epidemiology and Animal Health, Animal and Plant Health Inspection Service, Fort Collins, Colorado, USA.

USEPA (Environmental Protection Agency). 1995. Environmental Monitoring and Assessment Program (EMAP): Laboratory methods manual—Estuaries. volume 1. Biological and physical analyses. EPA/620/R-95/008. U.S. Environmental Protection Agency, Office of Research and Development, Narragansett, Rhode Island, USA.

U. S. Fish and Wildlife Service. 1995. San Marcos/Comal (revised) recovery plan. Albequerque, New Mexico, USA.

Vincent, E. R. 1996. Whirling disease and wild trout: The Montana experience. Fisheries 21(6):32–33.

Walker, P. G. and R. B. Nehring. 1995. An investigation to determine the cause(s) of the disappearance of young wild rainbow trout in the upper Colorado River, in Middle Park, Colorado. Colorado Division of Wildlife, Denver, Colorado, USA.

Waltman, W. D., E. B. Shotts, and V. S. Blazer. 1985. Recovery of *Edwardiella ictaluri* from danio (*Danio devario*). Aquaculture 46:63–66.

Wedemeyer, G. A., N. C. Nelson, and C. A. Smith. 1978. Survival of salmonid infectious hematopoietic necrosis (IHNV) and infectious pancreatic necrosis (IPNV) in ozonated, chlorinated, and untreated waters. Journal of the Fisheries Research Board of Canada 35:875–879.

Weston, D. P. 1991. The effects of aquaculture

on indigenous biota. Pages 534–567 *in* D. E. Brune and J. R. Tomasso, editors. Aquaculture and water quality. Advances in world aquaculture, volume 3. World Aquaculture Society, Baton Rouge, Louisiana, USA.

Whoriskey, F. 2000. Infectious salmon anaemia: A review and the lessons learned from wild salmon on Canada's east coast. Pages 46–51 *in* P. Gallaugher and C. Orr, editors. Aquaculture and the protection of wild salmon. Simon Fraser University, Burnaby, British Columbia, Canada.

Williams, I. and D. F. Amend. 1976. A natural epizootic of infectious hematopoietic necrosis in fry of sockeye salmon (*Oncorhynchus nerka*) at Chilko Lake, British Columbia. Journal of the Fisheries Research Board of Canada 33:1564–1567.

Wobeser, G. A. 1981. Diseases of wild waterfowl. Plenum Publishing, New York, New York, USA.

Wolf, K. 1988. The viruses and viral diseases of fish. Cornell University Press, Ithaca, New York, USA.

Wolf, K. and M. E. Markiw. 1984. Biology contravenes taxonomy in the Myxozoa: New discoveries show alterations of the invertebrate and vertebrate hosts. Science 225:1449–1452.

Wolf, K., M. C. Quimby, L. L. Pettijohn, and M. L. Landolt. 1973. Fish viruses: isolation and identification of infectious hematopoietic necrosis in eastern North America. Journal of the Fisheries Research Board of Canada 30:1625–1627.

Yamamoto, T. and J. Kilistoff. 1979. Infectious pancreatic necrosis virus: quantification of carriers in lake populations during a 6-year period. Journal of the Fisheries Board of Canada 36:562–567.

Yoder, W. G. 1972. The spread of *Myxosoma cerebralis* into natural trout populations in Michigan. Progressive Fish-Culturist 43:103–106.

Yukhimenko, S. S. 1970. On the occurrence of *Bothriocephalus gowkongenesis* Yeh, 1955 (Cestoda, Pseudophyllidea) in young of cyprinidae from the Amour river. Parasitologia 4:480–483.

Zelikoff, J. T. 1994. Fish immunotoxicology. Pages 71–95 *in* J. H. Dean, M. I. Luster, A. E. Munson, and I. Kimber, editors. Immunotoxicology and immuno-pharmacology. Raven Press, New York, USA.

Ecological Impacts of Escaped Organisms

CHRISTOPHER A. MYRICK

Department of Fishery and Wildlife Biology, Colorado State University,

Fort Collins, Colorado 80523-1474 USA

ABSTRACT

The escape of organisms, especially fish, from aquaculture operations has the potential to impact the ecology of the surrounding area. The purpose of this chapter is to examine some of the ecological impacts of escaped organisms and to propose possible strategies for minimizing these impacts. Potential areas where escaped organisms can have an impact are disease transmission, genetic interaction with wild organisms, competition for limiting resources, predation on wild organisms, and colonization of novel environments. The first two topics are addressed in other chapters in this volume. The degree to which a group of escaped organisms influences the remaining three areas depends on: 1) the number of escaped organisms, as larger numbers will have larger impacts; 2) the species of the escaped organisms, as species with wide environmental tolerances and requirements (especially for reproduction) will have greater impacts than those with narrow tolerances, and; 3) the biological and physical characteristics of the receiving environment. Strategies for minimizing the ecological impacts of escape events include the use of containment or isolation technologies, the use of inventory control systems, the adoption of local or regional monitoring and notification programs, and careful site and species selection. Escapes from aquaculture operations will continue to occur, but it is in the best interest of the operators and stakeholders to work together to minimize the impacts of these escapes.

The culture of aquatic organisms for food has been an important part of food production in some regions for centuries. In the last 40 years, the role of aquaculture as a source of aquatic food production has increased rapidly, with close to 30 million metric tons produced worldwide in 1997 (Stickney 2000). Aquaculture's contribution in both economic and nutritional terms is undisputed and well-recognized, however, the aquaculture industry has recently been criticized for the potentially negative impacts it may have on the environment (Naylor et al. 2000). Considerable efforts have been made by private and public stakeholders to document the potential ecological effects of aquaculture and come up with viable management solutions that address all sides of the issue (e.g., The Salmon Aquaculture Review of the Environmental Assessment Office of British Columbia).

A potential source of ecological impact that may result from an aquaculture system, with the notable exception of a completely closed recirculating system, is that of escaped organisms. As the number of aquaculture operations and the number of organisms produced increases, so too increases the number of escaped organisms (Carr et al. 1997). This increase in escaped organisms may increase the risk of local, and perhaps regional impacts on biological systems. Concern over the potential impacts of escaped organisms exists throughout the world, including Asia (Nelson and Eldredge 1991), Africa (Bruton and van As 1986), Australia (Thresher 1999), Europe (Lura et al. 1993), and the Americas (McMichael et al. 1999). Recent research on potential ecological impacts has been centered in the latter two regions, with a majority of the focus on the effects of escaped salmonids.

For the purpose of this chapter, an escaped organism is defined as one that escapes directly from an aquaculture system like a net-pen or raceway, and expressly excludes non-cultured organisms that may co-occur with cultured organisms. Additionally, this chapter addresses only the unintentional releases of cultured organisms, as an extensive body of literature exists on the effects of intentional releases (i.e., stocking).

The goals of this chapter are as follows:

1) Provide an overview of the sources and mechanisms of escape from aquaculture operations.

2) Discuss the potential ecological effects of these escapes.

3) Discuss some options for managing and/or preventing the escape of cultured organisms.

4) Identify areas where more research or development activities are required.

Sources and Mechanisms of Escaped Organisms

Despite the best efforts of culturists, some escapes are to be expected from all but the most physically isolated culture systems. The causes of these escapes can be grouped into four broad classes:

1) A natural event (e.g., weather) compromises the integrity of the culture system, allowing the escape of some or all of the animals contained therein. Examples of such events include the loss of pond-reared fishes when floods over-top pond levees, and the loss of cage-reared fishes when wave and current action are strong enough to physically damage or destroy nets and support structures.

2) Operator error compromises the integrity of the culture systems and/or inadvertently releases fish into the natural environment. Examples of such events include net damage caused by boat handling errors, and the loss of fish during transfers between culture units.

3) Acts of vandalism intentionally compromise the integrity of culture systems. These acts generally involve damage to culture units and may be associated with poaching attempts.

4) Predators may also cause escapes by damaging holding structures in an effort to reach the fish inside. This is particularly a problem in marine environments where large, powerful predators such as seals are present.

Fish culturists have no direct control over the first of these categories, but can control the latter three. A moderate degree of control can be exerted over weather-related risks by constructing facilities to withstand expected environmental stresses. The degree of control that can be exerted is a function of financial, technological, regulatory, and personnel resources—rarely would sufficient resources be available to allow the creation of a completely escape-proof culture system and environment.

A wide variety of fish species have escaped from culture operations. In the United States, the majority of the escapes have been fishes cultured for the tropical aquarium fish trade. Courtenay (1995) reported that at least 70 exotic species are now found in the U.S.; an additional 200 native species now occur outside of their native ranges. Benson (1999) listed 50 species that are known to have escaped from culture operations in the U.S. Of the 50, 15 species have established populations but only a handful of these are species that are cultured for human consumption. Table 1 provides a partial listing of native and exotic freshwater fishes that have been introduced outside of their native range in the United States. Fishes that escape from aquaculture operations can be subdivided into two groups, the first of which is composed of exotics that did not previously

TABLE 1. *Partial listing of native and exotic freshwater fishes that have been introduced outside of their native ranges in the United states. Classes of fish are native (N) and exotic (E). Exotic species that have established populations in the U.S. are marked with (*). Sources of the fishes are classified as known (K), suspected (S), aquaculture escape (A), and intentionally transplanted for reintroduction (T). Data are from Benson (1999), Courtenay (1995), Moyle et. al (Moyle et al. 1995), Herbold et al. (1992), Brown and Moyle (1991), Moyle and Williams (1990), and Moyle (1976).*

Family	Scientific name	Common name	Class	Source
Petromyzontidae—lampreys	*Petromyzon marinus*	sea lamprey	N	K
Acipenseridae—sturgeons	*Acipenser transmontanus*		N	K
Amiidae—bowfins	*Amia calva*	bowfin	N	K
Anguiliidae—freshwater eels	*Anguilla anguilla*	European eel	E	A
	Anguilla australis	Shortfin eel	E	A
Clupeidae—herrings	*Alosa pseudoharenga*	alewife	N	K
	Alosa sapdissima	American shad	N	K
	Dorosoma cepedianum	gizzard shad	N	K
	Dorosoma petense	threadfin shad	N	K
Engraulidae—anchovies	*Anchoa mundeoloides*	anchovy	E	K
Cyprinidae—carps and minnows	*Agosia chrysogaster*	longfin dace	N	K
	Brachydanio rerio	zebra danio	E*	A
	Campostoma anomalum	central stoneroller	N	K
	Campostoma oligolepis	largescale stoneroller	N	K
	Carassius auratus	goldfish	E*	K
	Ctenopharyngodon idella	grass carp	E*	K
	Cyprinella galactura	whitetail shiner	N	K
	Cyprinella lutrensis	red shiner	N	K
	Cyprinella spiloptera	spotfin shiner	N	K
	Cyprinella venusta	blacktail shiner	N	K
	Cyprinus carpio	common carp	E*	K
	Dionda episcopa	roundnose minnow	N	S
	Exoglossum maxillingua	cutlips minnow	N	S
	Gila atraria	Utah chub	N	K
	Gila bicolor	tui chub	N	K
	Gila coerulea	blue chub	N	K
	Gila copei	leatherside chub	N	K
	Gila orcutti	arroyo chub	N	K
	Gila pandora	Rio Grande chub	N	K
	Gila purpurea	Yaqui chub	N	T
	Hesperoleucus symmetricus	California roach	N	K
	Hybognathus hankinsoni	brassy minnow	N	K
	Hybognathus placitus	plains minnow	N	K
	Hypophthalmichthys molitrix	silver carp	E*	K
	Hypophthalmichthys nobilis	bighead carp	E*	K
	Leuciscus idus	ide	E*	K
	Luxilus cerasinus	crescent shiner	N	S
	Luxilus chrysocephalus	striped shiner	N	K
	Luxilus coccogenis	warpaint shiner	N	S
	Luxilus zonistus	bandfin shiner	N	K
	Lythrurus ardens	rosefin shiner	N	S
	Lythrurus atrapiculus	blacktip shiner	N	K
	Macrohybopsis storeiana	silver chub	N	K
	Mylopharyngodon piceus	black carp	E	A

TABLE 1 (continued). *Partial listing of native and exotic freshwater fishes that have been introduced outside of their native ranges in the United states. Classes of fish are native (N) and exotic (E). Exotic species that have established populations in the U.S. are marked with (*). Sources of the fishes are classified as known (K), suspected (S), aquaculture escape (A), and intentionally transplanted for reintroduction (T). Data are from Benson (1999), Courtenay (1995), Moyle et. al (Moyle et al. 1995), Herbold et al. (1992), Brown and Moyle (1991), Moyle and Williams (1990), and Moyle (1976).*

Family	Scientific name	Common name	Class	Source
	Nocomis biguttatus	hornyhead chub	N	K
	Nocomis leptocephalus	bluehead chub	N	S
	Nocomis micropogon	river chub	N	K
	Nocomis raneyi	bull chub	N	S
	Notemigonus crysoleucas	golden shiner	N	K
	Notropis amoenus	comely shiner	N	S
	Notropis atherinoides	emerald shiner	N	K
	Notropis baileyi	rough shiner	N	K
	Notropis bairdi	Red River shiner	N	K
	Notropis bifenatus	bridle shiner	N	S
	Notropis buccula	smalleye shiner	N	K
	Notropis buchanani	ghost shiner	N	K
	Notropis chiliticus	redlip shiner	N	S
	Notropis dorsalis	bigmouth shiner	N	K
	Notropis girardi	Arkansas River Shiner	N	K
	Notropis hudsonius	spottail shiner	N	K
	Notropis leuciodus	Tennessee shiner	N	S
	Notropis lutipinnis	yellowfin shiner	N	K
	Notropis oxyrhynchus	sharpnose shiner	N	K
	Notropis ozarcanus	Ozark shiner	N	K
	Notropis procne	swallowtail shiner	N	S
	Notropis rubellus	rosyface shiner	N	K
	Notropis rubricroceus	saffron shiner	N	S
	Notropis schumardi	silverband shiner	N	K
	Notropis spectrunculus	mirror shiner	N	S
	Notropis telescopus	telescope shiner	N	K
	Notropis volucellus	mimic shiner	N	K
	Notropis winchelli	clear chub	N	K
	Orthodon microlepidotus	Sacramento blackfish	N	K
	Phenacobius mirabilis	suckermouth minnow	N	K
	Phoxinus oceas	mountain redbelly dace	N	K
	Pimephales notatus	bluntnose minnow	N	K
	Pimephales promelas	fathead minnow	N	K
	Ptychocheilus grandis	Sacramento pikeminnow	N	K
	Puntius conchonius	rosy barb	E	A
	Puntius filamentosus	blackspot barb	E*	K
	Puntius gelius	dwarf barb	E	A
	Puntius schwanenfeldii	tinfoil barb	E	A
	Puntius semifasciolatus	green barb	E*	K
	Puntius tetrazona	tiger barb	E	A
	Rhinichthys cataractae	longnose dace	N	K
	Rhinichthys osculus	speckled dace	N	K
	Richardsonius balteatus	redside shiner	N	K
	Scardinius erythrophthalmus	rudd	E*	K

TABLE 1 (continued). *Partial listing of native and exotic freshwater fishes that have been introduced outside of their native ranges in the United states. Classes of fish are native (N) and exotic (E). Exotic species that have established populations in the U.S. are marked with (*). Sources of the fishes are classified as known (K), suspected (S), aquaculture escape (A), and intentionally transplanted for reintroduction (T). Data are from Benson (1999), Courtenay (1995), Moyle et. al (Moyle et al. 1995), Herbold et al. (1992), Brown and Moyle (1991), Moyle and Williams (1990), and Moyle (1976).*

Family	Scientific name	Common name	Class	Source
	Semotilus atromaculatus	creek chub	N	K
	Semotilus corporalis	fallfish	N	K
	Tinca tinca	tench	E*	K
Catostomidae—suckers	*Carpoides carpio*	river carpsucker	N	K
	Catostomus commersoni	white sucker	N	K
	Catostomus fumeiventris	Owens sucker	N	K
	Catostomus platyrhynchus	mountain sucker	N	K
	Catostomus plebeius	Rio Grande sucker	N	K
	Catostomus santaanae	Santa Ana sucker	N	K
	Catostomus tahoensis	Tahoe sucker	N	S
	Erimyzon sucetta	lake chubsucker	N	S
	Hypentelium etowanum	Alabama hog sucker	N	K
	Hypentelium nigricans	northern hog sucker	N	K
	Ictiobus bubalus	smallmouth buffalo	N	K
	Ictiobus cyprinellus	bigmouth buffalo	N	K
	Minytrema melanops	spotted sucker	N	K
	Moxostoma duquesnei	black redhorse	N	S
	Moxostoma erythrurum	golden redhorse	N	K
	Moxostoma macrolepidotum	shorthead redhorse	N	K
	Moxostoma rhothoecum	torrent sucker	N	S
	Moxostoma rupiscartes	striped jumprock	N	S
Cobitidae—loaches	*Misgurnus angillicaudatus*	Oriental weatherfish	E*	K
	Misgurnus mizolepis	Chinese fine-scaled loach	E	K
Characidae—characins	*Aphyocharax anisitsi*	bloodfin tetra	E	A
	Astyanax mexicanus	Mexican tetra	N	K
	Colossoma macropomum	black pacu	E	K
	Gymnocorymbus ternetzi	black tetra	E	K
	Hemigrammus ocellifer	head-and-tailight tetra	E	A
	Hoplias malabaricus	trahira	E	A
	Hoplosternum littorale	brown hoplo	E	S
	Hyphessobrycon serpae	serpae tetra	E	A, K
	Moenkhausia sanctaefilomenae	redeye tetra	E	A, K
Locariidae—armored catfishes	*Hypostomus plecostomus*	suckermouth catfish	E*	A, K
Corydoridae—S. American catfish	*Corydoras aeneus*	green corydoras	E*	K
	Pterygoplichthys disjunctivus	vermiculated sailfin catfish	E*	A, K
	Pterygoplichthys multiradiatus	Orinoco sailfin catfish	E*	A, K
	Pterygoplichthys pardalis	Amazon sailfin catfish	E	K
Ictaluridae—bullhead catfishes	*Ameiurus catus*	white catfish	N	K
	Ameiurus melas	black bullhead	N	K
	Ameiurus natalis	yellow bullhead	N	K
	Ameiurus nebulosus	brown bullhead	N	K
	Ictalurus furcatus	blue catfish	N	K
	Ictalurus punctatus	channel catfish	N	K
	Noturus gyrinus	tadpole madtom	N	K

TABLE 1 (continued). *Partial listing of native and exotic freshwater fishes that have been introduced outside of their native ranges in the United states. Classes of fish are native (N) and exotic (E). Exotic species that have established populations in the U.S. are marked with (*). Sources of the fishes are classified as known (K), suspected (S), aquaculture escape (A), and intentionally transplanted for reintroduction (T). Data are from Benson (1999), Courtenay (1995), Moyle et. al (Moyle et al. 1995), Herbold et al. (1992), Brown and Moyle (1991), Moyle and Williams (1990), and Moyle (1976).*

Family	Scientific name	Common name	Class	Source
	Noturus insignis	margined madtom	N	K
	Pylodictis olivaris	flathead catfish	N	K
Clariidae—air-breathing catfishes	*Clarias batrachus*	walking catfish	E*	A
	Clarias fuscus	whitespotted clarias	E*	K
Esocidae—pikes	*Esox americanus americanus*	redfin pickerel	N	K
	Esox americanus vermiculatus	grass pickerel	N	K
	Esox lucius	northern pike	N	K
	Esox masquinongy	muskellunge	N	K
	Esox niger	chain pickerel	N	K
Umbridae—mudminnows	*Dallia pectoralis*	Alaska blackfish	N	K
	Umbra limi	central mudminnow	N	K
Osmeridae—smelts	*Hypomesus nipponensis*	wakasagi	N	K
	Osmerus mordax	rainbow smelt	N	K
Salmonidae—trouts	*Coregonus albula*	vendace	E	K
	Coregonus artedi	cisco	N	K
	Coregonus clupeaformis	lake whitefish	N	K
	Coregonus lavarettus	powan	E	K
	Oncorhynchus apache	Apache trout	N	T
	Oncorhynchus clarki	cutthroat trout	N	K
	Oncorhynchus gilae	Gila trout	N	K
	Oncorhynchus gorbuscha	pink salmon	N	K
	Oncorhynchus keta	chum salmon	N	K
	Oncorhynchus kisutch	coho salmon	N	K
	Oncorhynchus masou	cherry salmon	E	K
	Oncorhynchus mykiss	rainbow trout	N	A, K
	Oncorhynchus mykiss aguabonita	golden trout	N	K
	Oncorhynchus nerka	sockeye salmon or kokanee	N	K
	Oncorhynchus tshawytscha	chinook salmon	N	K
	Salmo salar	Atlantic salmon	N	A, K
	Salmo trutta	brown trout	E*	K
	Salvelinus alpinus	Arctic char	N	K
	Salvelinus fontinalis	brook trout	N	K
	Salvelinus namaycush	lake trout	N	K
	Thymallus arcticus	Arctic grayling	N	K
Amblyopsidae—cavefishes	*Chologaster agassizi*	spring cavefish	N	K
Gadidae—cods	*Lota lota*	burbot	N	K
Aplocheilidae—rivulines	*Aplocheilus lineatus*	striped panchax	E	S
Cyprinodontidae—killifishes	*Belonesox belizanus*	pike killifish	E*	K
	Crenichthys nevadae	Railroad Valley springfish	N	T
	Cyprinodon bovinus	Leon Springs pupfish	N	T
	Cyprinodon diabolis	Devils Hole pupfish	N	T
	Cyprinodon elegans	Comanche springs pupfish	N	T
	Cyprinodon macularius	desert pupfish	N	T
	Cyprinodon rubrofluviatilis	Red River pupfish	N	K

TABLE 1 (continued). *Partial listing of native and exotic freshwater fishes that have been introduced outside of their native ranges in the United states. Classes of fish are native (N) and exotic (E). Exotic species that have established populations in the U.S. are marked with (*). Sources of the fishes are classified as known (K), suspected (S), aquaculture escape (A), and intentionally transplanted for reintroduction (T). Data are from Benson (1999), Courtenay (1995), Moyle et. al (Moyle et al. 1995), Herbold et al. (1992), Brown and Moyle (1991), Moyle and Williams (1990), and Moyle (1976).*

Family	Scientific name	Common name	Class	Source
	Empetrichthys latos	Pahrump pupfish	N	T
	Fundulus diapharus	banded killifish	N	K
	Fundulus grandis	gulf killifish	N	K
	Fundulus sciadicus	plains topminnow	N	K
	Fundulus seminolis	Seminole killifish	N	K
	Fundulus stellifer	southern studfish	N	K
	Fundulus waccamensis	Waccamaw killifish	N	K
	Fundulus zebrinus	plains killifish	N	K
	Leptolucania ommata	pygmy killifish	N	K
	Lucania goodei	bluefin killifish	N	K
	Lucania parva	rainwater killifish	N	K
Poeciliidae—livebearers	*Gambusia affinis*	western mosquitofish	N	K
	Gambusia gaigei	Big Bend gambusia	N	T
	Gambusia geiseri	largespring gambusia	N	K
	Gambusia holbrooki	eastern mosquitofish	N	K
	Gambusia nobilis	Pecos gambusia	N	T
	Heterandia formosa	least killifish	N	K
	Poecilia formosa	Amazon molly	N	K
	Poecilia latipinna	sailfin molly	N	K
	Poecilia latipunctata	broadspotted molly	E	A, K
	Poecilia mexicana	shortfin molly	E*	A, K
	Poecilia petenensis	swordtail molly	E	K
	Poecilia reticulata	guppy	E*	A, K
	Poecilia sphenops	liberty molly	E*	A, K
	Poeciliopsis gracilis	porthole livebearer	E	A, K
	Poeciliopsis occidentalis	Gila topminnow	N	T
	Xiphophorus helleri	green swordtail	E*	K
	Xiphophorus maculatus	southern platyfish	E*	A, K
	Xiphophorus variatus	variable platyfish	E*	K
Atherinidae—silversides	*Atherinops regius*	silverside	E	K
	Labidesthes sicculus	brook silverside	N	K
	Menidia beryllina	inland silverside	N	K
Gasterosteidae—sticklebacks	*Culaea inconstans*	brook stickleback	N	K
	Gasterosteus aculeatus	threespine stickleback	N	K
	Pungitius pungitius	ninespine stickleback	N	K
Cottidae—sculpins	*Cottus bairdi*	mottled sculpin	N	K
Percichthyidae—temperate basses	*Morone americana*	white perch	N	K
	Morone chrysops	white bass	N	K
	Morone mississippiensis	yellow bass	N	K
	Morone saxatilis	striped bass	N	K
Centrarchidae—sunfishes	*Ambloplites cavifrons*	Roanoke bass	N	K
	Ambloplites contellatus	Ozark bass	N	K
	Ambloplites rupestris	rock bass	N	K
	Archoplites interruptus	Sacramento perch	N	K

TABLE 1 (continued). *Partial listing of native and exotic freshwater fishes that have been introduced outside of their native ranges in the United states. Classes of fish are native (N) and exotic (E). Exotic species that have established populations in the U.S. are marked with (*). Sources of the fishes are classified as known (K), suspected (S), aquaculture escape (A), and intentionally transplanted for reintroduction (T). Data are from Benson (1999), Courtenay (1995), Moyle et. al (Moyle et al. 1995), Herbold et al. (1992), Brown and Moyle (1991), Moyle and Williams (1990), and Moyle (1976).*

Family	Scientific name	Common name	Class	Source
	Enneacanthus gloriosus	bluespotted sunfish	N	K
	Lepomis auritus	redbreast sunfish	N	K
	Lepomis cyanellus	green sunfish	N	K
	Lepomis gibbosus	pumpkinseed	N	K
	Lepomis gulosus	warmouth	N	K
	Lepomis humilis	orangespotted sunfish	N	K
	Lepomis macrochirus	bluegill	N	K
	Lepomis megalotis	longear sunfish	N	K
	Lepomis microlophus	redear sunfish	N	K
	Lepomis punctatus	spotted sunfish	N	K
	Micropterus coosae	redeye bass	N	K
	Micropterus dolomieu	smallmouth bass	N	K
	Micropterus punctulatus	spotted bass	N	K
	Micropterus salmoides	largemouth bass	N	K
	Micropterus treculi	Guadalupe bass	N	K
	Pomoxis annularis	white crappie	N	K
	Pomoxis nigromaculatus	black crappie	N	K
Percidae—perches	*Etheostoma blennioides*	greenside darter	N	S
	Etheostoma caeruleus	rainbow darter	N	K
	Etheostoma chlorosomum	bluntnose darter	N	K
	Etheostoma edwini	brown darter	N	K
	Etheostoma fusiforme	swamp darter	N	K
	Etheostoma olmstedi	tesselated darter	N	K
	Etheostoma zonale	banded darter	N	K
	Gymnocephalus cernuus	ruffe	E*	K
	Perca flavescens	yellow perch	N	K
	Percina caprodes	logperch	N	K
	Percina macrolepida	bigscale logperch	N	K
	Percina roanoka	Roanoke darter	N	K
	Stizostedion canadense	sauger	N	K
	Stizostedion lucioperca	zander	E	K
	Stizostedion vitreum	walleye	N	K
Haemulidae—grunts	*Anisotremus davidsoni*	sargo	N	K
Sciaenidae—drums	*Aplodinotus grunniens*	freshwater drum	N	K
Cichlidae—cichlids	*Aquidens pulcher*	blue acara	E*	A
	Astronotus occellatus	oscar	E*	A
	Cichla ocellaris	peacock cichlid	E*	K
	Cichla temensis	speckled pavon	E	K
	Cichlasoma beani	green guapote	E	K
	Cichlasoma bimaculatum	black acara	E*	A
	Cichlasoma citrinellum	Midas cichlid	E*	K
	Cichlasoma cyanoguttatum	Rio Grande cichlid	N	K
	Cichlasoma labiatum	red devil	E	K
	Cichlasoma managuense	jaguar guapote	E*	K

TABLE 1 (continued). *Partial listing of native and exotic freshwater fishes that have been introduced outside of their native ranges in the United states. Classes of fish are native (N) and exotic (E). Exotic species that have established populations in the U.S. are marked with (*). Sources of the fishes are classified as known (K), suspected (S), aquaculture escape (A), and intentionally transplanted for reintroduction (T). Data are from Benson (1999), Courtenay (1995), Moyle et. al (Moyle et al. 1995), Herbold et al. (1992), Brown and Moyle (1991), Moyle and Williams (1990), and Moyle (1976).*

Family	Scientific name	Common name	Class	Source
	Cichlasoma meeki	firemouth cichlid	E*	A, K
	Cichlasoma nigrofasciatum	convict cichlid	E*	K
	Cichlasoma octofasciatum	Jack Dempsey	E*	K
	Cichlasoma salvini	yellowbelly cichlid	E*	K
	Cichlasoma spilurum	blue-eyed cichlid	E*	K
	Cichlasoma trimaculatum	threespot cichlid	E	K
	Cichlasoma urophthalmus	Mayan cichlid	E*	A, K
	Geophagus brasiliensis	pearl eartheater	E	A
	Geophagus surinamensis	redstriped eartheater	E	A, K
	Hemichromis letourneauxi	African jewelfish	E*	K
	Hemicrhomis elongatus	banded jewelfish	E*	K
	Heros severus	banded cichlid	E	K
	Oreochrmois macrochir	longfin tilapia	E*	S
	Oreochromis aureus	blue tilapia	E*	A, K
	Oreochromis mossambicus	Mozambique tilapia	E*	K
	Oreochromis niloticus	Nile tilapia	E	A
	Oreochromis urolepis	Wami tilapia	E*	K
	Oreochromis zilli	redbelly tilapia	E*	A, K
	Sarotherodon melanotheron	blackchin tilapia	E*	A
	Symphysodon discus	red discus	E	A
	Telmatochromis bifrenatus	Lake Tanganyika dwarf cichlid	E	A, K
	Tilapia mariae	spotted tilapiae	E*	A, K
	Tilapia rendalli	redbreast tilapia	E	K
	Tilapia sparrmannii	banded tilapia	E	A, K
Gobiidae—gobies	*Acanthogobius flavimanus*	yellowfin goby	E*	K
	Gillichthys mirabilis	longjaw mudsucker	N	K
	Gillichthys seta	mudsucker	E	K
	Neogobius melanostomus	round goby	E*	K
Anabantidae—climbing perch	*Anabas testudineus*	climbing perch	E	A
	Ctenopoma nigropannosum	twospot climbing perch	E	A
Belontiidae—gouramis	*Betta splendens*	Siamese fighting fish	E	A
	Colisa fasciata	banded gourami	E	K
	Colisa labiosa	thicklip gourami	E	A, K
	Colisa lalia	dwarf gourami	E	A, K
	Helostoma temminckii	kissing gourami	E	A, K
	Trichogaster leeri	pearl gourami	E	A
	Trichogaster trichopterus	blue gourami	E	A
	Trichopsis vitatta	croaking gourami	E*	A
Channidae	*Channa micropeltes*	giant snakehead	E	K
	Channa striata	chevron snakehead	E*	K

occur in the region. Examples of such species include Atlantic salmon on the Pacific coast of North America (McKinnell and Thomson 1997; Thomson and McKinnell 1997), tilapia *Oreochromis spp.* in sub-tropical and tropical regions of the U.S. (including U.S. possessions), and grass carp *Ctenopharyngodon idella* in several regions of the United States (Stanley et al. 1978). The second group is comprised of native species that are cultured in the same area as existing wild populations. Examples of species in this group include Atlantic salmon (*Salmo salar*) in Norway (Sægrov et al. 1997) and steelhead *Oncorhynchus mykiss* on the North American west coast (McMichael et al. 1999). Both groups have the potential to impact ecological systems in different manners.

Potential Impacts of Escaped Cultured Organisms

Whenever organisms are added to a community, regardless of source or whether they are exotic or native, there will be some type of impact (Begon et al. 1990). An escapee from aquaculture will impact on the receiving ecosystem (Courtenay 1995), although the impact will probably be slight for a single escapee. As the number of escaped organisms increases, the probability of a significant ecological impact increases concurrently. The ecological impacts from these escapes vary in nature and severity. The direct and indirect impacts of the escapes belong to one of the following categories.

1) <u>Genetic impacts</u>. If native populations of a given species are sympatric with conspecific or congeneric escaped organisms, genetic impacts may occur as the two groups interbreed (Crozier 1993). These impacts will include genetic dilution, loss of genetic diversity, and hybridization.

2) <u>Disease impacts</u>. Escaped organisms may serve as hosts or vectors for pathogens that have adverse effects on wild organisms (McVicar 1997).

3) <u>Competition</u>. Escaped organisms may compete with wild organisms for habitat, food, or mates (McMichael et al. 1999). This competition may be interspecific or intraspecific in nature.

4) <u>Predation.</u> Escaped organisms may affect local ecological systems by preying on wild organisms. Theoretically, the increase in local abundance following an escape event may also attract predators, thereby increasing the overall predation pressure in the immediate vicinity.

5) <u>Habitat alteration</u>. Escaped organisms may directly or indirectly alter the habitats into which they escape (Crutchfield 1995; Pflieger 1997). If they do so, then wild organisms may not be capable of adapting to the altered habitat conditions. Habitat alterations are normally manifested over several years (Crutchfield 1995).

6) <u>Colonization</u>. Perhaps the most widely feared impact of escapes from aquaculture (at least by the general public) is that they will successfully colonize local ecosystems to the detriment of the existing biota.

Genetic and disease impacts are discussed in chapters 9 and 10 of this volume. Habitat alteration will not be discussed in further detail because impacts from competition, predation, and colonization are likely to be apparent. Spatial scale is important in regards to the impacts of aquaculture escapes. Impacts from escape events decrease in severity and significance as the distance from the point of escape increases. It is important, therefore, to determine whether the primary concern is the ecological health of a small-scale local system or that of a larger-scale system.

Competition

Competition between escaped and wild organisms arises over access and utilization of habitat resources, food resources, or mates. Competition for these resources is not necessarily mutually exclusive. For example, successfully competing for limited habitat resources may increase access to food and or mates (Griffith 1988; Alanärä and Brännäs 1997). The type and intensity of the competition varies, and largely depends on factors such as the species involved, relative populations sizes, and the degree to which the resource is limited.

Competition for Habitat. Upon entry to a novel environment, an escaped organism is immediately subjected to some level of habitat competition. The intensity of competition is a function of the species of the escaped organism, the rearing history of the escaped organism, and the biotic and physical composition of the receiving environment. Cultured fish often show poor adaptation to natural environments by selecting less advantageous or suitable habitats (Mesa 1991). Escaped organisms that are native to the area should have a higher probability of short-term survival than exotic organisms, especially if wild populations are small relative to the number of escapees. If, however, the environment is notably different than the culture environment, the escapees may be incapable of acclimating and adjusting to their new surroundings. When local environmental conditions are within the tolerance limits of the escaped organisms, the outcome of competitive interactions will be influenced by other factors.

Competitive encounters between an escaped organism and a wild organism have three possible outcomes. The first outcome is that the two organisms partition the resource and remain in close proximity to one another. The second outcome is that the wild organism is dominant and remains in the suitable habitat while the escaped organism is displaced. Wild organisms may spend a lot of energy dominating or defeating the escaped cultured organisms, resulting in a net loss of energy reserves that may have otherwise been used for somatic growth (Busacker et al. 1990; Jobling 1994). Weiss and Schmutz (1999) documented

such a reduction in growth rates for wild brown trout *Salmo trutta* that were interacting with cultured brown trout. A reduction in energy reserves may also increase the predation risk and susceptibility to pathogens (Wald and Wilzbach 1992).

In the opposite case, the escaped organism is dominant and the wild organism displaces into less suitable habitat (Jonsson 1997; McMichael et al. 1999). This reduces the wild organism's feeding opportunities and protection from predators or harsh environmental conditions, with the net result of decreasing its fitness. The escaped organism may derive an energetic advantage from this situation because it now occupies prime habitat, but it may also experience a net loss of energy, as mentioned above. The net effect of competitive interactions is often negative, with both wild and escaped organisms experiencing losses in energy reserves and overall ecological fitness.

A large body of literature examining the interactions between wild salmonids and intentionally released cultured salmonids exists e.g., (Schramm and Piper 1995). The results from these studies may not be directly applicable to the issue of escaped salmonids because intentional releases of juvenile and adult salmonids are timed to give the cultured fish a high probability of survival, whereas accidental releases can occur at any time during the culture process. Interactions among non-salmonid species have also been documented, but to a lesser extent. Cichlids, including the spotted tilapia *Oreochromis mariae*, like most cichlids, aggressively defend territories and are capable of driving away other fishes, including centrarchids (Moyle 1976; Courtenay and Hensley 1979). Without careful field and laboratory studies of inter- and intraspecific interactions, similar to the large body of work conducted on salmonids, the outcomes of such interactions are difficult to predict.

If the escaped organisms are adults, or if

they mature and return to habitats with resident populations of wild organisms, then the potential for competition for spawning habitats and mates exists (Crozier 1993; Lura et al. 1993; Hansen et al. 1997). Successful reproduction in salmonids requires habitat suitable for redd construction (Moyle 1976; Groot and Margolis 1991). Suitable spawning habitat is often limited (Saunders and Smith 1965; Vogel 1992), thus providing the trigger for direct or indirect competition. Direct competition occurs when both escaped and wild organisms are present simultaneously on the spawning grounds. Under these circumstances, individual size, aggression, and past experience appear to determine success. Wild fish have been reported to hold a competitive edge over cultured fish (Berejikian et al. 1997). This superior competitive ability will result in a preponderance of wild x wild matings or an increase in the number of wild x cultured matings (wild males mating with cultured females) if the escaped population is larger than the wild population. Indirect competition results when the wild and escaped populations are temporally segregated. Whichever population occupies the habitat last is likely to superimpose their redds on redds left by earlier spawnings (Taniguchi et al. 2000). When the second population excavates their redds, they may destroy or disturb (physically shock) eggs during the sensitive pre-pigmentation period (Post et al. 1974; Dwyer et al. 1993), causing mass mortality. The digging activities also can displace eggs or developing embryos, exposing them to egg predators such as sculpins, salmonids, and other creatures (Moyle 1976). Sægrov et al. (1997) reported that escaped Atlantic salmon spawned earlier than wild fish in the River Vosso, Norway. In this case wild fish, particularly males, dominated spawning habitats and activities. If spawning habitat in the River Vosso is limiting, wild females (55% of the total number of females captured in the river) may have destroyed the majority of the embryos produced by the earlier-spawning escaped salmon.

As reviewed by Jonsson (1997), however, some escaped Atlantic salmon spawn later than wild stocks; similar findings have been reported for cultured Pacific salmon, including coho salmon *Oncorhynchus kisutch* (Fleming and Gross 1992, 1993). Jonsson attributes the delayed spawning to a lack of freshwater experience. The spawning delay could also result from the widespread use of non-resident stocks for culture organisms, or from manipulation of temperature and photoperiod regimes in aquaculture operations (Bromage et al. 1984; Pavlidis et al. 1992).

Competition for food. Cultured organisms represent a broad array of feeding guilds, including omnivores (e.g., channel catfish *Ictalurus punctatus*), filter-feeding plank-tivores (e.g., Pacific oysters *Crassostrea gigas*), and predators (e.g., Atlantic salmon). Unlike their wild counterparts, most cultured organisms are trained to accept prepared diets and are naïve with respect to natural food items. Because they are unfamiliar with natural foods, escaped organisms may jeopardize their immediate fitness by sampling a variety of edible and inedible items. Additionally, escaped predators may lack the skills needed to capture and subdue prey. Two recent studies on brown trout demonstrated that captive rearing does not eliminate the ability to learn to identify and capture natural food items (and therefore survive), although such learning does not occur immediately. Weiss and Schmutz (1999) reported that 3-mo survival of juvenile cultured brown trout was similar (48–80%) to that of wild trout (49–90%) when both groups were present in the same streams. Johnsen and Ugedal (1986) compared the food habits of cultured brown trout with those of wild brown trout already present in the stream. The cultured trout displayed learning behavior with respect to suitable natural foods, and the diets of the two groups were essentially the same after 1 wk.

When an organism escapes into the natural environment, the intensity of competition for food depends on the local abundance of food items. Abundant food resources will limit the intensity of competition, whereas limited food resources will generate fairly intense competitive pressures. Limited food resource situations are particularly serious where the food resource has already been impacted by other factors, such as habitat degradation (Moyle 1994). Perhaps the most important factor in determining the intensity of food resource competition is the total demand for those resources. If the combined populations of escaped and wild organisms exceed the limits set by the size of the resource, competition will be intense and an overall reduction of fitness is likely to be observed.

Omnivorous- or generalist-cultured species (e.g., *Oreochromis zilli* and *Oreochromis aurea*) may have an advantage over more specialized wild species because of their ability to exploit a wide variety of food resources. Unfortunately, few studies comparing the relative success of food competition between escaped and wild organisms have been conducted. When investigating competition for food resources, it is particularly important to pay attention to competition during larval and juvenile stages, as many species are particularly vulnerable to insufficient or intermittent food availability at these stages.

Competition for mates. Unlike competition for habitat or food, which can be inter- or intraspecific in nature, competition for mates is almost exclusively intraspecific (Verspoor 1988). As with competition for food and habitat, the critical variable is the number of escaped organisms relative to the number of wild organisms present. In cases where the escaped cultured organisms outnumber the wild organisms, flooding the spawning areas with either males or females will likely have severe impacts. Because most aquaculture escape events involve small numbers of organisms, competition for mates should not be overly prevalent.

Predation

The second major type of ecological interaction is predation (Begon et al. 1990). When dealing with aquatic species, it is important to consider the implications of life-history omnivory (Polis 1991) and other complex trophic interactions such as ontogenetic changes in feeding guilds. For example, bluegill *Lepomis macrochirus* may feed on largemouth bass *Micropterus salmoides* eggs and larvae, only to be in turn fed upon by juvenile and adult largemouth bass. Additionally, as is the case with grass carp, many larval and juvenile fish are zooplanktivorous, even when the adults are herbivorous (Stanley et al. 1978). Predation impacts can be direct, where escaped cultured organisms feed directly on a wild organism (e.g., coastal cutthroat trout *Oncorhynchus clarki clarki*, preying on juvenile sockeye salmon, *Oncorhynchus nerka* fry, Cartwright et al. 1998), or indirect where escaped organisms feed on a prey item used by wild organisms. The latter case represents competition for food resources, and is discussed above.

Studies of the impacts of predation by escaped organisms on wild organisms are limited. Crutchfield (1995) noted that *Oreochromis zilli* and *Oreochromis aurea* were known predators upon the eggs and fry of native clupeids and centrarchids. In a study of the effects of the introduction of these tilapia species on the fish community of a power plant cooling reservoir, Crutchfield reported declines in golden shiner *Notemigonus crysoleucas*, eastern mosquitofish *Gambusia holbrooki*, and green sunfish *Lepomis cyanellus* populations. These declines may have resulted from the combined effects of predation on early life stages, competition for food resources, and habitat alteration.

Without pre-escape baseline data and post-escape monitoring efforts, the full impact of predation by escaped organisms cannot be known. If naïve wild prey species are present (Zaret and Paine 1973; Zaret 1975) or if large numbers of predators escape into a system, predation can have a significant impact on the prey species (Zaret 1975). As with any change in ecosystem function, predator-prey interactions can have wide-ranging impacts throughout the ecosystem (Johnson and Goettl 1999). These impacts are hard to predict without a sound understanding of the linkages and energy flow through the ecosystem (Cohen and Brian 1984; Cohen and Newman 1988).

Colonization

Despite the popular public perception that organisms escaping from aquaculture will establish viable populations, few instances of successful colonizations exist, relative to the number of escape events. Colonizing a new environment or location requires that the following criteria be met:

1) Environmental conditions such as temperature extremes, annual flow regimes, and physical habitat types must meet the requirements of the escaped organisms, not only for survival but for successful reproduction. Fish reproduction tends to be constrained by a narrow set of physicochemical tolerance limits (Combs 1965; Beacham and Murray 1990; USFWS 1999); conditions outside these limits produce high mortality rates in embryonic and juvenile fishes.

2) The receiving environment must provide the necessary conditions for the completion of the life cycle of the escaped organisms. For example, a catadromous organism like an eel *Anguilla spp.* requires access to seawater to successfully complete its lifecycle. If the environment doesn't provide all the necessary components, then reproduction will not occur.

3) Enough escaped organisms must be present to form a viable breeding population. Effective population sizes vary, but the small size of most aquaculture escape events combined with the low return rates of adult organisms reduces the probability that sufficient numbers will be present. If any one escape event does not provide enough recruits for the new population, then a continuous series of events, serving as a metapopulation source, is required.

These assumptions have been met for a few species in recent history. One region that seems particularly hospitable for colonization by aquaculture escapes is southern Florida (Courtenay and Robins 1973; Courtenay and Hensley 1979; Benson 1999). The combination of favorable environmental and biological conditions has led to an exotic fish fauna with a surprising number of representatives from the tropical aquarium fish trade (Table 1). Of the fish species cultured for food, Crutchfield (1995) reported that two tilapia species successfully colonized a lake where winter temperatures did not drop below 10 C. Escaped Atlantic salmon are reported to have entered streams on the North American west coast (McKinnell and Thomson 1997; Thomson and McKinnell 1997), presumably to spawn, but the author was unable to find documentation of any established populations. Successful efforts to introduce food or game fish to new areas generally release hundreds of thousands or even millions of individuals into the new environment—most escapes from aquaculture occur on a much smaller scale and are less likely to succeed.

Management Options

Commercial aquaculture, like any other agricultural venture, is ultimately driven by the need for profitability. Aquaculture profit margins are generally low, so the loss of even a small number of animals produces an unfavorable shift in the bottom line. In order to maximize the profitability of their operations, most culturists actively try to prevent escapes. This leads to a willingness to adopt new methods of safeguarding their stocks, either through the use of new innovative equipment or culture practices.

The primary causes of escapes are weather, predation, equipment failure, and operator error. Any modification of culture systems, or regulatory and culture practices that reduces the incidence of these causes will help reduce the numbers of escaped fish. Some possible options are listed below.

Isolation or Containment of Stock

In the most extreme form, fish culturists could use systems that are completely isolated from the environment (e.g., recirculating systems). While theoretically attractive, such systems are impractical for a number of culture situations. A more realistic approach would be the adoption of partial containment systems (e.g., double-walled nets for net-pens, isolation from water sources, etc.) that would prevent escapes caused by anything short of gross vandalism, operator error, or massive structural failures.

Inventory Control

Fish culturists should use some form of inventory control to mark and track their fish so that any escaped fish can be distinguished from wild fish. Various marking techniques have been successfully used on large batches of fish, including fin clips, coded wire tags, VI-alpha tags and PIT tags. In order for an inventory control system to be effective, unique tags that can at least identify fish by source (e.g., Farm A vs. Farm B), and preferably by lot or even individual fish, are needed. The use of farm-unique tags would be important for regulatory and management purposes, as it would allow government officials to determine the source of an escaped fish. Tags that can be read without specialized equipment are

required if the detection and reporting of escaped fish by untrained personnel is considered important (e.g., the Atlantic Salmon Watch program in the Pacific Northwest).

Re-capture Contingency Plans

All aquaculture operations should develop a series of contingency plans that addresses escape events. Where possible, such plans should be approved by or coordinated with the appropriate agricultural or fisheries agency. Each plan should provide a threshold number of escaped organisms that generates a full-scale response. This number will vary depending on the species, the type of culture operation, and the location of the aquaculture operation. Contingency plans for species that have wild populations or species whose environmental tolerances and requirements match those of the receiving environment should emphasize rapid, direct action. Plans for species that are unlikely to thrive for an extended period in the receiving environment (e.g., tropical fish released in a temperate reservoir during the summer) can also emphasize direct action, though it may be acceptable to simply monitor the population of escaped fishes.

Under certain situations, the use of re-capture contingency plans may be feasible. Perhaps the simplest systems for which contingency plans could be developed are small freshwater lakes. If a large escape occurs, chemical treatment of the lake with an ichthyocide would suffice. In the case of anadromous species, it may be possible to remove mature adults from spawning areas and to destroy any redds that have been created. This method would be labor intensive, but it has the advantage of targeting those few escaped individuals that have survived long enough to return to spawn. Other possible methods of concentrating and capturing escaped organisms include the use of feeds, scent-based homing behavior, etc. This is clearly an area that warrants further research.

Monitoring/Notification

Carefully designed monitoring and notification systems are the nuclei of any escape control and mitigation system. Regular monitoring programs should be established in the vicinity of culture operations so that rapid detection of the marked fish can occur. In the event of an escape, culturists should notify the proper authorities immediately so that the contingency plans can be activated. Finally, the general public should be kept appraised of the situation. This may generate some initial backlash, but the long-term benefits of an up-front and frank exchange of information should outweigh any initial resentment. Aquaculture operations are viewed negatively in some areas (especially in western North America) and any action that demonstrates a "good neighbor" attitude will reflect favorably on the aquaculture operations.

Species Selection

One attractive proactive option for preventing long-term ecological impacts from escaped organisms is the culture of species or strains that are incapable of surviving in the natural environment. The most feasible means of adopting this approach is to culture tropical species in a temperate environment or marine species in inland areas. As more domesticated strains of fish are developed, it may be possible to produce fish that are wholly dependent on artificial diets or habitats for survival.

Another proactive option is the culture of sterile organisms. Even if sterile organisms escape, their ecological impacts will not be long-lived. Short-term impacts from competition and predation may be significant, but these will only persist as long as the organisms do. The sterile culture approach has already been adopted by some resource agencies with regards to the use of grass carp *Ctenopharyngodon idella*. In Colorado, for example, only triploid grass carp are legal for release in waters west of the Continental Divide. Techniques for producing sterile cultures (e.g., triploids or hybrids) are being developed for commonly cultured species like rainbow trout (Suresh and Sheehan 1998), brook trout *Salvelinus fontinalis* (Stillwell and Benfey 1997), and Atlantic salmon (Cotter et al. 2000). Early experiments prove promising, and the use of hybrid organisms may be attractive to aquaculture producers because of their high growth rates (Cheng et al. 1987; Hayward et al. 1997).

Conclusions

Escapes of organisms from aquaculture systems will continue to occur. The ecological impact of these events is a function of the number of organisms escaping, the species escaping, the proximity to any wild populations, and the biological and physical characteristics of the receiving ecosystem. Two of the most important factors are the number and species of the organisms escaping. Small numbers of organisms or exotic species that cannot survive or reproduce are less cause for concern than large numbers and/or native species that can successfully adapt to the receiving ecosystem.

Escaped organisms that manage to survive in the receiving environment will have ecological impacts. Predation on wild organisms by the escaped organisms will be local in nature, but will probably be minimal. Competition for food, habitat, and mates will depend largely on the total number of organisms present and the scarcity of the contested resource. Outcomes of competitive encounters range from little effect on the wild organisms to a loss of fitness for both wild and escaped organisms. Colonization of novel environments by escaped organisms does occur, but appears to be most common in smaller species that have broad reproductive and juvenile environmental tolerances and requirements. Both large and small species do colonize novel environments, particularly when environmental conditions are favorable and large numbers of escaped organisms are present.

The frequency and magnitude of escape events can be reduced through careful risk management practices. Careful site and species selection are proactive measures that will reduce the likelihood and impact of escape events, as will careful training and equipment selection. All culturists should prepare contingency plans for dealing with a variety of escape scenarios. Contingency plans should be integrated into local and regional monitoring and notification systems so that the proper authorities and stakeholder groups are made aware of escaped events as quickly as possible. Some level of government-mandated risk management of aquaculture escapes is warranted, however, the aquaculture industry should be given the opportunity to develop industry-standard procedures before strict government controls and regulations are imposed. Aquaculture operations stand to lose as much from the escape of their culture organisms as does the environment. By working with regulatory agencies and stakeholders, it should be possible to develop strategies to prevent escapes and/or reduce their impacts in the early stages.

Literature Cited

Alanärä, A. and E. Brännäs. 1997. Diurnal and nocturnal feeding activity in Arctic char (*Salvelinus alpinus*) and rainbow trout (*Oncorhynchus mykiss*). Canadian Journal of Fisheries and Aquatic Sciences 54:2894–2900.

Beacham, T. D. and C. B. Murray. 1990. Temperature, egg size and development of embryos and alevins of five species of Pacific salmon: a comparative analysis. Transactions of the American Fisheries Society 119:927–945.

Begon, M., J. L. Harper, and C. R. Townsend. 1990. Ecology, 2nd edition. Blackwell Scientific Publications, Cambridge, Massachusetts, USA.

Benson, A. J. 1999. Documenting over a century of aquatic introductions in the United States. Pages 1–31 *in* R. Claudi and J. H. Leach, editors. Nonindigenous freshwater organisms. CRC Press, Boca Raton, Florida, USA.

Berejikian, B. A., E. P. Tezak, S. L. Schroder, C. M. Knudsen, and J. J. Hard. 1997. Reproductive behavioral interactions between wild and captively reared coho salmon (*Oncorhynchus kisutch*). ICES Journal of Marine Science 54:1040–1050.

Bromage, N. R., J. A. K. Elliot, J. R. C. Springate, and C. Whitehead. 1984. The effects of constant photoperiods on the timing of spawning in the rainbow trout. Aquaculture 43:213–223.

Brown, L. R. and P. B. Moyle. 1991. Changes in habitat and microhabitat partitioning within an assemblage of stream fishes in response to predation by Sacramento squawfish (*Ptychocheilus grandis*). Canadian Journal of Fisheries and Aquatic Sciences 48:849–856.

Bruton, M. S. and J. van As. 1986. Faunal invasions of aquatic ecosystems in southern Africa, with suggestions for their management. L. A. W. Macdonald, F. J. Kruger, and A. A. Ferrar, editors. The ecology and management of biological invasions in southern Africa: proceedings of the National Synthesis Symposium on the Ecology of Biological Invasions. Oxford University Press, Cape Town, South Africa.

Busacker, G. P., I. R. Adelman, and E. M. Goolish. 1990. Growth. Pages 363–388 *in* C. B. Schreck and P. B. Moyle, editors. Methods for fish biology. American Fisheries Society, Bethesda, Maryland, USA.

Carr, J. W., J. M. Anderson, F. G. Whoriskey, and T. Dilworth. 1997. The occurrence and spawning of culture Atlantic salmon (*Salmo salar*) in a Canadian River. ICES Journal of Marine Science 54:1064–1073.

Cartwright, M. A., D. A. Beauchamp, and M. D. Bryant. 1998. Quantifying cutthroat trout (*Oncorhynchus clarki*) predation on sockeye salmon (*Oncorhynchus nerka*) fry using a bioenergetics approach. Canadian Journal of Fisheries and Aquatic Sciences 55:1285–1295.

Cheng, K. M., I. M. McCallum, R. I. McKay, and B. E. March. 1987. A comparison of survival and growth of two strains of chinook salmon (*Oncorhynchus tshawytscha*) and their crosses reared in confinement. Aquaculture 67:301–311.

Cohen, J. E. and F. Brian. 1984. Trophic links of community food webs. Proceedings of the National Academy of Sciences 81:4105–4109.

Cohen, J. E. and C. M. Newman. 1988. Dynamic basis of food web organization. Ecology 69:1655–1664.

Combs, B. D. 1965. Effect of temperature on the development of salmon eggs. Progressive Fish-Culturist 27:134–137.

Cotter, D., V. O'Donovan, N. O'Maoiléidigh, G. Rogan, N. Roche, and N. P. Wilkins. 2000. An evaluation of the use of triploid Atlantic salmon (*Salmo salar* L.) in minimising the impact of escaped farmed salmon on wild populations. Aquaculture 186:61–75.

Courtenay, W. R., Jr. 1995. The case for caution with fish introduction. Pages 413–424 *in* H. L. Schramm, Jr. and R. G. Piper, editors. Uses and effects of cultured fishes in aquatic ecosystems. American Fisheries Society, Bethesda, Maryland, USA.

Courtenay, W. R., Jr. and D. A. Hensley. 1979. Range expansion in southern Florida of the introduced spotted tilapia, with comments on its environmental impress. Environmental Conservation 6:149–151.

Courtenay, W. R. and C. R. Robins. 1973. Exotic organisms in Florida with emphasis on fishes: A review and recommendations. Transactions of the American Fisheries Society 102:1–12.

Crozier, W. W. 1993. Evidence of genetic interaction between escaped farmed salmon and wild Atlantic salmon (*Salmo salar* L.) in a Northern Irish river. Aquaculture 113:19–29.

Crutchfield, J. U., Jr. 1995. Establishment and expansion of redbelly tilapia and blue tilapia in a power plant cooling reservoir. Pages 452–461 *in* H. L. Schramm, Jr. and R. G. Piper, editors. Uses and effects of culture fishes in aquatic ecosystems. American Fisheries Society, Bethesda, Maryland, USA.

Dwyer, W. P., W. Fredenberg, and D. A. Erdhal. 1993. Influence of electroshock and mechanical shock on survival of trout eggs. North American Journal of Fisheries Management 13:839–843.

Fleming, I. A. and M. R. Gross. 1992. Reproductive behavior of hatchery and wild coho salmon (*Oncorhynchus kisutch*): Does it differ? Aquaculture 103:101–121.

Fleming, I. A. and M. R. Gross. 1993. Breeding success of hatchery and wild coho salmon (*Oncorhynchus kisutch*) in competition. Ecological Applications 3:230–245.

Griffith, J. S. 1988. Review of competition between cutthroat trout and other salmonids. Pages 134–140 *in* R. E. Greswell, editor. Status and management of interior stocks of cutthroat trout. American Fisheries Society, Bethesda, Maryland, USA.

Groot, C. and L. Margolis. 1991. Pacific salmon life histories. UBC Press,

Vancouver, British Columbia, Canada.

Hansen, L. P., M. L. Windsor, and A. F. Youngson. 1997. Interactions between salmon culture and wild stocks of Atlantic salmon: the scientific and management issues. ICES Journal of Marine Science 54:963–964.

Hayward, R. S., D. B. Noltie, and N. Wang. 1997. Use of compensatory growth to double hybrid sunfish growth rates. Transactions of the American Fisheries Society 126:316–322.

Herbold, B., A. D. Jassby and P. B. Moyle. 1992. San Francisco estuary project: status and trends report on aquatic resources in the San Francisco estuary. San Francisco Estuary Project, San Francisco, California, USA.

Jobling, M. 1994. Fish bioenergetics. Chapman and Hall, London, England.

Johnsen, B. O. and O. Ugedal. 1986. Feeding by hatchery-reared and wild brown trout, *Salmo trutta* L., in a Norwegian stream. Aquaculture and Fisheries Management 17:281–287.

Johnson, B. M. and J. P. Goettl, Jr. 1999. Food web changes over fourteen years following introduction of rainbow smelt into a Colorado reservoir. North American Journal of Fisheries Management 19:629–642.

Jonsson, B. 1997. A review of ecological and behavioural interactions between cultured and wild Atlantic salmon. ICES Journal of Marine Science 54:1031–1039.

Lura, H., B. T. Barlaup, and H. Sægrov. 1993. Spawning behaviour of a farmed escaped female Atlantic salmon (*Salmo salar*). Journal of Fish Biology 42:311–313.

McKinnell, S. and A. J. Thomson. 1997. Recent events concerning Atlantic salmon escapees in the Pacific. ICES Journal of Marine Science 54:1221–1225.

McMichael, G. A., T. N. Pearsons, and S. A. Leider. 1999. Behavioral interactions among hatchery-reared steelhead smolts and wild *Oncorhynchus mykiss* in natural streams. North American Journal of Fisheries Management 19:948–956.

McVicar, A. H. 1997. Disease and parasite implications of the coexistence of wild and cultured Atlantic salmon populations. ICES Journal of Marine Science 54:1093–1103.

Mesa, M. G. 1991. Variation in feeding, aggression, and position choice between hatchery and wild cutthroat trout in an artificial stream. Transactions of the American Fisheries Society 120:723–727.

Moyle, P. B. 1976. Inland fishes of California. University of California Press, Berkeley, California, USA.

Moyle, P. B. 1994. The decline of anadromous fishes in California. Conservation Biology 8:869–870.

Moyle, P. B. and J. E. Williams. 1990. Biodiversity loss in the temperate zone-decline of the native fish fauna of California. Conservation Biology 4:275–284.

Moyle, P. B., R. M. Yoshiyama, J. E. Williams, and E. D. Wikramanayake. 1995. Fish species of special concern in California. Department of Wildlife & Fisheries Biology, University of California, Davis, California, USA.

Naylor, R. L., R. J. Goldburg, J. H. Primavera, N. Kautsky, M. C. M. Beveridge, J. Clay, C. Folke, J. Lubchenco, H. Mooney, and M. Troell. 2000. Effect of aquaculture on world fish supplies. Nature 405:1017–1024.

Nelson, S. and L. G. Eldredge. 1991. Distribution and status of introduced cichlid fishes of the genera *Oreochromis* and *Tilapia* in the islands of the South Pacific and Micronesia. Asian Fisheries Science 4: 11–22.

Pavlidis, M., V. Theochari, J. Paschos, and A. Dessypris. 1992. Effect of six photoperiod protocols on the spawning time of two rainbow trout, *Oncorhynchus mykiss* (Walbaum), populations in northwest Greece. Aquaculture and Fisheries Management 23:431–441.

Pflieger, W. L. 1997. The fishes of Missouri, revised edition. Missouri Department of Conservation, Jefferson City, Missouri, USA.

Polis, G. J. 1991. Complex trophic interactions in deserts: an empirical critique of food-web theory. American Naturalist 138:133–155.

Post, G., D. V. Power, and T. M. Kloppel. 1974. Survival of rainbow trout eggs after receiving physical shocks of known magnitude. Transactions of American Fisheries Society 103:711–716.

Sægrov, H., K. Hindar, S. Kålås, and H. Lura. 1997. Escaped farmed Atlantic salmon replace the original salmon stock in the River Vosso, western Norway. ICES Journal of Marine Science 54:1166–1172.

Saunders, J. W. and M. W. Smith. 1965. Changes in a stream population of trout associated with increased silt. Journal of the Fisheries Research Board of Canada 22:395–404.

Schramm, H. L., Jr. and R. G. Piper, editor. 1995. Uses and effects of cultured fishes in aquatic ecosystems. American Fisheries Society, Bethesda, Maryland, USA.

Stanley, J. G., W. W. Miley, Jr., and D.L. Sutton. 1978. Reproductive requirements and likelihood for naturalization of escaped grass carp in the United States. Transactions of the American Fisheries Society 107:119–128.

Stickney, R. R., editor. 2000. Encyclopedia of aquaculture. John Wiley and Sons, Inc., New York, New York, USA.

Stillwell, E. J. and T. J. Benfey. 1997. The critical swimming velocity of diploid and triploid brook trout. Journal of Fish Biology 51:650–653.

Suresh, A. V. and R. J. Sheehan. 1998. Biochemical and morphological correlates of growth in diploid and triploid rainbow trout. Journal of Fish Biology 52:588–599.

Taniguchi, Y., Y. Miyake, T. Saito, H. Urabe, and S. Nakano. 2000. Redd superimposition by introduced rainbow trout (*Oncorhynchus mykiss*) on native charrs in a Japanese stream. Ichthyological Research 47:149–156.

Thomson, A. J. and S. McKinnell. 1997. Summary of reported Atlantic salmon (*Salmo salar*) catches and sightings in British Columbia and adjacent waters in 1996. Department of Fisheries and Oceans, Nanaimo, British Columbia, Canada.

Thresher, R. E. 1999. Diversity, impacts and options for managing invasive marine species in Australian waters. Australian Journal of Environmental Management 6:137–145.

USFWS. 1999. Effect of temperature on early-life survival of Sacramento River fall-and winter-run chinook salmon. Northern Central Valley Fish and Wildlife Office, Red Bluff, California, USA.

Verspoor, E. 1988. Widespread hybridization between native Atlantic salmon, *Salmo salar,* and introduced brown trout, *S. trutta*, in eastern Newfoundland. Journal of Fish Biology 32:327–334.

Vogel, D. 1992. Sacramento winter chinook: problem and recovery. Northwest Science 66:141.

Wald, L. and M. A. Wilzbach. 1992. Interactions between native brook trout and hatchery brown trout: effects on habitat use, feeding, and growth. Transactions of the American Fisheries Society 121:287–296.

Weiss, S. and S. Schmutz. 1999. Performance of hatchery-reared brown trout and their effects on wild fish in two small Austrian streams. Transactions of the American Fisheries Society 128:302–316.

Zaret, T.M. 1975. The ecology of introductions—a case study from a Central American lake. Environmental Conservation 1:308–309.

Zaret, T. N. and R. T. Paine. 1973. Species introductions in a tropical lake. Science 182:449–455.

The Economics of Environmental Impacts of Aquaculture in the United States

CAROLE R. ENGLE AND

DIEGO VALDERRAMA

Aquaculture/Fisheries Center, Mail Stop 4912, University of Arkansas at Pine Bluff,

Pine Bluff, Arkansas 71601 USA

ABSTRACT

The economics of environmental impacts of aquaculture in the United States is a complex topic. The field of environmental economics has developed a theoretical framework for analysis of these issues. The concept of pollution as a production externality is central to discussions of why producers should internalize waste disposal costs. The economics revolve around balancing the costs of effluent controls with the costs to society of poor water quality. The costs of effluent controls include the farm costs of installing treatment systems in addition to impacts on local, state, regional, and national economies of decreased acreage in production. Molluscan aquaculture provides positive externalities in terms of water quality. Pond aquaculture internalizes much of the waste disposal cost. Imposing additional treatment on the infrequent and dilute discharges is likely to result in high cost as well as strongly negative impacts on the economy. Production systems such as raceways that discharge more concentrated wastes may create greater water quality problems and higher costs to society. Imposing treatment costs on raceway farmers may be economically advisable if the treatment costs are less than the costs of impaired water quality. In all cases, strong economic incentives should be built into programs to ensure maximum voluntary adoption by farmers.

This chapter first will outline briefly how economists approach and define environmental conflicts. The types of conflicts and impacts to the environment that can occur from different aquaculture production systems are then discussed from an economics perspective. Costs and benefits to society of adopting effluent controls are contrasted. The chapter then presents specific, farm-level costs associated with effluent treatment alternatives.

Environmental Economics

Economics is often considered to be the study of costs, revenues, and profits of businesses. However, economics encompasses much broader themes. To an economist, the essential problem is that resources are scarce and human wants and needs are unlimited. Thus, it is necessary to establish priorities in order to allocate resources to meet the wants and needs of human beings. Economists study how resources are allocated to produce goods for human beings.

In many situations, supply and demand are the forces that operate to distribute resources to the production of goods for purchase by consumers. Fluctuating prices serve as the primary catalyst for changes in both the quantity supplied (produced) and the quantity demanded (purchased). Thus, consumers who most want and need a particular product will be willing to pay more. Producers looking to make profits will produce the more profitable goods and will purchase more resources. Thus, the price system and market forces result in resources being directed towards production of those goods most wanted and needed by consumers.

The above description of how resources are allocated in a market economy is simplistic.

One underlying assumption is that all resources used in the production of a particular good are accounted for by the business. Businesses account for resource use by monitoring the costs associated with the acquisition and use of that resource.

What happens if there is no direct cost that a business had to pay for using a particular resource? In this case, the price of the product will not reflect the cost of using that particular product. The product's price will be lower and consumers are likely to purchase a greater quantity. This will result in greater production of that product and greater use of the resource than if its use had entailed a direct cost. If the quantity available of the resource being used is unlimited, then there would be no problem. However, if the resource exists in a limited quantity and if there are many users of that resource, conflicts will arise.

Conflicts commonly occur over use of resources that change its quality, particularly if it is a common-property resource. Common-property resources are those not owned by any one individual but used by many individuals. In these cases, other users suffer if one user causes the quality of the resource to deteriorate.

An example of this concept might be a factory that discharges its wastes into a stream. The stream is a common-property resource not owned by the company and its water is used by a town located downstream from the factory. The factory in this case is "using" the stream resource for waste disposal without incurring any cost. It sells its product at a cost that does not include waste disposal costs. If waste disposal costs were included in its costs, it would likely produce less because the quantity demanded by consumers at the higher price would be less. If it produced less, it would discharge less and reduce its "use" of the stream.

Moreover, the town downstream will need to treat the water before using it. The town's costs will be higher due to the treatment and the town might use less water because its costs are higher. In this case, resources are not being allocated efficiently because the factory's low product price is being subsidized by the town downstream. Resources will be allocated efficiently only if costs are paid directly by the consumers of the product, not by others in the society.

Economists view pollution as a negative "externality" because the cost of the good whose production generates pollution is not borne by the consumers of that good. Pollution costs are considered "spillover costs" because these costs spill over to segments of the economy other than the market for that particular good. Moreover, the consumers affected by spillover costs from an externality are not compensated for these costs. The polluters have lower costs and non-polluters have higher costs associated with cleaning up the resource.

When do negative externalities occur? They typically occur when users of resources bear only part of the cost of its use. Price systems and market forces will work to allocate resources to their best use only if full costs (long and short-run costs) are charged to the consumers of that product. An example might be that of fossil fuels. Full costs of fossil fuel use would include not only the costs of extracting fossil fuels but also of replacing them. This is similar to the use of depreciation to account for the use of capital goods in production. The full cost of a tractor to a farm includes its operating costs, but also the cost of replacing it when it wears out. If the business does not earn enough over time to replace it when it wears out, then that business is not profitable.

Negative externalities also occur when the private costs (marginal costs) are less than the costs to society. If costs of production are not internalized (paid for by the firm), the firm is likely to overproduce and generate greater profits than if it had paid for its waste disposal.

If all costs of production, including pollution costs, are charged to the business, its product will sell at a higher price and fewer units will be sold. Alternative products will be produced with the additional resources made available.

Government intervention is often called upon to develop rules that force producers to internalize costs such as those of waste disposal. Regulations that limit discharge or require producers to treat wastes prior to disposal are justified in terms of the improvements to the quality of natural resources. The rationale is that the costs of waste disposal would be passed on to consumers in the form of higher prices. Those individuals in society most interested in the particular products being produced would be those to pay for the total costs of production, including the costs of waste disposal.

Aquaculture and Environmental Externalities

Negative externalities tend to attract the most attention. Nevertheless, externalities can be either positive or negative. For example, aquaculture generates a positive externality by reducing pressure on wild stocks of fish, some of which are endangered or threatened. In many instances, consumers prefer farm-raised seafood products to wild-caught products. Thus, aquaculture provides a means of substituting farm-raised fish and seafood products for wild-caught seafood. Aquaculture production further offers the level of control over the product that provides for assurances of quality and food safety. A safe food supply generates clear and positive externalities for society.

Other externalities that may be generated are specific to various aquaculture production systems. The following discusses some of the externalities, positive and negative, generated by raceways, cages, recirculating aquaculture systems, ponds, and shellfish culture systems. One distinction should be kept in mind

throughout. While aquaculture has been defined as a point source of pollution by the U.S. Environmental Protection Agency, its effluents differ significantly from those of manufacturing industries. Effluents from aquaculture have dilute concentrations of nutrients. Thus, as is the case with effluents from agriculture and forestry, these effluents are not easily treatable.

Raceways

The water source for raceway systems is typically a free-flowing spring or stream that enters a series of raceway units. The primary cost of water supply is for pipes or canals to perhaps filter and convey water to the raceways. In such a continuous flowing system, wastes are carried away by the flowing water into a stream or other public waterway. This scenario often presents a classic example of a negative externality. Disposing wastes into a publicly-owned waterway results in little or no cost to the producer, but additional costs may be incurred downstream by other users.

A raceway farmer who constructs a waste treatment unit (quiescent zone or off-line settling basin) would incur additional investment and operating costs. Waste disposal costs would be internalized and production costs would increase. Most trout-producing states require trout raceway farms to treat wastes before disposal, thereby internalizing the costs of waste treatment.

Raceway production facilities may also generate positive externalities. Trout wastes discharged into a nutrient-poor stream may increase the number, size, and biomass of fish in the stream by increasing the base of the food pyramid. Enhancement of the fishery will increase recreational benefits for the area. Increased recreational activities will provide economic benefits to local communities and to the region in the form of increased expenditures for food, lodging, transportation, fishing gear, and incidentals.

Cages

Most cage operations are sited in public waters without either water supply or waste treatment costs. Fish feed is converted to fish flesh. Waste products that disburse through the mesh of the cages into public waters may constitute a negative externality that imposes costs to other potential users of the public water resource. Kautsky et al. (1997) examined intensive cage aquaculture in Lake Kariba, Zimbabwe, showing that an area 10,000 times the size of the cage area would be needed to treat wastes from the operation.

Other chapters in this book address specific biological impacts that constitute negative externalities from cage production of fish. These may include nutrient enrichment of water bodies, the potential for escape of non-indigenous fish species and unintended effects on native fish communities or habitat, and the potential for increased parasite loading in areas with high concentrations of cages.

As in the case of raceways, positive externalities may occur in the case of cages sited in nutrient-poor, oligotrophic waters. Increasing nutrient levels of the impoundment or water body will increase the yield of fish. Enhanced fisheries will lead to additional economic activity in the form of increased fishing, commercial or recreational, and the subsequent economic activity and value that will be generated.

Recirculating systems

Recirculating systems in theory recycle water, but some discharge is inevitable. In practice, many recirculating systems discharge some percentage of water, particularly when problems are experienced. Even in the absence of any direct discharge, the recirculating system produces sludge that must be disposed of at a public treatment facility (discussed in Chapter 6). Wastes removed in the sludge are liquefied and represent a type of discharge. This disposal method imposes some cost to society in the form of subsidies received by public treatment facilities. Discharge standards for public treatment facilities appear to be more lenient than those used for private businesses (USEPA 1996). Specifically, public treatment facilities are allowed to discharge a higher level of nutrients than is allowed for many other types of businesses.

Recirculating aquaculture systems include explicit waste treatment processes. Thus, waste disposal costs are internalized to a large degree in recirculating systems (with the exception of sludge processed at public treatment facilities). The difference between waste treatment in ponds and in recirculating systems is that ponds take advantage of naturally-occurring processes fueled by sunlight whereas recirculating systems use a technological process fueled by electricity.

Other negative externalities associated with recirculating systems include an increased reliance on energy through increased heating, pumping, and other costs associated with running the system. While these are costs internalized by the firm, there is no cost to the firm of the additional demand for non-renewable resources that is created by the use of energy by recirculating systems.

Ponds

Ponds are the most common form of aquaculture production unit in the United States. Most commercial aquaculture growers rely on ground water resources, pumped into ponds from wells to supply ponds with water. Unlike the previous cases, pond producers incur costs associated with supplying water to production units. Furthermore, most pond systems are managed as static, not continuous flow, systems. While an argument may be made that the cost to the farmer does not include a cost for increased demands on the aquifer, this is mitigated to varying degrees by the fact that surface aquifers do recharge.

In static water systems, the nutrients (primarily nitrogen and phosphorus) applied to the pond in feed for fish are not flushed regularly from the system. After a percentage of feed is converted to fish flesh, the remaining nitrogen and phosphorus is released into the pond in the form of waste products. Without a continuous flow of water, these waste products remain in the pond and stimulate growth of phytoplankton (discussed in Chapter 3).

Nutrient cycles in ponds are complex, but natural processes that involve plankton and bacteria process wastes from fish production. However, the plankton biomass also limits fish production through its demand for oxygen. Thus, the waste treatment process (plankton and bacteria) in a pond also limits the kg of fish that can be produced. By not flushing wastes away, a type of opportunity cost of production foregone is incurred. This is in distinct contrast to waste disposal from raceways and cages in which the wastes leave the area of fish production.

Catfish ponds are self-sustaining units that typically are left in production for periods of 7–10 yr. When catfish ponds are drained, they are drained to be able to re-work the levees and bottoms of ponds or to catch the large fish that have evaded capture by seining. If there were no levee erosion and if all fish could be harvested by seining, ponds would not need to be drained. Ponds are not drained to dispose of waste materials. Organic material in the bottom of ponds is decomposed through natural processes and much of what accumulates in the pond bottoms is really soil particles from erosion of levees (Boyd 1995).

When ponds are drained, even though infrequently, nutrients are released in the discharge. These nutrients, while dilute and at lower levels than municipal treatment plant discharges, still contribute to overall nutrient loading and represent a negative externality.

Catfish ponds are also aerated. Catfish farmers aerate primarily to keep fish alive, but the aeration is as much a waste treatment activity as it is related to fish production. The fish uses only about 18% of the oxygen in a catfish pond (Boyd 1985); the remainder of the oxygen demand is from plankton and bacterial respiration.

Ponds also provide positive externalities to the environment in a variety of ways. They provide resting areas and a source of food for a wide variety of wading, diving, and migratory birds. While some species of birds actively prey on the fish being raised and cause serious damages, other species hunt only incidental species such as frogs, aquatic plants, or insects. Many fish farmers tolerate a wide variety of birds on their farms, and only become concerned with scaring those species (cormorants, pelicans) that significantly damage their fish crops.

Many fish farms have been constructed in what formally had been row crop land. The switch to fish farming from row crops generates a positive externality in the form of reduced soil loss and pesticide use.

Molluscan Shellfish

Shellfish growers raise an animal that filters water to remove nutrients. An oyster, for example, can filter 189 L of water each day (Canzoneri 1996). Because shellfish can concentrate relatively low levels of water pollution, the quality and cleanliness of water where shellfish are grown is a major factor in consumer acceptance of shellfish products.

Shellfish are highly susceptible to water pollution. Shellfish growers in various parts of the United States strongly support enforcement of water quality regulation (Canzoneri 1996). Failing septic systems have contributed to 80% of the marketing restrictions in shellfish growing areas. Productive molluscan aquaculture industries provide an economic incentive for maintaining high water quality. Even shoreline owners (who are opposed to the visual impact of floating aquacultural facilities) benefit from the incentives molluscan aquaculture provides for enhancing water quality.

Costs to Society of Effluent Controls

From an economics perspective, there is a trade-off between the cost to society of poor environmental quality resulting from pollution and the costs of controlling or reducing pollution to maintain a high level of environmental quality. Fig. 1 shows that the costs to society of poor environmental quality (i.e., hypoxia in the Gulf of Mexico, pfiesteria outbreaks in the northeast, etc.) are high and decrease as environmental quality

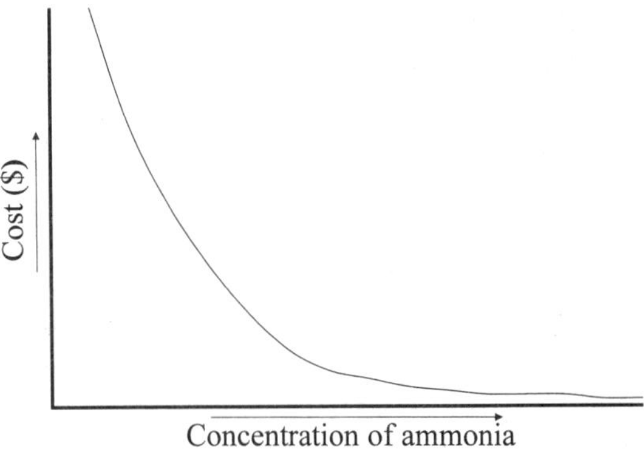

FIGURE 2. *Theoretical cost per unit of ammonia reduction at varying concentrations of ammonia (Adapted from Hailstones and Mastrianna 1986).*

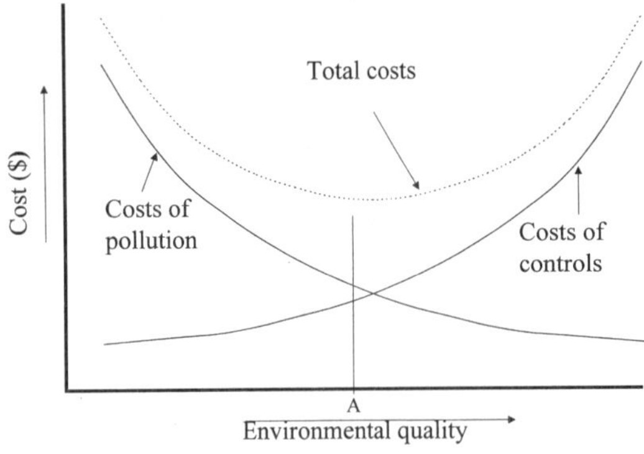

FIGURE 1. *Economic approach to evaluating the trade-offs between the cost to society of pollution and the cost to society of maintaining high environmental quality. "A" represents the optimal level of environmental quality given costs to society of pollution versus costs of pollution control (adapted from Hailstones and Mastrianna 1986).*

improves. On the other hand, costs of controlling pollution are low if environmental quality is poor because few controls are in place (Sharp and Leftwich 1986). In order to improve environmental quality, pollution control costs would increase. To maintain a very low level of pollution, it would be necessary to invest heavily in pollution-control equipment and to make other economic sacrifices. Fig. 2 presents a theoretical schematic of per-unit costs of

removing a "pollutant" such as ammonia. At high levels of ammonia, it is relatively inexpensive to remove a high percentage of ammonia. However, it is very costly to remove the last bits. Pond wastes are very dilute as compared to other types of pollutants, and the cost to remove such dilute amounts will be very high.

Pollution control costs are passed on to consumers in the form of higher prices. Higher-income families tend to allocate a larger share of their income for services, which are not affected greatly by pollution control costs. Hence, lower income families may be penalized more severely by increasing pollution control costs than will higher-income families. For example, costs of treating effluents from crawfish ponds might increase crawfish prices. Since food budgets constitute a high percentage of the household budget in low-income families, the higher crawfish prices would represent a higher percentage increase of the family budget. On a percentage basis, these cost increases would have a lesser effect on high-income households.

Polluted water and pollution control are both costly choices for society. From Fig. 1, it is clear that a goal of zero pollution is not rational. The pollution control costs would greatly exceed the cost to society of a minimal level of pollution. The sensible goal is to minimize the sum of the costs of pollution and the costs of controlling pollution (Mansfield 1986). This minimum could be viewed as a socially (not necessarily biologically) "optimal" level of environmental quality.

The conceptual framework illustrated in Fig. 1 is used in a general sense by economists to analyze environmental impacts and quality. Specific situations require data on the specific costs to society associated with different types of pollutants. For example, the costs to society of a given level of ammonia discharge would be much less than an equivalent amount of plutonium discharged. Plutonium would cause much greater public health problems than ammonia and would persist much longer in the environment. Those types of "pollutants" that can be assimilated readily in receiving streams will have lower costs to society than those that persist for a long time. Different receiving systems will also have differing capacities to assimilate nutrients and other types of pollutants.

Since pollution control (a cleaner environment) has costs, society must make a choice between the level of goods and services its resources will be used to produce and the degree of cleanliness of its environment. If the society experiences a level of pollution that is distasteful to it, it will be willing to sacrifice some quantity of goods and services for some level of pollution control. If the benefits of additional control (what cleaner water is worth to the citizens of the society) exceed the acceptable costs of additional control, then pollution control should be increased.

International Costs

The net effect on the global environment of increased regulation of U.S. aquaculture should be considered. Regulations may create unintended incentives for aquaculture production to move off-shore. Countries with fewer bureaucratic hurdles and fewer restrictions on environmental discharges can produce at costs lower than those in highly regulated economies. The net global environmental effect could be negative if production were displaced to less highly regulated countries. Similarly, if restrictions placed on aquaculture production result in increased demand for wild-caught fish, the additional pressure on natural fisheries could result in even greater damages to natural fish stocks. Populations of threatened fish species or stocks of fish that are already overfished could be driven to disastrous levels.

National Costs

If additional effluent controls were to be imposed on the aquaculture industry in the United States, it is likely that costs of production would increase. Economic theory indicates that increasing costs of production would result in a decrease in supply of aquaculture products. The decrease in supply would result in either less production from existing farms or a reduction in the overall number of aquaculture businesses. In either case, lower production levels would result in fewer jobs, less economic growth, and less economic impact.

Economic impacts are generated through a multiplier effect. As dollars are spent by an aquaculture business for feed, equipment, labor, pond construction, etc., the dollars spent constitute revenue to feed and equipment companies, to workers, and to companies that construct ponds, etc. As these companies hire additional workers and purchase additional equipment and supplies, these dollars are spread even further. These additional workers spend more money at local grocery stores, retailers, restaurants, and entertainment centers, etc. The more money spent in these sectors, the more other jobs and spending is generated that continues to reverberate throughout the economy. The increased circulation of money generates a multiplier effect on the economy.

Capital-intensive industries such as aquaculture generate higher economic multipliers than industries that use lower levels of capital. The reduction in supply of aquaculture products that would be created through increasing costs of production with effluent controls would reduce the amount of capital spent throughout the industry. A reduction in the amount of capital spent would in turn result in the reduction of money multiplying throughout local and regional economies. The result would be fewer jobs, less economic development, and less economic growth. The losses would be both immediate and over time from lost employment and economic growth that would have occurred had the controls not been put in place.

Moreover, regulations on aquaculture businesses that increase costs would affect the global competitiveness of U.S. aquaculture businesses. Major international competitors for aquaculture and seafood products have different policies regarding environmental issues, government subsidies, and labor conditions. Increased U.S. production costs may enhance the competitive position of other companies that use practices more damaging to the environment than those used in the U.S.

Edible fish and seafood products are one of the leading contributors to the U.S. trade deficit. Given the status of many of the world's fisheries, aquaculture is one of the few viable alternatives available to reduce the U.S. trade deficit in edible seafood products.

A safe, quality, nutritious food supply has implications for all U.S. citizens. One advantage of aquaculture in producing seafood products is its control over its product. This control ensures that aquaculture products meet standards for food safety and nutrition.

The vast majority (90%) of aquaculture businesses in the U.S. are classified as small businesses (< $500,000 in sales; USDA 1998).

This includes 84% of catfish farms, 95% of trout farms, 92% of ornamental fish farms, and 93% of baitfish farms. Families operating single farms appear most vulnerable to costs of pollution control (Hailstones and Mastrianna 1986). Economies of scale, access to capital, and other factors indicate that small businesses will be disproportionately affected by pollution-control regulations. Large corporate businesses have access to legal, accounting, research and development departments, and other support services that facilitate their compliance with regulations. The transformation of the agriculture sector from a family-farm based industry to a corporate-dominated one will be hastened with increasing amounts of regulation.

Local and Regional Costs

Much aquaculture production in the United States occurs in rural areas such as the Mississippi Delta region. These areas are characterized as impoverished areas with high unemployment, low income levels, poor economic growth, and little hope for economic development. The economy of the Mississippi Delta region has been further depressed in recent years. The phase-out of government programs for agricultural commodities has reduced payments to row crop farmers for low prices or low incomes. As a result, grain prices have fallen, and the rural economies dependent on these types of crops have experienced even greater economic challenges than before. Aquaculture is one of the few viable alternative enterprises in many of these communities. Aquaculture has played a major role in creating jobs and contributing to long-term economic growth in many of the most economically depressed regions of the United States. This contribution to economic growth has occurred without federal crop subsidies for aquaculture commodities.

In some rural communities, local banks have developed portfolios that depend heavily on loans to aquaculture businesses. This is particularly true of counties in the Mississippi Delta regions. Contractions in the catfish industry would have devastating financial effects on these banks that would reverberate throughout all sectors of these communities.

Economic impacts occur throughout the economy when any one major sector is damaged. The first impacts to be seen are the direct impacts that would result from a change in aquaculture output (Dicks et al. 1996). Indirect impacts then follow that include changes in the output of the industries that supply inputs such as feed, equipment, custom harvesting, fish hauling, net companies, and the processing and marketing companies that sell the commodities produced. Induced impacts follow that result from changes due to the spending levels of the farmers, their families, and workers in the upstream and downstream industries, and the effect of changes in spending levels on food, durable and non-durable goods, recreation, and private and public services. Expenditures on goods and services create additional employment, which in turn generates further economic activity. Dicks et al. (1996) estimated that, in the southern United States, aquaculture accounted for more than 121,000 domestic jobs and nearly 2.5 billion dollars of income. This is likely conservative because there is little information on production and processing expenditures on more than 30 species (such as baitfish, which is the fourth largest aquaculture industry in the U.S.) currently being produced by aquaculture.

Industry groups estimated that in 1993 each job in the oyster industry supported 1.13 jobs elsewhere in the economy of the state of Washington (Canzoneri 1996). These jobs were a significant employment base for rural counties, and shellfish aquaculture is one of the few resource-based industries that demonstrates long-term, sustainable economic growth in distressed communities. Shellfish growers bring revenues into the state. About 80% of oysters are sold out of state while 80% of the purchases are made within the state.

On-Farm Micro-Level Costs Associated with Effluent Control

Pollution can be reduced by either reducing the volume of the source of pollution or by reducing the concentration of "pollutants" in the discharges. In the case of aquaculture, the primary concern is with the discharge of nutrients. Methods to either reduce the volume of discharge (source reduction) or to reduce the concentration of nutrients can include either a formal discharge treatment, Best Management Practices (BMP's), or a combination of both. However, aquaculture discharges differ from most others in that the "pollutants" are nutrients in dilute concentrations. This makes it difficult to reduce concentrations. Furthermore, the primary form of the biological oxygen demand (BOD) or nutrients in pond systems is the plankton. Since plankton, by definition, floats, it is not subject to settling without the use of chemical treatments. The following discussion will begin with an estimation of the costs associated with the naturally occurring in-pond treatment of wastes in earthen ponds and will be followed by a general examination of economic aspects of BMP's. Estimates of costs associated with construction and use of settling basins, production/storage ponds, and constructed wetlands will be presented. Aspects of land application of aquaculture discharges also will be reviewed.

Ponds

In-Pond Treatment Costs

Much of the focus of the discussion of aquaculture effluents has focused on treatment to remove nutrients in discharge. However, earthen ponds actually treat the majority of nutrients that are added to the pond. In fact, as much as 57% of the nitrogen and 55% of the phosphorus added to catfish ponds in the feed

TABLE 1. *Pond production costs ($) partitioned into costs of waste treatment and costs of fish production (Valderrama and Engle 2001).*

Cost	64 ha (160 acre)			256 ha (640 acre)		
	Fish production	Waste treatment	Total pond	Fish production	Waste treatment	Total pond
Repairs and maintenance	120	11,880	12,000	360	35,640	36,000
Pond renovation	72	7,128	7,200	230	22,770	23,000
Aeration	3,029	13,799	16,828	12,312	56,088	68,400
All fuel	1,870	-	1,870	7,600	-	7,600
Chemicals	485	-	485	2,000	-	2,000
Telephone	2,000	-	2,000	3,100	-	3,100
Water quality	-	-	-	1,000	1,000	2,000
Fingerlings	42,000	-	42,000	136,560	-	136,560
Feed	215,600	-	215,600	876,260	-	876,260
Labor	14,664	21,996	36,660	73,185	109,778	182,963
Management	21,000	-	21,000	60,000	-	60,000
Harvesting/hauling	28,000	-	28,000	85,350	-	85,350
Accounting/legal	1,800	-	1,800	3,500	-	3,500
Bird scaring	500	500	1,000	2,000	2,000	4,000
Int. operating cap.	27,319	4,563	31,882	104,235	18,750	122,985
Total variable costs	358,459	59,866	418,325	1,367,692	246,026	1,613,718
Total fixed costs	17,113	78,188	95,301	49,268	305,948	355,216
Total costs	375,572	138,054	513,626	1,416,960	551,974	1,968,934
Cost/kg	1.17	0.44	1.61	1.10	0.42	1.52

is either taken up by plankton, volatilized into the atmosphere, or adsorbed onto mud particles (Boyd 1985, Chapter 3). The remaining percentages (43% nitrogen and 45% phosphorus) are assimilated by the fish. Yields (kg/ha) of fish produced in an earthen pond are limited by its waste treatment properties. The feed fed acts as fertilizer as the waste products from the fish provide nutrients for plankton production. The plankton represents a demand for oxygen that far exceeds the oxygen demand of fish in the pond. Moreover, it is the plankton (in addition to bacterial populations) that serves as the primary waste treatment unit in the pond. Thus, the waste treatment unit (plankton and bacteria) is also what limits fish production in pond production systems.

Table 1 presents an enterprise budget in which the fish production costs are separated from the waste treatment cost components of an earthen pond. In a tank in which the wastes are simply flushed out instead of processed, approximately 0.06 kg fish/L of water can be produced (Hargreaves 2000). At this rate, approximately 378,500 L of water would be needed to hold the 22,727 kg of fish that would be produced in a 4-ha earthen pond. This volume of water (378,500 L) represents approximately 1% of the water volume of a 4-ha pond that has an average depth of 1.3 m. Thus, in terms of the pond costs, 1% of the costs associated with the pond are for fish costs; 99% are for waste treatment. In the enterprise budget in Table 1, 1% of the costs associated with repairs and maintenance of the pond, and pond renovation are charged to fish production.

All feed, fuel for trucks for marketing and regular business activities, chemicals, telephone, fingerlings, feed, management, harvesting, hauling, accounting and legal expenses are charged at 100% to fish production. Bird scaring is charged at 50% to fish (to compensate for any bird netting that would have to be charged for an intensive containment). Labor would be reduced if fish were to be confined to one locale and is budgeted at 40% to fish production. Water quality monitoring costs are charged at 50% to fish production and 50% to waste treatment.

Aeration costs are more complex. Since fish account for approximately 18% of the oxygen demand in the pond, 18% of aeration costs were charged to fish, with the remainder charged to waste treatment. In the absence of the oxygen produced by phytoplankton during the day, additional aeration would be required to supply all the oxygen needed by the fish. To calculate the costs associated with the oxygen demand of the fish alone, oxygen consumption values of 0.8–0.45 mg O_2/g fish/per/h were taken from Andrews and Matsuda (1975). For a 4-ha pond with 22,727 kg of fish, the oxygen needed for a 210-d growing period with 24-hr/d demand for oxygen would be 70,875-kg O_2, based on the average values from Andrews and Matsuda (1975). A 10-hp floating electric paddlewheel aerator transfers approximately 21-kg O_2 (Boyd and Ahmad 1987). Thus, to meet the oxygen demand of fish in this pond would require 3,375 h of aeration. At 9.7 Kw/h and a price of $0.075/Kw-h, the cost would be $2,455. For the entire 64-ha farm, the total aeration cost would be $34,370. However, plankton also creates off-flavor problems that cost $0.08/kg of catfish produced, on average (Engle et al. 1995). This would generate an additional expense of waste treatment. These costs nearly cancel each other out, but add slightly more weight to fish production costs.

To estimate fixed costs associated with just the fish production, 1% of the depreciation and interest on investment of the ponds, water supply, and land were charged to the fish alone.

All of the fixed costs associated with the office building and the feed storage units were charged to fish production. Equipment costs were separated out into those items that are used exclusively for fish and those more related to the pond system. Taxes and insurance were charged at 25% to fish production.

Table 1 shows that waste treatment functions in the pond cost $0.42–$0.44/kg, while the fish production costs ranged from $1.10 to $1.17/kg. Thus, 27–28% of the cost of producing fish in ponds is related to the waste treatment processes.

It should be pointed out that, in spite of the treatment costs of the pond unit, pond production is still the least expensive way to raise fish commercially. Technologically-based waste treatment costs in intensive recirculating systems result in overall costs of production that are two to three times higher than pond production costs (Lutz 2000).

Best Management Practices

One proposed alternative for minimizing environmental impacts from aquaculture is to develop a set of Best Management Practices that would minimize the volume of discharge, the concentration of nutrients in the discharge, and that would be economically beneficial to farmers. Adoption of BMP's by farmers is most successful when there are strong economic incentives to do so. Several specific management practices that have been proposed for pond production will be discussed as an example of BMP's that also offer economic incentives for farmers. BMP's that would reduce the volume of discharge include: the 6/3 water management rule, harvesting without discharging water, and re-using water where possible. Many catfish farmers already practice the 6/3-water management rule (Pote and Wax 1993). Under this guideline, farmers allow pond water levels to drop (through evaporation) by 15 cm (6 in.)

prior to filling and then re-fill 7.5 cm (3 in.). In this manner, there is always at least 7.5 cm (3 in.) of rainwater storage capacity in the ponds. This reduces both overflow discharge and groundwater use. Farmers practice this rule primarily to minimize pumping costs, but the practice also minimizes overflow from ponds. Harvesting without discharging water is also practiced by farmers primarily due to the high cost of pumping water from groundwater wells to replace water discharged. Baitfish farms tend to re-use water drained from ponds to minimize pumping costs and to reduce demand on groundwater resources. Utility costs are the fifth largest cost in catfish production. Water conservation practices such as the 6/3 rule and water re-use minimize pumping costs and reduce utility costs in catfish production.

BMP's to reduce the concentration of nutrients in discharges focus principally on feeding practices. Careful feeding reduces waste feed and reduces nitrogen and phosphorus additions to the water. In catfish farming, feed is the single greatest cost of production, representing from 42–44% of total annual costs and from 52–54% of total operating costs. Reductions in waste feed through careful feeding will make more efficient use of the feed input. More efficient use of feed will reduce feed costs per kg of fish produced.

Fingerling cost is the second greatest cost in catfish production. Losinger et al. (2000) showed that maximum profits occur at stocking and feeding rates that are less than common industry standards that produce maximum yields. Thus, those farmers who are attempting to maximize yield could reduce both feed and fingerling costs by producing at economically efficient levels. However, restricting feeding and stocking rates at levels lower than the profit-maximizing levels would impose additional costs to catfish farmers.

Best Management Practices to reduce both the source and concentration of nutrients in pond aquaculture can possibly reduce fingerling, feed, and pumping costs if these levels of input use would be set at the economically efficient levels. Thus, it would appear to be possible to design BMP's for pond aquaculture that provide economic incentives for their adoption. However, if BMP's would be developed that entail input use at levels less than the economically efficient ones, economic disincentives would be created and adoption by farmers would not be likely.

Settling Basins

The use of settling basins for treatment of effluents from ponds has been under investigation (Tucker et al. 2000). However, their use for treating effluents from ponds has several disadvantages. The first major disadvantage is that settleable solids form only a portion of effluents and, for much of the time, are a minor constituent. Most of the organic suspended solids in pond effluents is associated with live phytoplankton cells that do not settle out readily (Boney 1975). Since a good portion of nutrients are tied up in the plankton, settling basins do little to reduce nutrients in pond effluents. The second major disadvantage is that land must either be available for construction of basins or existing production ponds would have to be retrofitted. In either case, conveyance of farm effluents into these basins is a complex problem and will require either careful engineering design or substantial pumping investment and operating costs.

Costs associated with settling basins are dependent on the size and number of basins, and whether sufficient land is available for basin construction or if existing production ponds must be retrofitted and taken out of production. Sizing of settling basins is controlled by factors such as the type of effluent to be treated (draining or storm overflow), layout of ponds, size of the largest foodfish pond, the number of drainage canals, and the scope of regulations governing the release of aquacultural effluents. The number of settling basins is affected by the hydraulic residence time (HRT) which is calculated from Stoke's Law (Boyd 1995). HRT, in turn, is affected by the size of particles in suspension. Fine clay particles (1 mm) may require a minimum HRT of 20 d, while larger, 5-mm particles may only require an HRT of 20 h. While Boyd et al. (1998) found that an HRT of 8 h achieved significant removal of critical parameters, Alabama soils have a much lower content of fine clays than Mississippi delta soils. The number of settling basins is also affected by the number of drainages on a farm. Some farms may drain in four to five different directions.

Furthermore, farms that have ponds that are not contiguous would need a greater number of basins.

Three farm size scenarios (Keenum and Waldrop 1988; Engle and Killian 1997) were considered in an analysis of settling basin costs: a 64-ha farm with approximately 56-ha of water, a 128-ha farm with 114 water ha, and a 256-ha farm with 228-ha of water (Valderrama and Engle 2001). Average sizes of foodfish ponds in this analysis were assumed to be 4-ha and 6-ha, while fingerling ponds were 2-ha each. The total area of fingerling ponds was assumed to occupy 10% of the farm area.

Scenarios were developed that included farms with ponds that drain all to one side of the farm and those that drain to two different sides of the farm (Fig. 3). Options to either

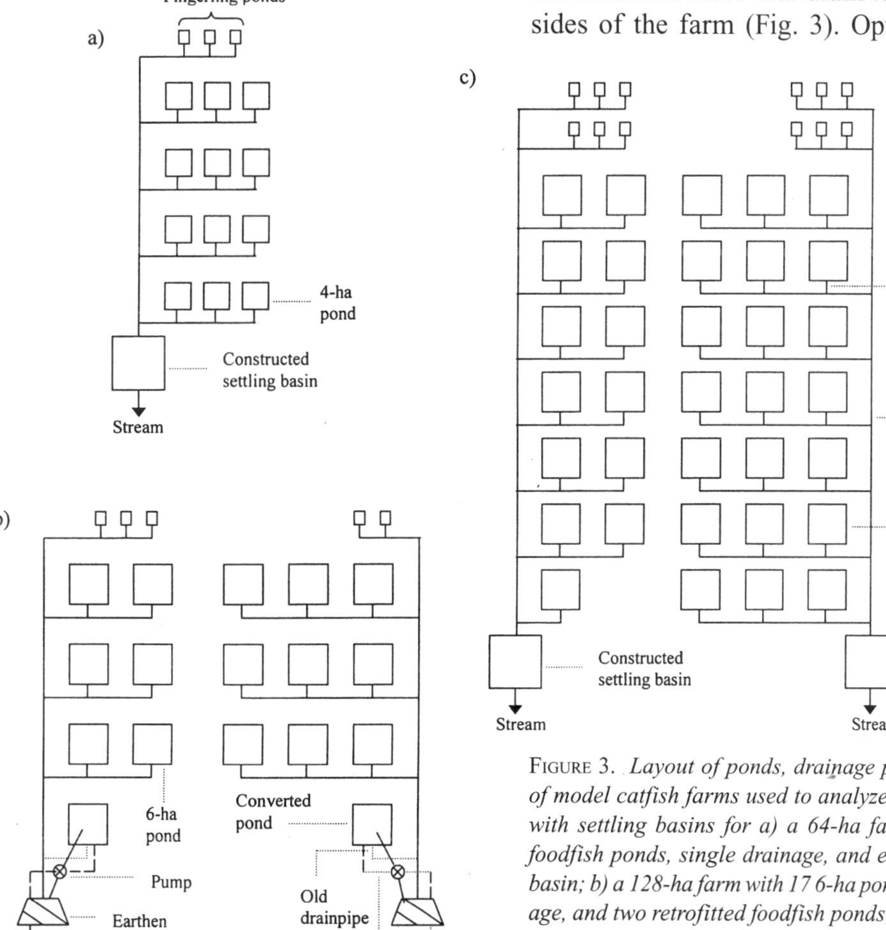

FIGURE 3. *Layout of ponds, drainage pipes, and canals of model catfish farms used to analyze costs associated with settling basins for a) a 64-ha farm with 12 4-ha foodfish ponds, single drainage, and excavated settling basin; b) a 128-ha farm with 17 6-ha ponds, double drainage, and two retrofitted foodfish ponds; and c) a 256-ha farm with 34 6-ha ponds, double drainage, and two excavated settling basins (Valderrama and Engle 2001).*

TABLE 2. *Total investment and annual costs associated with construction of settling basins on new land and retrofitting existing foodfish ponds for draining and storm overflow effluents for a 64-ha farm (Hydraulic Residence Time = 1–20 d) (Valderrama and Engle 2001)*

Foodfish pond size	Layout	Portion of effluents/ overflow[a]	Required settling area	Total investment costs[b]	Total annual operating costs[c]	Total annual fixed costs[d]	Total annual cost[e]	Unit cost
(ha)			(ha)	($)	($)	($)	($)	($/kg)
Construction of settling basins on new land/draining (HRT=1–20)								
4	Single	last 20%	0.8–1.2	28,648–40,932	443–560	3,442–4,828	3,885–5,388	$0.012–$0.017
4	Single	all	4–6	127,194–188,916	1,381–1,967	14,569–21,540	15,950–23,507	$0.050–$0.074
4	Double	last 20%	0.8	57,296	677	6,884	7,561	$0.024
4	Double	all	4	254,387	2,553	29,138	31,691	$0.10
6	Single	last 20%	1.2	40,932	560	4,828	5,388	$0.017
6	Single	all	6	188,916	1,967	21,540	23,507	$0.074
6	Double	last 20%	1.2	81,864	912	9,657	10,569	$0.033
6	Double	all	6	377,832	3,725	43,081	46,806	$0.147
Retrofitting foodfish production ponds/draining (HRT=1–20)								
4	Single	last 20%	0.8–1.2	5,837	1,042	3,679	4,721	$0.015
4	Single	all	4–6	5,837–8,656	1,042–1,921	3,679–6,659	4,721–8,580	$0.015–$0.027
4	Double	last 20%	0.8	10,891	1,918	7,358	9,276	$0.029
4	Double	all	4	10,891	1,918	7,358	9,276	$0.029
6	Single	last 20%	1.2	6,413	1,485	5,169	6,654	$0.021
6	Single	all	6	6,413	1,485	5,169	6,654	$0.021
6	Double	last 20%	1.2	12,039	2,801	10,338	13,139	$0.041
6	Double	all	6	12,039	2,801	10,338	13,139	$0.041
Construction of settling basins on new land/overflow (HRT=1)								
4	Single	41,634	3	96,356	1,102	11,086	12,189	$0.038
4	Double	41,634	3	100,314	1,102	11,740	12,842	$0.040
6	Single	41,634	3	96,356	1,102	11,086	12,189	$0.038
6	Double	41,634	3	100,314	1,102	11,740	12,842	$0.040

[a]Overflow from an 8-cm storm event in m^3.
[b]Includes construction costs of basins and acquisition of pumps.
[c]Includes copper sulphate and pumping costs.
[d]Includes depreciation and interest on investment.
[e]Equal to the sum of total annual operating and fixed costs.

purchase land to construct settling basins or to retrofit existing ponds in situations in which no additional land is available were included. In both cases, ponds were assumed to drain by gravity flow, although a re-lift pump would be required to pump either from the ditch to the basin or back. Since evidence shows that total suspended solids in the last 20% of draining effluents from watershed ponds are much higher than in the initial drawdown, scenarios of treating only the last 20% versus all effluents were considered (Boyd et al. 2000). It should be noted that this may not be the case for levee ponds because levee ponds are not drained during harvest and less sediment is stirred up in the seining process.

Table 2 presents total investment costs and annual operating and fixed costs for each scenario for a 64-ha farm. Larger farm sizes will result in higher and more variable costs. Investment costs included excavation of settling basins and the installation of stationary re-lift pumps to drain effluents from excavated basins. Annual operating costs consisted of copper sulfate applications (to promote sedimentation of phytoplankton cells), the annual cost of pumping, and levee mowing and maintenance, whereas annual fixed costs refer to depreciation of basins and pumps, interest on investment, and the opportunity costs associated with land taken out of production for the settling basin. Estimates of lost revenue due to production foregone from retrofitted production ponds were: $745, $855, and $1,185/ha ($298, $342, and $474 per acre) for the 64, 128, and 256-ha (160, 320, and 640-acre) farms, respectively (Engle 1998; Engle and Killian 1997). Pond yields of 5,682 kg/ha were assumed in each case. Treatment of overflow effluents would require construction of new settling basins because operating retrofitted and linked production ponds requires a degree of supervision that may not be available during rainfall events.

Large investments are needed for the construction of settling basins. This investment cost depends heavily on the drainage layout of the farm and the scope of regulations governing the release of effluents. For instance, investment costs in a 64-ha (640-acre) catfish farm may range from $28,648 to over $375,000 (Table 2). Total annual operating costs are relatively low because the analysis assumed a design based on gravity flow. On farms where pumping would be required to move effluents to the basin, operating costs would be much higher. Total annual fixed costs are high due to interest on the investment. Retrofitting production ponds entails a lower investment cost, but the system must be designed carefully and monitored closely, especially when several retrofitted ponds are interconnected. If pumping rates from the drainage canal exceed the conveyance capacity of the system, overflow in the first basin will be unavoidable.

Given the high costs associated with excavation of basins and the technical difficulties that may arise from retrofitting and linking production ponds, it appears that a much more logical approach is simply to use each production pond as its own settling basin, as suggested by Tucker et al. (2000). Certain good management practices, such as harvesting

ponds without ever draining the last 20% of water, or holding this volume of water in existing farm ditches prior to final discharge, may be implemented to achieve reductions in solid and nutrient loads comparable to what could be accomplished with the construction of independent and expensive structures.

Production/Storage Ponds

The use of production/storage (P & S) ponds for the reduction of effluent discharge and groundwater use in catfish farming has been examined by researchers at Mississippi State University (Cathcart et al. 1999). This approach is based on the simultaneous use of certain production ponds for water storage and fish production. One pond (the production/ storage pond), in a group of linked ponds, is deepened to increase storage capacity. While all ponds in the group collect rain water, the unmodified production ponds drain overflow into the production/storage pond via small culverts. Drainage from the linked system only occurs when the storage capacity is exceeded, and it is always through the production/storage pond. This system also results in reduced groundwater use because water is taken first from the storage pond when water is needed in the production ponds. Only when the production/storage pond depth reaches a minimum level (about 1 m), is groundwater pumped to the system.

Cathcart et al. (1999) obtained estimates of annual groundwater use and overflow release from the system assuming several pond configurations, production/storage pond depths, and infiltration rates. This information was used to develop farm-level estimates of groundwater use surface release, as well as costs of implementation for three different farm sizes (64, 128, and 256 ha; 160, 320, and 640 acre) (Valderrama and Engle 2001) (Table 3).

Two configurations (1:1 and 1:3), based on the number of production ponds served by each production/storage pond, were assumed

a)

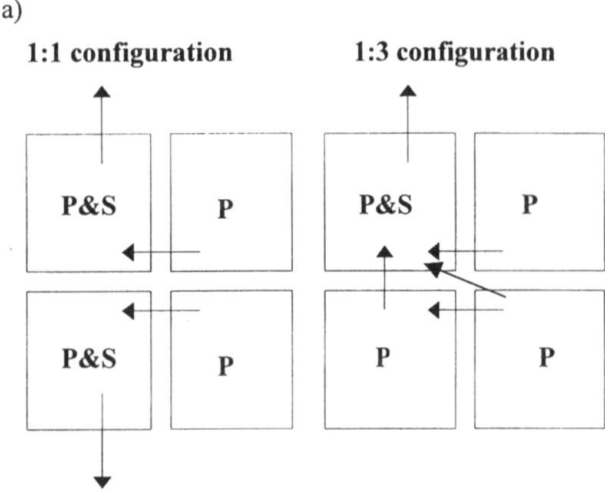

b)

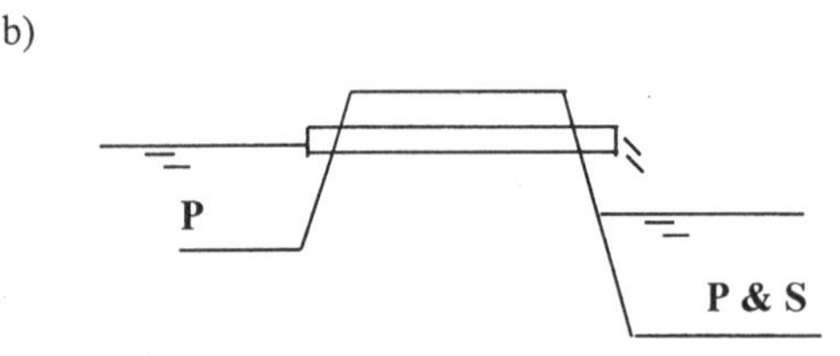

FIGURE 4. *Plan and section view of production and production/storage ponds. a) plan view showing two pond configurations: one production/storage pond per production pond (left), and one production/storage pond per three production ponds (right). b) section view of production and production/storage ponds showing gravity draining. P & S represent production and storage pond. P represents a production pond (taken from Cathcart et al. 1999).*

(Fig. 4). Increased depth of the P & S pond increases the storage capacity of the system but higher earthmoving costs are incurred. Three additional depths were considered: 30, 60, and 90 cm (12, 24, and 36 in., respectively). Seepage tends to be low in soils where pond aquaculture occurs. Seepage values of 0.0 and 1.0 mm/d were assumed. The water use and release estimates from Cathcart et al. (1999) were based on the "6/3" scheme of water management (Pote and Wax 1993). Cost estimates were developed for farms with average foodfish pond sizes of 4 and 6 ha (10 and 15 acre). Fingerling ponds were not linked.

In total, 24 scenarios were defined for each farm size. Table 3 presents total investment costs and total annual operating and fixed costs for each farm scenario. Total investment costs are the summation of excavation costs for P & S ponds, installation costs of culverts for each 1:1 and 1:3 pond system, and acquisition of tractor PTO-driven portable pumps to move water from the P & S ponds back into the production ponds. Because the use of P & S ponds results in reduced groundwater requirements, total operating costs include the annual cost of pumping using the portable pumps as well as the savings resulting from reduction in well pumping costs. For all farm scenarios considered, savings in well pumping surpassed PTO driven pumping costs, which resulted in a net reduction in annual operating costs. Finally, total annual fixed costs were based on pump depreciation and interest on investment.

Depth of storage pond and pond configuration are the two most important factors affecting implementation costs of this technology. For instance, estimated total investment costs for a 64-ha (160-acre) farm with 4-ha (10-acre) foodfish ponds, 1:1 configuration, and 0-mm/d infiltration rate, ranged from $76,123 to $215,088 as the additional depth of storage ponds was increased from 30 to 90 cm (Table 3). However, if the configuration is 3:1, investment costs decreased (the respective range goes from $44,782 to $115,947).

Despite the savings in groundwater use and the reduction in total annual operating costs, large amounts of investment are involved in the modification of regular production ponds into production/storage ponds. Depending on the scenario, this technology may be more expensive to implement than the construction of settling basin units. Also, practical knowledge of farmers indicates that disease and off-flavor episodes may become widespread among interconnected ponds. All these factors

TABLE 3. *Total investment and annual costs for production/storage ponds for reduction of effluent discharge and groundwater use on catfish farms, 64-ha farm (Valderrama and Engle 2001). Numbers in parentheses are negative. Ranges are for additional storage depths of 30–90 cm.*

Pond configuration	Infiltration rate	Total investment cost	Total annual operating cost	Total annual fixed cost	Total annual cost	Unit cost
	(mm/d)	($)	($)	($)	($)	($/kg)
			4-ha ponds			
1:1	0	76,123–215,088	(12,223)–(21,845)	5,125–14,820	(7,098)–(6,025)	$0.022-$0.019
1:1	1	76,163–215,088	(10,089)–(16,155)	5,125–14,820	(4,965)–(1,335)	$0.016-$0.004
3:1	0	44,782–115,947	(6,214)–(14,749)	2,839–7,964	(3,375)–(6,785)	$0.011-$0.021
3:1	1	44,782–119,352	(5,697)–(11,603)	2,839–8,517	(2,858)–(3,086)	$0.009-$0.010
			6-ha ponds			
1:1	0	75,366–214,291	(12,223)–(21,845)	5,125–14,820	(7,098)–(6,025)	$0.022-$0.019
1:1	1	75,366–214,291	(10,089)–(16,155)	5,125–14,820	(4,965)–(1,335)	$0.016-$0.004
3:1	0	43,587–114,752	(6,214)–(14,749)	2,839–7,964	(3,375)–(6,785)	$0.011-$0.021
3:1	1	43,587–118,157	(5,697)–(11,603)	2,839–8,517	(2,858)–(3,086)	$0.009-$0.010

need to be considered in the selection of this strategy as a technology option for effluent reduction.

Constructed Wetlands

A number of studies have demonstrated that the use of constructed wetlands to treat catfish pond effluents significantly reduces the concentrations of nutrients and suspended solids (Schwartz and Boyd 1995; Posadas 1998). Depending on the size of the wetland and the HRT, concentrations of total ammonia-nitrogen, total phosphorus, and total suspended solids can be reduced by up to 80–90%. Wetlands have been shown to significantly improve water quality even during late fall and winter when vegetation is dormant.

The problem with using constructed wetlands to treat catfish pond effluents is that a very large area would need to be devoted in large catfish farms to provide adequate HRTs. Several studies have evaluated the economic feasibility of constructed wetlands as an effluent treatment option on catfish farms (Hebicha 1989; Cerezo 1993; Kouka and Engle 1996; Posadas 1998). The conclusion in each case was that, although effective in improving water quality, constructed wetlands do not appear to be a feasible approach for the treatment of all catfish effluents due to the large cost outlay associated with their implementation on a commercial scale.

Hebicha (1989) estimated total investment costs and total annual operating and fixed costs for wetlands of several sizes. He conducted regression analyses to estimate the percentage of BOD removed as a function of initial BOD concentration and the loading rate of effluents. Assuming a constant initial BOD concentration, the removal percentage of BOD increases with lower loading rates. However, if lower loading rates are used, HRT increases and size of the wetland must be increased to keep treatment time within a reasonable limit.

Kouka and Engle (1996) developed cost estimates for surface and subsurface flow wetland systems for catfish farms of various sizes (65, 130, and 260 ha). Wetlands were assumed to treat only 5% of the pond volume and equaled 20% and 10% of pond area for surface and subsurface flow systems. Table 4 presents investment and operating costs of constructed wetlands (surface flow) for each farm-size scenario. Estimates are presented for both unlined and lined systems, as well as for the current and an expanded (wetlands built on additional acreage) land base. Expressed in 1989 dollars, these annual cost estimates

TABLE 4. *Investment and annual operating costs for constructed wetlands for the treatment of effluents from catfish ponds. Adapted from Kouka and Engle (1994), Posadas (1998), and Hebicha (1989).*

Farm size (ha)	Wetland size (ha)	Investment cost			Operating cost	
		Unlined		Lined		
		($)	($/ha of farm)	($)	($)	($/ha)
Kouka and Engle (1996)						
			Current land base			
65	11	800,832	12,320	1,123,249	1,654	25
130	22	1,655,103	12,732	2,311,254	3,065	24
260	44	3,301,355	12,698	4,595,558	5,827	22
			Expanded land base			
65	13	984,036	15,139	1,378,170	1,961	30
130	26	1,959,903	15,076	2,734,510	3,559	27
260	52	3,910,213	15,039	5,440,495	6,790	26
Posadas (1998)						
19.2	0.5	21,250	1,107	-	2,796	146
19.2	0.8	33,284	1,734	-	3,204	167
19.2	1.1	45,318	2,360	-	4,235	221
Hebicha (1989)						
4.1	1.2	3,706	904	-	69	17
4.1	2.4	7,051	1,710	-	136	33
4.1	4.1	11,565	2,821	-	230	56

ranged from $904–$2,821/ha. Kouka and Engle (1996) concluded that the construction of wetlands would represent an average increase in the production cost of catfish of $0.11/kg across farm-size scenarios, meaning that small farmers would likely be driven out of business and new farmers would face additional barriers to entry.

Posadas (1998) evaluated the use of constructed wetlands for the treatment of catfish pond effluents from multi-enterprise farms in the Mississippi Black Belt area. Although significant reductions in the concentrations of nitrogen and phosphorus compounds and total suspended solids were achieved, the economic analysis demonstrated that the use of constructed wetlands is prohibitively costly.

Land Application

It has been suggested that effluents from catfish ponds could be used as a water source for irrigation of row crops. Some early reports indicated that water from catfish ponds may actually contribute to enhanced rice yields, at least near the point of discharge (Burtle and Morrison 1987). Based on this assumption, Kouka and Engle (1996) developed cost estimates for rice irrigation with pond water. They found that the revenue from the additional crop yield compensated for the costs incurred in the system (mostly piping and plumbing, and an additional pumping cost), which makes this practice a much more attractive option than expensive technologies such as constructed wetlands or settling basins. However, the increased crop yield was due to reducing effects of iron toxicity of groundwater in the first 100 m from the well. This benefit would only accrue to farmers using groundwater with high iron content. Otherwise, there would not likely be additional crop yield.

Catfish pond effluents are in fact quite dilute with respect to the major plant nutrients (Tucker et al. 2000); thus the benefit to the irrigated crop is doubtful. Furthermore, water demand for irrigation is higher during the summer months, when there is virtually no overflow from fish ponds. In addition, water for irrigation may be needed quickly and at relatively large volumes, and pumping may have to cover great distances, all of which implies that it may not be feasible to use crop irrigation as a treatment for pond effluents.

Other Treatment Options

In addition to the treatment technologies already mentioned, there are several other options for effluent treatment that have been examined and then disregarded because of their prohibitive costs or simple impracticability. Hebicha (1989) evaluated the use of intermittent sand filters in catfish pond effluents and concluded that this is not an economically viable approach to improve effluent quality. For instance, in order to reduce the BOD concentration of the effluents from a 4.1-ha (10-acre) catfish pond from 30 to 5 mg/L, a 0.3-ha sand filter was required. The construction cost of this filter was nearly $50,000 (1988 costs) while annual fixed cost was over $5,000. In comparison, the cost of construction of a 2.4-ha wetland for the same pond was only $7,051.

Hebicha (1989) also evaluated the feasibility of using groundwater to dilute harvest effluents. Although less costly than the construction of intermittent sand filters and wetlands, using dilution as a method of pollution control is unacceptable given the potential for rapid depletion of aquifers if a large number of farmers followed this practice.

Several approaches for water recycling within fish farms have been analyzed. Cerezo (1993) proposed the construction of additional ponds on existing catfish farms that could be used for temporarily holding draining effluents from production ponds. Once harvesting was completed, water could be pumped back from the recycling ponds. The problem with this approach is that large portions of farm land would have to be converted to recycling ponds, thereby significantly reducing production levels, and large pumping and plumbing costs would be incurred to convey pond water over great distances.

Kouka and Engle (1996) evaluated a strategy of paired ponds on catfish farms, whereby water from a catfish pond is drained into another pond with filter-feeding fish and is pumped back. This practice combines recycling with filtration. However, this method requires removing one pond from catfish production for every pond remaining in production. The lost revenue resulting from such a system constitutes a major cost to this alternative. Kouka and Engle (1996) estimated that this system would increase production costs by $0.07/kg, which would mostly affect small catfish farmers given the small profit margins under which they operate.

Raceways

Trout raceway farms that sell more than 9,000 kg/yr and use more than 2,270 kg of feed in a month are required to have a permit to discharge effluents (State of Idaho 1997). Most trout farms have some type of waste management system and plan in place. The most common element of waste management is a settling unit. For small-scale raceway farms, a settling pond is commonly used, while for larger farms, combinations of quiescent zones and off-line settling ponds are most commonly used (Fornshell 2001).

The size of quiescent zones and off-line

TABLE 5. *Annual costs of adding quiescent zones and off-line settling basins for different sizes of trout raceway farms.*

Farm size	Quiescent zone costs		Off-line settling basin costs		Total cost	Cost/kg
	Fixed	Variable	Fixed	Variable	$	$/kg
Small						
400 gpm	76	169	15	31	291	0.160
	76	169	30	31	306	0.168
	76	169	45	31	321	0.177
600 gpm	120	169	15	31	335	0.122
	120	169	30	31	350	0.127
	120	169	45	31	365	0.133
800 gpm	157	169	15	31	372	0.102
	157	169	30	31	387	0.106
	157	169	45	31	402	0.110
Large						
3500 gpm	1493	338	75	210	1906	0.080
	1493	338	105	210	1936	0.081
	1493	338	135	210	1966	0.082
4000 gpm	1717	338	75	240	2130	0.078
	1717	338	105	240	2160	0.079
	1717	338	135	240	2190	0.080
4500 gpm	1916	338	75	270	2329	0.076
	1916	338	105	270	2359	0.077
	1916	338	135	270	2389	0.078

settling ponds, as well as the volume of trout produced, varies with the flow rates of water. The costs involved in adding quiescent zones and off-line settling basins consist primarily of the additional investment to construct the two units, as well as additional labor costs involved with sludge removal.

Construction costs of trout raceways were estimated by adapting costs from Dunning and Sloan (no date) and Hinshaw et al. (1990). Flow rates of 400–800 gpm for small farms and 3,500–4,500 gpm for larger farms were specified. Sizes of quiescent zones and off-line settling basins were calculated based on Fornshell (2001) and State of Idaho (1997) using these flow rates as a basis.

The additional costs are higher per kg of trout produced on smaller farms (Table 5). For smaller farm sizes, the additional costs imposed by quiescent zones and off-line settling basins ranged from $0.102/kg to $0.177/kg. On larger farms, the additional costs of effluent treatment and sludge removal ranged from $0.076 to $0.082/kg. Sludge removal costs alone can represent as much as $0.0088/kg of trout produced. However, this cost figure is underestimated because it does not include the depreciation, interest on investment, or opportunity costs that compose the annual fixed costs. These fixed costs tend to be higher than operating costs for many waste treatment systems.

Conclusions

Economists view pollution as a production externality. Negative externalities that increase costs to other users may require government intervention. Regulations can force producers to internalize waste disposal costs and pass these on to their respective consumers. The economically optimum level is determined by balancing the costs of effluent controls with the costs of poor water quality. Selecting a level of water quality at which the pollution control costs are far higher than the benefits to society of clean water does not make economic sense.

The costs of effluent controls include the direct farm-level costs of installing treatment options. It also includes the indirect effects on local, state, and national economies that result from decreases in supply. Unintended environmental impacts of imposing effluent controls must be evaluated likewise.

Molluscan shellfish culture provides significant positive externalities to aquatic resources. The economic importance of shellfish industries in the U.S. has resulted in pressure on other sources of pollution due to the sensitivity of shellfish to environmental quality.

The majority of pond aquaculture poses little risk for contributing to environmental degradation and creates positive environmental impacts in many areas. Fish farmers have internalized waste disposal costs as 27–28% of production costs due to the waste treatment properties of static fish culture in earthen ponds. The cost of treating the remaining dilute amount of nutrients will be high with little environmental benefit. Adding cost to catfish products will reduce supply and will result in serious economic consequences in impoverished rural areas in the delta. Settling basins, production/storage ponds, constructed wetlands, and filtration units are impractical and too costly to justify the minimal benefit that would be obtained. Land application may be possible in some areas. BMP's may be feasible, but it is not possible to evaluate these until one definitive set is defined.

The use of quiescent zones and off-line settling basins on trout farms has increased trout production costs. These costs are lower per unit of product on larger farms as a result of economies of scale in construction and operation. Effluent treatment regulations may result in cost structures that favor larger trout farms and impose higher costs on smaller farms.

Aquaculture is a broad term that covers a wide variety of aquatic species and production systems. There clearly is no one economically optimum solution to the conflict over aquaculture production and environmental impacts. Policies, programs, and regulations should be tailored to deal with specific environmental problems caused by aquaculture operations. However, care should be taken to avoid creating unintended economic or environmental problems through lack of careful and comprehensive scientific scrutiny of all components of the problem and all proposed solutions.

Aquaculture production systems that rely upon direct discharge of wastes need to be assessed carefully. The cost of treating discharges must be weighed against the cost to society of negative environmental impacts. Comprehensive, in-depth studies will be required to develop policies that will improve environmental quality with minimal effects on local economies. Care must also be taken to avoid unintended negative economic and environmental effects. These can result from regulations developed from insufficient data and analysis. In those cases in which the cost to society is greater than the costs of controlling pollution, then economically efficient treatment alternatives should be developed. Strong economic incentives should be built into programs to ensure maximum voluntary adoption by farmers.

Literature Cited

Andrews, J. W. and Y. Matsuda. 1975. The influence of various culture conditions on the oxygen consumption of channel catfish. Transactions of the American Fisheries Society 104:322–327.

Boney, A. D. 1975. Phytoplankton. Edward Arnold Publishers, London, United Kingdom.

Boyd, C.E. 1985. Chemical budgets for channel catfish ponds. Transactions of the American Fisheries Society 11:291–298.

Boyd, C. E. 1995. Bottom soils, sediment, and pond aquaculture. Chapman and Hall, New York, USA.

Boyd, C. E. and T. Ahmad. 1987. Evaluation of aerators for channel catfish farming. Bulletin 584, Alabama Agricultural Experiment Station, Auburn University, Alabama, USA.

Boyd, C. E., A. Gross, and M. Rowan. 1998. Laboratory study of sedimentation for improving quality of pond effluents. Journal of Applied Aquaculture 8(2):39–48.

Boyd, C. E., J. Queiroz, J. Lee, M. Rowan, G. N. Whitis, and A. Gross. 2000. Environmental assessment of channel catfish *Ictalurus punctatus* farming in Alabama. Journal of the World Aquaculture Society 31:511–544.

Burtle, G. L. and J. R. Morrison. 1987. Circulation of water in catfish ponds. American Chemistry Society, Annual Meeting, New Orleans, Louisiana, USA.

Canzoneri, D. 1996. The economic value of shellfish aquaculture in Puget Sound: a case study prepared for People for Puget Sound. Seattle, Washington, USA.

Cathcart, T. P., J. W. Pote, and D. W. Rutherford. 1999. Reduction of effluent discharge and groundwater use in catfish ponds. Aquacultural Engineering 20:163–174.

Cerezo, G. A. 1993. Catfish pond effluent control: an economic analysis into wetlands, water recycling, and the effect of taxation of effluents. Master's thesis. Auburn University, Alabama, USA.

Dicks, M. R., R. McHugh, and B. Webb. 1996. Economy-wide impacts of U.S. aquaculture. Oklahoma Agricultural Experiment Station P-946, Oklahoma State University, Stillwater, Oklahoma, USA.

Dunning, R. and D. Sloan. No date. Aquaculture in North Carolina, rainbow trout: inputs, outputs, and economics. North Carolina Department of Agriculture and Consumer Services, Division of Aquaculture and Natural Resources, Franklin, North Carolina, USA.

Engle, C. R. 1998. Annual costs and returns of raising bighead carp in commercial catfish ponds. FSA9078-3M-9-98N. Cooperative Extension Program, University of Arkansas at Pine Bluff, Pine Bluff, Arkansas, USA.

Engle, C. and H. S. Killian. 1997. Costs of producing catfish on commercial farms in levee ponds in Arkansas. ETB 252. Cooperative Extension Program, University of Arkansas at Pine Bluff, Pine Bluff, Arkansas, USA.

Engle, C., G. L. Pounds, and M. van der Ploeg. 1995. The cost of off-flavor. Journal of the World Aquaculture Society 26:297–306.

Fornshell, G. 2001. Settling basin design. Western Regional Aquaculture Center. WRAC Publication No. 106. University of Washington, Seattle, Washington, USA.

Hailstones, T. J. and F. V. Mastrianna. 1986. Contemporary economic problems and issues. South-Western Publishing Co., Livermore, California, USA.

Hargreaves, J. A. 2000. Tilapia culture in the southeast United States. Pages 60–81 in B. A. Costa-Pierce and J. E. Rakocy, editors. Tilapia aquaculture in the Americas. Volume 2. The World Aquaculture Society, Baton Rouge, Louisiana, United States.

Hebicha, H. 1989. Catfish pond harvesting pollution control and costs for West Central Alabama. Doctoral dissertation. Auburn University, Alabama, USA.

Hinshaw, J. M., L. E. Rogers, and J. E. Easley. 1990. Budgets for trout production: estimated costs and returns for trout farming in the south. Southern Regional Aquaculture Center Publication No. 221, Stoneville, Mississippi, USA.

Kautsky, N., H. Berg, C. Folke, J. Larsson, and M. Troell. 1997. Ecological footprint for assessment of resource use and development limitations in shrimp and tilapia aquaculture. Aquaculture Research 28:753–766.

Keenum, M. and J. Waldrop. 1988. Economic analysis of farm-raised catfish production in Mississippi. Mississippi Agricultural and Forestry Experiment Station Technical Bulletin 155, Mississippi State, Mississippi, USA.

Kouka, P. J. and C. R. Engle. 1996. Economic implications of treating effluents from catfish production. Aquacultural Engineering 15:273–290.

Losinger, W., S. Dasgupta, C. Engle, and B. Wagner. 2000. Economic interactions between feeding rates and stocking densities in intensive catfish *Ictalurus punctatus* production. Journal of the World Aquaculture Society 31:491–502.

Lutz, G. 2000. Production economics and potential competitive dynamics of commercial tilapia culture in the Americas. Pages 119–132 *in* B. A. Costa-Pierce and J.E. Rakocy, editors. Tilapia aquaculture in the Americas, volume 2. The World Aquaculture Society, Baton Rouge, Louisiana, USA.

Mansfield, E. 1986. Principles of microeconomics. W.W. Norton & Company, New York, USA.

Posadas, B. C. 1998. Evaluating the use of constructed wetlands in producing catfish in multi-enterprise farming in Mississippi Black Belt Area. Doctoral dissertation. Mississippi State University, Mississippi, USA.

Pote, J. W. and C. L. Wax. 1993. Modeling the climatological potential for water conservation in aquaculture. Transactions of the American Society of Agricultural Engineers 36:1343–1348.

Schwartz, M. F. and C. E. Boyd. 1995. Constructed wetlands for treatment of channel catfish pond effluents. Progressive Fish-Culturist 57:255–266.

Sharp, A. M. and R. H. Leftwich. 1986. Economics of social issues. Business Publications, Inc., Plano, Texas, USA.

State of Idaho. 1997. Idaho waste management guidelines for aquaculture operations. Division of Environmental Quality, Boise, Idaho, USA.

Tucker, C., C. Boyd, J. Hargreaves, N. Stone, and R. Romaire. 2000. Effluents from channel catfish aquaculture ponds. Report from the Technical Subgroup for Catfish Production in Ponds, Joint Subcommittee on Aquaculture, Washington, D.C., USA.

USDA (United States Department of Aquaculture). 1998. Census of aquaculture. 1997 Census of agriculture volume 3, part 3. National Agricultural Statistics Service, United States Department of Agriculture,

Washington, D.C., USA.

USEPA United States Environmental Protection Agency). 1996. USEPA. NPDES permit writers manual. EPA-833-B-96-003, USEPA, Washington, D.C., USA.

Valderrama, D. and C. Engle. 2001. Economic analysis of treatment alternatives for catfish pond effluents. Annual report to Southern Regional Aquaculture Center, Stoneville, Mississippi, USA.

INDEX